"Accomplished naturopathic physician Dr. Jill Stansbury brings her vast clinical experience, deep knowledge, commitment, and passion for herbs to this exciting modern-day formulary. *Herbal Formularies for Health Professionals, Volume 1* is an extensive reference guide to the creation of classic herbal formulas using traditional herbal healing wisdom as well as modern science. This text should be a foundational part of herbal education. There is no other formulary as comprehensive or as accessible to the clinician; this book truly fills an empty space in the list of current herbal clinical education publications."

—DR. MARY BOVE, ND, author of *Encyclopedia of Natural Healing for Children and Infants*

"*Herbal Formularies for Health Professionals, Volume 1* delves deeply into herbal formulas for digestive, liver, urinary, and skin problems, along with descriptions and, where known, causes of these problems. In a chapter on the art of formulation, Jill Stansbury outlines example cases and provides appropriate formulas for each. Her considerations are fascinating. This refreshing work will be highly useful for both serious students and practitioners."

—HENRIETTE KRESS, herbalist; founder of Henriette's Herbal Homepage

"Dr. Jill Stansbury's *Herbal Formularies for Health Professionals* is a formidable accomplishment and historic contribution to the rapidly developing field of botanical medicine. . . . *Herbal Formularies* is practical and immediately usable for anyone interested in the applications of herbal medicine. Its depth will satisfy senior clinicians, academic researchers, and medical educators, while its accessibility will greatly further the education of students and the serious lay person."

—DAVID CROW, LAc, founder of Floracopeia Aromatic Treasures

"Reading Dr. Stansbury's formulary has quickly supported us in updating and filling in gaps in our dispensary stock to better serve our customers. I believe this book will similarly inform the dispensing practices of practitioners of all experience levels looking to glean insight from one of the most experienced herbalist physicians in the United States.

"I'm thrilled to see this first of five volumes of herbal internal medical formularies being released to a wide audience. . . . Clinicians will be surprised to discover how much groundwork has already been laid for future generations of health care professionals. This book offers mature templates for improvisation to treat a myriad of complex health conditions with botanical medicines."

—BENJAMIN ZAPPIN, LAc, herbalist; cofounder of Five Flavors Herbs

"Formulation is one of the key elements of a clinical herbalist's practice. It is also an aspect that intimidates many practitioners, especially those newer to herbal medicine. Jill Stansbury's book is a valuable guide to formulation that will lend a strong guiding hand for herbalists of all skill levels. Her language is clear, instructive, and easy to read. She gives multiple examples drawn from her extensive practice, and her explanations offer direction for working with individual patients. *Herbal Formularies for Health Professionals, Volume 1* is a comprehensive and practical book that I will recommend to my herbal students."

—7SONG, director of Northeast School of Botanical Medicine; Director of Holistic Medicine, Ithaca Free Clinic

HERBAL FORMULARIES FOR HEALTH PROFESSIONALS

VOLUME 2

CIRCULATION AND RESPIRATION

INCLUDING THE CARDIOVASCULAR, PERIPHERAL VASCULAR, PULMONARY, AND RESPIRATORY SYSTEMS

DR. JILL STANSBURY, ND

Chelsea Green Publishing
White River Junction, Vermont
London, UK

Project Manager: Patricia Stone
Editor: Fern Marshall Bradley
Editorial Assistance: Mikailah Grover and Patryk Madrid
Copy Editor: Nancy N. Bailey
Proofreader: Deborah Heimann
Indexer: Shana Milkie
Designer: Melissa Jacobson

Printed in the United States of America.
First printing September, 2018.
10 9 8 7 6 5 4   24 25 26 27

ISBN: 978-1-60358-798-3

Library of Congress Cataloging-in-Publication Data
Names: Stansbury, Jill, author.
Title: Herbal formularies for health professionals. Volume I, Digestion and elimination, including the gastrointestinal system,
    liver and gallbladder, urinary system, and the skin / Dr. Jill Stansbury.
Other titles: Digestion and elimination, including the gastrointestinal system, liver and gallbladder, urinary system, and the skin
Description: White River Junction, Vermont : Chelsea Green Publishing, [2017] | Includes bibliographical references and index.
Identifiers: LCCN 2017044410| ISBN 9781603587075 (hardcover) | ISBN 9781603587082 (ebook)
Subjects: | MESH: Formularies as Topic | Phytotherapy—methods | Digestive System Diseases—drug therapy | Urologic Diseases
    —drug therapy | Skin Diseases—drug therapy
Classification: LCC RM666.H33 | NLM QV 740.1 | DDC 615.3/21—dc23
LC record available at https://lccn.loc.gov/2017044410

Chelsea Green Publishing
White River Junction, Vermont, USA
London, UK
www.chelseagreen.com

To the Standing Green Nation—
all plants everywhere as living, breathing, sentient beings.
May we learn to treat them with respect and care.

And to all the traditional peoples, cultures, and nations
who have lovingly entered into sacred relationship with
the earth and the plants, and contributed to the art
and craft of herbal wisdom shared in these pages.

— CONTENTS —

INTRODUCTION
Honoring Traditional Knowledge   1

CHAPTER 1
The Art of Herbal Formulation   9

CHAPTER 2
**Creating Herbal Formulas for
Cardiovascular, Peripheral Vascular,
and Pulmonary Conditions**   25

CHAPTER 3
**Creating Herbal Formulas for
Respiratory Conditions**   107

Acknowledgments   *185*
Scientific Names to Common Names   *187*
Common Names to Scientific Names   *194*
Glossary of Therapeutic Terms   *202*
Notes   *205*
Index   *221*

Honoring Traditional Knowledge

Formularies are a long-standing tradition in herbal medicine, the history of which is discussed in Volume I of this set. Formularies grew from folkloric empiricism in which the recipes and remedies were as cherished as any other homesteading tool. They were also essential in times and places where doctors were few and far between. European, Asian, and other immigrants to early America brought along their herbal traditions, if not the plants and seeds themselves, from their homelands, but it must be recognized that Native American healing and plant knowledge contributed greatly to the survival of the colonizers, and thereby to herbalism of present day United States.

At the time of this writing, the cultural leitmotif is one of upheaval and conflict, where long-standing wounds are gaining greater voice and the faulty facade of old thought structures are beginning to crumble. Monuments honoring racist figures are being torn down, and sports teams using monikers offensive to Native Americans are changing their names in response to societally sanctioned shame. It is clearly offensive to wear a native headdress if you are not native and did not earn the right, but where does appreciation and assimilation turn into appropriation? A recent heated public discussion focused on whether the African American population has the sole right to sport dreadlocks. Following this logic, may only Asian women rock a kimono and hold a chignon in place with Japanese Kanzashi hair sticks? Is it crass to adorn yourself with a henna tattoo, just for fun, when mehndi holds deep spiritual significance in India? While these fashions may seem innocent at first glance, the fact is that an African or Asian person may be criticized, or even limited in job opportunities, simply for choosing their natural hairdo, while the dominant white culture is able to adopt the fashion with no consequences. Such cultural "borrowing" makes light of the real struggle that others face each day.

To bring the discussion back to herbalism, what about a New Age white man or woman teaching classes on herbs that open the chakras, herbs for ceremony, or entheogens as a spiritual path? Again, these endeavors can be pure of heart and may seem to harm no one, until one realizes that an Indian immigrant, an African American single mom, or a Latina may have difficulty in finding an audience or making a profit for the very same offerings. It is all too easy to overlook the oppression, the struggle, and the suffering that others face when trying to start a business, receive financing, rent a community venue, be positively reviewed in the media, or be invited into a networking group, when one has never faced such challenges. What seems like an innocent pursuit becomes an unjust ability to profit from someone else's culture, while those from the featured culture itself are unable to do so.

When I was in grade school, I learned the romantic definition of the term "melting pot" to be the happy result of different races, peoples, and cultures coming together, appreciating one another, and enriching one another's lives as a result. But in reality, only some members of the community can skim the cream from the top, while others may only avail themselves to dregs at the bottom of the pot. When some members of society are able to exploit the intellectual or cultural property of others for personal or financial gain, the pot has not melted into mutually beneficial sharing—it has stratified. Commercializing and monetizing that appropriation adds a further degree of injustice and outrage.

In light of the present struggle to address racial inequality, gender bias, and gross infringements of human rights in the world at large, I would be remiss not to mention the lineage holders who contributed to modern herbalism and to my own personal knowledge. In fact, the very term *Western herbalism* co-opts long-standing indigenous knowledge, whereby native wisdom is owned or claimed in giving it a "Western" identity when, in fact, some of this knowledge reached the West only by way of being incubated, developed, and matured within non-Western cultures. The field of herbal medicine is not exempt from the taint of cultural appropriation, where the conquerors claimed ownership and profited from something they usurped. While the term *Western herbalism* intends to differentiate the discipline from Traditional Chinese Medicine (TCM) and its form of herbalism, it is not hard to see that Western herbalism is a misnomer, given its broad underlying contributions.

I hold a particular appreciation of the First Nations of North America who have been the lineage holders of the beloved plants with whom I live and breathe. And I feel compelled to ask myself: Am I standing and making my living on land that was brutally taken from the first nations? Am I, or are you, gathering medicine whose virtues were lovingly coaxed from the plants by generations of people linked to this land, whose toil, whose observance, and whose generations of experience matured into healing wisdom? I say the answer is yes. Although European immigrants found some familiar cosmopolitan species in North America, we have First Nations knowledge to thank for introducing *Echinacea*, goldenseal, California poppy, American ginseng, black cohosh, gravel root, *Asclepius*, and many other important medicines into our *materia medica*—another term from folklore referring to collected information on individual medical materials, in this case, the plants themselves. First Nation peoples also selectively cultivated certain wild plants and modified entire plant species over the centuries—for example, Great Lakes region peoples, the Chippewa, Ojibwa, and Menominee, cultivated *Zizania aquatica*, a native grass, into "wild rice." The Inca and Amayra peoples of the high Andes cultivated hundreds of varieties of potatoes (*Solanum tuberosum*), enabling them to thrive in various elevations and ecosystems. By recognizing such millennia-long time frames, we can appreciate the generations-deep connection and the land-based identity that exists and anchors a people to their ways of life. Ethnobotany is the study of such connections a people hold with the plants of their surroundings.

Ethnobotany has fascinated me from a young age, and I began checking out books on Native American lore from the library when I was in elementary school. But to my credit, when I started my study of Amazonian plants, I didn't wish to read about the plants of the Bora, Matsigenka, and Wachiperi people in a book written by a nonnative scholar—I wanted to meet, live, and learn directly from the people themselves and from the forest itself. It has been my privilege and good fortune to do so, and I have gained knowledge about much more than ethnobotany. My most powerful lessons have come while sitting around the bonfire, talking with and listening to the feelings and frustrations of the community members. For example, as I described all the places I had traveled to in their fascinating country and through their magical jungle and told of all the people and tribes with whom I had lived and worked, I realized that they were not in the same economic position to easily do the same. "You know our country better than we do," I was told. As I listened, I learned how their own government had betrayed them. Land supporting the biggest trees, the highest waterfalls, and the most magnificent scenery had been sold to foreign enterprises in profit-driven trade agreements. Expensive lodges and fancy restaurants were built on land that only foreign tourists could afford and that they could never enjoy. North Americans, Europeans, and others with grant funding or research institute backing have come to study flora, fauna, climate, and natural history. The resulting books have been published in English, German, and French—languages the native people cannot read. Those books include photos of their ancestral land, of their sacred sites, and even of the people themselves, but no compensation was offered them, not even simple acknowledgment. Time after time, I have been made increasingly aware of my privilege, my "affluenza." I was infected, and I didn't even know it.

Many native communities have been forcibly displaced from their ancestral lands and thereby separated from the plants known best to them. And worse, many communities have been subjected to environmental terrorism—the forced placement of pollution-generating mining operations, oil fields, and pipelines—negatively impacting food and game and corrupting health and traditional ways of life. Actions that threaten health and safety and destroy the environmental resources of a selected population is the very definition of ecoterrorism, and the majority of the indigenous peoples of the world have been subjected to unfair, if not atrocious, and violent seizure of property and other transgressions for corporate gain, which poverty, racism, and power dynamics make difficult for the targeted populations to oppose. One need look only as far as the placement of the Dakota Access pipeline to see how the Sioux were held at gunpoint (or water cannons in freezing temperatures, as the case may be) and forced to accept the oil pipelines they vehemently opposed. While the process was heartbreaking, the Standing Rock Sioux gave us all a lesson in courage and peaceful resistance, and I hold them in great respect.

While conquerors often write history books to spin tales that suit, they have also written herb books. We can see the arrogance of early nonnative botanists in the naming of some Amazonian plant species in their honor. Bit by bit, those in power start to own the wisdom and control the narrative, while those whose ancestors developed the knowledge struggle to survive. It can ring false when one simply restates long-won cultural wisdom stripped down to a square peg of information and forced it into the round hole of a completely different culture. After studying herbal medicine for most of my life, can I comfortably assert that my knowledge is truly "mine"? Have I appreciated the finer points and nuances of the various disciplines such that they can be effectively used? I raise such questions with the endgame of maturing the field of herbal medicine. Maturing a template for a new era will require appreciation rather than appropriation, sophistication rather than simplification, and refinement rather than reductionism—all actions necessitating creative, collaborative, and integrative thinking. This set of volumes represent my own effort to synthesize the various threads that I have tugged on, or perhaps more accurately that have tugged on me like a vigorous vine encircling my ankle and drawing me in.

When you learn to speak another language, the phrases and idiosyncrasies of the language force your brain to think in new ways. So, too, learning the uses of a plant within a certain cultural context offers more information than just the words, or the list of what the plant is "good for." As one learns how a certain culture uses and relates to various plants, one starts to gain insights into the whole system of thought—a paradigm in which to make sense of the information. When such information is decontextualized, its value is diminished, and when the knowledge base is fractured, there is obviously much that will be lost. For example, saying that an herb promotes *vata* outside of greater understanding of Ayurveda, that an herb has an affinity for the liver *meridian* outside of TCM, or that an herb reduces fluid stasis in those with *phlegmatic* constitutions, is useless information if one does not fully understand these philosophical paradigms. A single word (vata, meridian, phlegmatic) may involve a lifetime's worth of understanding. If all we have gleaned from various paradigms are the words, the dots without the ability to connect them, the medicine cannot be expected to be effective.

I do not believe that one needs to be a master in all such disciplines or in every culture's contribution to the field in order to be an effective herbalist. However,

in this information age, even herbal research requires policing to avoid wild extrapolation based on an isolated molecule delivered to an engineered cell culture. There is a great deal of research of this ilk, and some of it may evolve to be of great value, yet we also see such research fueling reductionistic thinking, rather than the vitalistic thinking that most herbalists consider an essential aspect of deep healing. Nonetheless, I scour the published research for the tiny glimpses of insight that it might afford, as well as to give myself permission to think creatively and boldly to synthesize the various energetic models I have studied. I have done so out of a sense of necessity and an earnest desire to help my patients, using all that I have learned or been exposed to. I aim to learn from direct experience what I may lack in broad cultural understanding because "life is short and art is long" (Hippocrates) and I don't have decades to continue studying—I have patients sitting in front of me. Preferable, however, is interaction, conversation, and meaningful sharing with other herbalists to mature the craft of the entire profession. Such interactions are a prime reason why attending herbal conferences is so soul-nourishing, and reading the books and following the work of others is so thought-provoking.

As the microcosm of herbal medicine reflects the macrocosm of the broader society, herbalists may have a role to play in helping to shift consciousness. Perhaps acknowledging the fact that Western herbalism is not entirely "Western" after all is one important step in moving the herbal profession forward. It would be easy to place blame for the cultural appropriation occurring in herbal medicine on the society at large. Far better, however, would be to engage in deeper soul searching to see the ways in which we have personally contributed to societal mores, whether deliberate and conscious, or undesired programmed reflexes that are within our power to become conscious of and change. As herbalists, we are privileged to work with people of all walks of life, and our work may also allow us to find ways to continue the conversation and enable cultural evolution.

The New Face of Cosmopolitan Political Herbalism

A "cosmopolitan" plant species is a term used in botany to refer to a plant, such as dandelion, that can be found around the entire globe. Perhaps then, a cosmopolitan herbalist is an herbalist who draws from numerous healing traditions of the world. Ancient China, India, Egypt, and Babylonia all had rich herbal traditions that contributed to present-day herbalism. The Iranian scholar and physician Ibn Sina, better known as Avicenna, was referred to as a prince of physicians and influenced Western herbalism of ancient Greece as early as the eleventh and twelfth centuries. The curanderos of Latin America and healing notions of the many African nations have also contributed spiritual and other ideas to present-day herbalism. Contemporary clinical herbalism has developed through the fusion of many traditions, and maintenance of the underlying philosophical framework is necessary to preserve and hopefully improve upon the efficacy of the medicine. Because of this, there is pressure on modern herbalists to act as interpreters of disparate cultures, while simultaneously needing to bridge the gap between maintaining the vitalistic heart of herbal medicine and the work in the herbal research arena, which is presently seeking to validate the use of herbs via "evidence-based medicine." This moniker aims to bring scientific rigor to the discipline of herbal medicine that fits the paradigm of the modern medical model. Unfortunately, in most cases, those doing the research and looking for the "evidence" are not culturally competent in herbs. That is to say, they don't know the traditional paradigms. And in many cases they are doing the wrong kind of research, attempting to remove a plant from the herbal philosophical tradition—and often a single molecule from the plant—and conducting research that fits the drug model. For example, *Echinacea* was emphasized in traditional indigenous folklore for snakebite, for severe tissue decay and gangrene, and for deep abcesses and disseminated infections. Due to media spin, the plant became popularized as therapy for common colds and as an all-purpose, run-of-the-mill infection herb. Sadly, when researchers investigated *Echinacea* for treatment of the common cold in expensive clinical trials, the results were less than stellar. Yet when herbalists prescribe *Echinacea* for specific situations, using it as the long-standing tradition has established, the results can indeed be stellar.

I have noticed that the media acts as a rather incompetent broker of herbal wisdom, and the populace appears to be a willing consumer. Running an herbal apothecary, I have witnessed the public's fickle buying whims, influenced by whatever Dr. Oz recently said or what a glossy magazine or website is currently featuring. The more simplistic the message and robust the marketing images, the better. I commonly field community requests for specific brands and product names because only those products will "oxygenate the blood," "work at a cellular level," "boost metabolism," and so on.

Behind such requests run the high expectations that the sought-after pill or potion will, itself, do the curing, and thereby removes any personal responsibility for creating meaningful human connection or for addressing lifestyle, diet, stress, or any other obstacles to health. I put a great deal of effort into stocking my apothecary with medicines of the highest quality and serving as a competent community resource available to educate people as to the physiologic reasons for choosing various courses of treatment and to steer them to appropriate medical care and options available to them. But time and again, the temptation of a miraculous medicine without a cultural context or healing paradigm whatsoever is too attractive to be assuaged. The mighty media has put forth an apple of vibrant health to be gleaned from a bottle of pills. It falls to herbalists to intervene in a manner that impacts general knowledge and awareness to maintain the sophistication and specificity of the medicine.

Most of our media streams are controlled by the highest bidder, and Big Pharma is one such bidder—a formidable empire, adept at fearmongering the public to entrust their innate healing capacities to medical authorities. Doctors can even be seen as a modern-day priestly caste, donning white robes, quoting revered texts, and blessing us with the all-important diagnostic code number so that we can receive good favor from the gods at the top of the established system. Empowering ourselves and others to take back power is no easy feat, but learning some rudimentary amount of self-care and herbal medicine skills is a worthy start. Herbalists can help the public to navigate this quagmire and see through the media maelstrom. Integrating the various threads of information and slices of truth falls to alternative medicine practitioners by default. There are no governmental or social agencies doing so while there are many agencies pushing us through the cattle drive of modern medicine.

A challenge for modern-day herbal practitioners is to reweave the failing fabric of the vitalistic paradigm (with whatever energetic or philosophic model that most suits you), while at the same time integrating modern science. In addition, herbalists must discern what plants are ecologically safe to use, and which herbs are medically safe to use in various patient populations (children, pregnant women, diabetics, those with renal or liver disease, etc.). They must understand a plant's molecular constituents and interpret drug-herb interaction data, discern how a plant affects different people and different constitutions, and what portion of the scientific research-based evidence is meaningful and how it may be applied for any given individual given their unique energetic, physical, social, and spiritual complexity. As the herbal profession continues to mature, we must make an effort to give everyone who wishes the capacity to contribute to the conversation and blend the various knowledge bases in culturally competent ways. While the medical model offers impressive procedures and technologies, herbalists and alternative practitioners are essential to the health of our communities by offering the vitalistic and deeply restorative therapeutic approaches that the mainstream lacks.

I believe that some of the environmental, political, and cultural disasters of our present era stem from an existential angst where the security of money and comfort soothes the disenfranchised soul. When one has no tribe, when the family is dysfunctional, where neighbors fence the yard and don't know or interact with one another, and where the media is paid to produce obedient, patriotic consumers, many thinking people experience soul pain and yearn for a more meaningful connection to the world. The lack of such connection can result in a willingness to value personal gain and financial power more than the health of our mother earth. Our collective failure to realize that long-term comfort and survival are entirely dependent on the health of the earth is hard to comprehend. It follows then that cultivating connection may be vital to shifting the societal consciousness. When we are separated from the earth, it makes it easier to value self over tribe and artificial riches over the truly precious riches of the earth.

Herbalists have a natural role in helping community members cultivate real and meaningful connections to the earth and in anchoring medical therapy in the riches that truly sustain us: air, water, food, and human and nature-based connections. And, although I have taken this introduction to an herbal formulary into strangely political arenas, I do so with the goal of contributing to a new philosophical roadmap for the next generations of herbalists.

About This Book

This text is the second in a set of five comprehensive volumes aimed at sharing my own clinical experience and formulas to assist herbalists, physicians, nurses, and allied health professionals create effective herbal formulas. The information in this book is based on the folkloric indications of individual herbs, fused with modern research and my own clinical experience.

I have organized this set of volumes by organ systems. Volume I features the organs of elimination—the gastrointestinal system, the liver and biliary system, the urinary system, and the skin. Herbalists know these organs are foundational to the health of the entire body. The treatment of many inflammatory, infectious, hormonal, and other complaints will be improved by optimizing digestion and elimination. With that information as the foundation, this volume takes a closer look at treating respiratory and vascular issues, including both cardiovascular and peripheral vascular complaints.

Each volume in this set offers specific herbal formulas for treating common health issues and diagnoses within the selected organ system, creating a text that serves as a user-friendly reference manual as well as a guide for budding herbalists in the high art of fine-tuning an herbal formula for the person, not just for the diagnosis. Each chapter includes a range of formulas to treat common conditions as well as formulas to address specific energetic or symptomatic presentations. I introduce each formula with brief notes that help to explain how the selected herbs address the specific condition. At the end of each chapter, I have provided a compendium of the herbs most commonly indicated for a specific niche, a concept from folklore simply referred to as *specific indications*. These sections include most herbs mentioned in the corresponding chapter and highlight unique, precise, or exacting symptoms for which they are most indicated. Please note that these listings do not encompass *all* the symptoms or indications covered by the various herbs, but rather only those symptoms that relate to that chapter—the indications for cardiac and peripheral blood vessels and those for pulmonary and respiratory complaints. You'll find certain herbs repeated in the specific indications section of both system chapters in the book, but in each instance, the description will feature slightly different comments. Readers are encouraged to refer back and forth between the chapters to best compare and contrast the information offered.

The Goals of This Book

My first goal in offering such extensive and thorough listings of possible herbal therapies is to demonstrate and model how to craft herbal formulas that are precise for the patient, not for the diagnosis. It is my hope that after studying the formulas in this book and following my guidelines for crafting a formula, readers will assimilate this basic philosophic approach to devising a clinical formula. As readers gain experience and confidence, I believe they will find that they rely less and less on this book and more and more on their own knowledge and insight. That's what happened to me over the years as I read the research and folkloric herb books and familiarized myself with the specific niche-indication details of a wide range of healing plants. I now have this knowledge in my head, and devising an herbal formula for a patient's needs has become second nature and somewhat intuitive. But from talking with my herbal students over several decades of teaching, I have come to understand that creating herbal formulas is one of the most challenging leaps between simply absorbing information and using it to treat real, live patients. Students often feel inept as they try to sift through all their books, notes, and knowledge and struggle to use "information" to devise a single formula that best addresses a human being's complexities. Thus, I felt that it was high time that I created a user-friendly book to help students refine their formulation skills and to help all readers develop their abilities to create sophisticated, well-thought-out formulas.

Another goal I aim to achieve through this set of herbal formularies is to create an easy-to-use reference that practitioners can rely on in the midst of a busy patient day. In this "information age," it is not hard to track down volumes of information about an herb, a medical condition, or even a single molecule isolated from a plant. The difficulty lies in remembering and synergizing it all. While this text doesn't pretend to synergize the "art" of medicine in one source, I believe it will help health professionals quickly recall and make use of herbal therapies they already know or have read about by organizing them in a fashion that is easy to access quickly.

Naturopathic physicians are a varied lot. Add in other physicians and allied health professionals, and the skill sets are varied indeed. I rely on my naturopathic colleagues to inform me about the latest lab tests, my allopathic colleagues to inform me about new pharmaceutical options, and my acupuncture colleagues to inform me on what conditions they are seeing good results in treating. This text allows me to share my own area of expertise. I have included a large number of sidebars that feature some of the more in-depth research on the herbs and individual molecular constituents, helping to provide an evidence-based foundation for the present era of medical herbalism.

I realize that not all clinicians specialize in herbal medicine, not even naturopathic physicians. I hope that this formulary will serve as a handy reference manual for those who can benefit from my personal experience, formulas, and supportive discussions.

Creating Energetically Fine-Tuned Formulas

Much like a homeopathic *materia medica*, this set of formularies aims to demonstrate to clinicians how to choose herbs based on *specific indications* and clinical *symptoms* and *presentations*, rather than on diagnoses alone. For example, I do not offer a one-size-fits-all hypertension formula. Instead, I share the many situations that I see in my own practice and offer a formula for each "type" of hypertension: Hypertension Associated with Stress, Hypertension Associated with Metabolic Syndrome, Hypertension Associated with Cardiomyopathy, and Hypertension Associated with Long-Term Smoking. I include supportive research on herbs that helps to explain why a particular herb is chosen for a particular formula as well as endnote citations that provide details of specific studies for those interested. I also provide findings from research on individual herbs that are essential to the treatment of the various conditions featured in a chapter. To make the text as useful as possible for physicians and other clinicians, I also offer clinical pearls and special guidance from my own experience and that of my colleagues—the tips and techniques that grab attention at medical conferences year after year.

The Information Sourced in This Book

The source of the information in these volumes is based on classic herbal folklore, the writings of the Eclectic physicians, modern research, and my own clinical experience. Because this book is designed as a guide for students and a quick reference for the busy clinician, the sources and research are not rigorously cited, but enough so as to make the case for evidence-based approaches. When I offer a formula based on my own experience, I say so. I also make note of formulas I've created that are more experimental, due to lack of research on herbs for that condition or my lack of clinical experience with it.

My emphasis is on Western herbs, but I also discuss and use some of the traditional Asian herbs that are readily available in the United States. In some cases, formulas based on TCM are featured due to a significant amount of research on the formula's usage in certain conditions. I readily admit that TCM creates formulas *not* for specific diagnoses, but rather for specific energetic and clinical situations. However, I have included such formulas, perhaps out of context but with the overall goal of including evidence-based formulas, with the expectation that readers and clinicians can seek out further guidance from TCM literature or experienced clinicians where possible. In reality, TCM is a sophisticated system that addresses specific presentations, and I have borrowed from this system where I thought such formulas might be of interest or an inspiration to readers. I admit that listing just one formula for a certain condition based on the fact there have been numerous studies on it is somewhat of a corruption of the integrity of the TCM system, which is aimed at precise patterns and energetic specificity. Nonetheless, I chose to do so with the goal of creating a textbook to help busy clinicians find information quickly, while still encouraging individualized formulas for specific presentations.

While I have endeavored to create herbal formulas to address as many different conditions and presentations as possible, this text purposefully avoids addressing specific types of lung cancer because to do the topic justice would require a textbook all its own. And frankly, there is not yet the evidence to cite, nor do I have the clinical experience in dozens of such cases to feel I could pose enthusiastic herbal formula suggestions. Research is also limited in the arena of treating heart failure with herbal medicine, and this text does not attempt to replace the advice of an expert cardiologist. However, there is much written about herbs for cardiomyopathy and the related symptoms in the folkloric tradition. The historic efficacy of *Digitalis* for treating heart failure, tachyarrhythmias, and dependent edema led to isolation of cardiac glycoside, which is still in use to this day. Alternative medicine seeks to recognize the early contributors to heart disease and intervene as soon as possible, but there are some formulas in this volume that can be effective heart tonics for those with early stage cardiomyopathy.

How to Use This Book

Each chapter in this book details herbal remedies to consider for specific symptoms and common presentations of various diagnoses. Don't feel that you must be a slave to following the recipes exactly. When good cooks create a food recipe, they are always at liberty to alter the recipe to create the flavor that best suits the intended meal—the big picture. A formula listed should not be thought of as *the* formula to make, but rather as a guide and an example, inviting the clinician to tailor a formula for each individual patient.

To create an herbal formula unique to a specific person, the clinician should first generate a list of actions that the formula should perform (respiratory antimicrobial, expectorant, bronchodilator, mast cell stabilizer, and so on), and then generate a list of possible herbal *materia medica* choices that perform the desired actions. If these ideas are new to you, you may want to begin by reading chapter 1, The Art of Herbal Formulation, before you start generating lists.

Look to the formulas in chapters 2 and 3 that address specific symptoms for guidance and inspiration. (These formulas are grouped within the chapter by a general diagnosis, such as "Formulas for Dyspnea" or "Formulas for Hyperlipidemia and Atherosclerosis.") Regard the lists and formulas I have provided as starting points and build from there. In my commentary on the individual formulas and in sidebars that focus on specific herbs, I offer further guidance as to whether the formula or

Unity of Disease (Totality of Symptoms)

The concept that any given health issues a person may experience are actually one disease, as opposed to a number of disparate diagnoses to be treated individually, is a core tenet of naturopathic medicine and the philosophical underpinning of holistic medicine in general. Any one symptom does not provide the full story, and just because you can label the symptoms with a Western diagnosis and offer the established therapy for that diagnosis does not mean you are really helping a person to *heal*. A careful consideration of the sum totality of all symptoms is important to reveal underlying patterns of organ strength or weakness, excess or deficiency states, nervous origins versus nutritional origins, and, of course, a complex overlap of all such issues. The most effective therapies will address *all* issues in their entirety and involve an understanding of the entire energetic, mental, emotional, nutritional, hereditary, situational, and other processes creating a complex web of cause and effect—the unity of any given individual's "dis-ease."

individual herbs are safe in all people, possibly toxic in large doses, intended for topical use only, or indicated only in certain cases of that particular symptom. Once herb and formula possibilities have been identified, the reader should then review "Specific Indications" at the end of the chapter to narrow in on choices of which herbs would be *most* appropriate to select and to learn more about how those herbs might be used. Herbalists can narrow down long lists of herbal possibilities to just a few *materia medica* choices that will best serve the individual. In many cases, the reader/clinician will be drawing upon herbal possibilities from a number of chapters and organ systems as the clinical presentation of the patient dictates. Thus you are not making a formula by throwing together all the herbs listed as covering that symptom or symptoms, but you are studying further and narrowing down the list of possibilities to consider based on the sum totality of all the symptoms. In some cases, you will rule out herbs on the list for a particular symptom after reading the specific description of that herb at the end of the chapter. In some cases, you might decide to put one herb in a tea and another in a tincture due to flavor considerations. In other cases, you might decide that you will prepare only a topical remedy. And in other urgent situations, you might come up with a topical, a pill, an herbal tea, *and* a tincture to address the situation as aggressively as possible. Aim to select the best choices, and avoid using too many herbs in one formula. Larger doses of just a few herbs tend to work better than smaller doses of many herbs, which can confuse the body with a myriad of compounds all at once. The use of three, four, or five herbs in a formula is a good place to start; this approach also makes it simpler to evaluate what works when the formula is effective as well as what is poorly tolerated, should a formula cause digestive upset or other side effect.

Learning from the Formulas in This Book

In reviewing the formulas in this book, notice how specific herbs are combined with foundational herbs to create different formulas that address a variety of energetic presentations. There are a handful of all-purpose immune modulators, all-purpose alterative herbs, and all-purpose anti-inflammatories that can be foundational herbs in many kinds of formulas. Such foundational herbs can be made more specific for various situations by combining with complementary herbs that are energetically precise. Notice how the herbs are formulated to be somewhat exacting to address specific symptoms and make a formula be warming, drying, cooling, or moistening and so on. Also, note how acute formulas may have aggressive dosages and include some strong herbs intended for short-term use, while formulas attempting to shift chronic tendencies are dosed two or three times a day and typically include nourishing and restorative herbs intended for long-term use. Also notice how some potentially toxic herbs are used as just a few milliliters or even a few drops in the entire 2-ounce (60 milliliter) formula. These dosages should not be exceeded, and if this is a clinician's first introduction to potentially toxic herbs, further study and due diligence are required to fully understand the medicines and how they are safely used. Don't go down the poison path without a good deal of education and preparation. I am able to prepare all of the formulas in these texts upon request, but I can only offer those containing the "toxic" or cardiac glycoside-containing herbs (*Atropa belladonna*, *Aconitum*, *Convallaria*, *Digitalis*, and so on) to licensed physicians.

It is my sincere hope that this book helps you in your clinical work and efforts to heal people.

DR. JILL STANSBURY

— CHAPTER ONE —

The Art of Herbal Formulation

Creating an effective and sophisticated herbal formula is somewhat of an art, and like all art, it is difficult to put into step-by-step directions; however, this is my attempt to do just that. This book aims to explain how to create specific formulas for *presentations* rather than *diagnoses*—rather than offering a single formula or two for a general condition such as hypertension or irritable bowel syndrome, this text offers exacting formulations to best address the precise presentation of the person. I have personally seen many *different kinds* of hypertension, dyspepsia, cystitis, and other conditions that conventional medicine tends to treat with one across-the-board medication or therapy. My aim is to coach readers on how to create numerous finely tuned herbal options for treating the person and not the diagnosis—a core tenet of natural medicine.

Creating effective herbal formulas and treatment plans requires many skills: knowledge of the herbs; herbal combinations best for a particular situation; the proportions to use in a formula; the starting dose, frequency of dose, and how long to dose; what form of medicine is best, such as a tincture, a tea, a pill, or a topical application; broader protocol options that may include diet, exercise, nutritional supplements, or referral to allied health professionals; and the follow-up plan.

Hippocrates said, "It is more important to know what kind of person has a disease than to know what kind of disease a person has." To know what sort of person has a disease requires careful listening and skilled and nuanced questioning in a safe and comfortable setting (see "Asking the Right Questions" on page 10). Listen for underlying causes, for overarching emotional tone, for what a person is able to do for themselves (sleep more, exercise more, eat better, brew a daily tea), and for what they are resistant to (sleep more, exercise more, eat better, brew a daily tea). Address underlying causes and start where people show some interest and capacity. Cheerlead to instill enthusiasm if required, so that people become better educated and thereby better motivated

to make important changes or adopt valuable healing practices. Giving the right medicine is only one aspect of doing healing work with a person; creating a sacred space that invites truth and sharing is key to getting to the point where you know what sort of person has a disease.

Healing also stems from understanding pain, suffering, challenges, and unique situations, from non-judgmental listening to the stories so often linked to our physical ailments. It comes from giving encouraging words and sympathy, congratulating and complementing people's efforts and accomplishments, and giving people the tools, resources, and support to succeed. This kind of true caring plus an earnest effort to provide real support are among our best medicines.

The Importance of Symptoms

Naturopathic medicine has a different philosophical stance on physical symptoms than allopathic medicine, especially infectious and eruptive symptoms, and that view is worth briefly describing here for those who may be unfamiliar.

The Biochemical Terrain

The terrain of the human body invites microbes specific to the chemical composition of the ecosystem, and when infectious microorganisms have consumed those specific chemical substrates, the disease-producing microbes are no longer supported. The microbes themselves change the chemical composition of the tissue by consuming the "food" that invited them, and when those nutritional resources are exhausted, the biochemical makeup of the tissues is changed for the better. Antibiotics are not the best way to treat chronic infections. Instead, optimize the ecosystem to support the desired beneficial flora. As when bacteria start to stink up the compost bucket on the kitchen counter, we don't spray the compost with a germicide—we clean the bucket! For example, some cases of chronic sinus or lung infections will be improved with dietary changes, such as reducing the consumption of dairy products, because milk and cheese can be mucous forming and offer a hospitable habitat for opportunistic infectious agents. In other cases, chronic allergic reactivity in the airways or in the blood vessels can be improved by optimizing digestion and bowel health and increasing nutrient intake, both of which can reduce hypersensitivity to allergens. Readers will notice the use of alterative herbs to improve digestion and elimination in formulas for chronic respiratory infections. Even though the airways may seem quite far removed from the gut, when the bowels are "leaky" due to disrupted barrier integrity, inflammatory pathways are triggered and can contribute to allergic airway disease.

The Role of a "Healing Crisis"

An acute infection, such as a simple cold, is a classic example of a "healing crisis"—meaning the acute symptoms are actually a part of the body's attempt to heal itself, allowing the infectious organisms to consume the "morbid matter" and thereby restore a healthier ecosystem. The symptoms of the cold—runny nose, fever, loss of appetite—are part of the body's process to heal itself.

Naturopathic philosophy embraces the symptoms of a healing crisis as a triumph of the vital force. Our symptoms serve us and call attention to the imbalances requiring changes. Be thankful when the body has the vitality to manifest a healing crisis. Be concerned when infections, eruptions, and discharges stop, and allergies, autoimmunity, joint pain, ulcers, blood pathology, and so on emerge. These are not healing crises, but rather signs that vitality is being damaged and pathology is becoming deeper and more serious.

Because the symptoms of a healing crisis are the way in which the body can heal itself, such symptoms should not be suppressed, but rather supported and made as tolerable as possible. Only when such symptoms are so severe as to threaten damage to the body should they be suppressed. Use herbs to help a fever do its job, use herbs to help reduce excessive respiratory secretions, or use herbs to quiet a problematic cough—in other words, use herbs to palliate the discomfort involved in righting the ecosystem and to open the emunctories (the organs of elimination).

The Harm in Suppressing Symptoms

Habitual suppression of symptoms over time can force the body to give up its struggle for health. For example, laxatives may relieve constipation temporarily, but ultimately damage normal peristalsis. Stimulants may initially provide energy, but ultimately exhaust the adrenals and nervous system. For example, a pot of coffee may jolt an exhausted person awake, but it will not improve core energy status in the long run and will only further exhaust it. Antibiotics can kill infectious pathogens, but unless the underlying ecosystem that supports such microbes is significantly improved, the same pathogens will be supported a second, third, and fourth time and become resistant to antibiotics if they are given repeatedly, such as for chronic otitis, cystitis, or sinusitis.

Asking the Right Questions

Learning how to ask questions that will elicit relevant information is as much an art form as creating an herbal formula. Aim your questions to gather information in several categories. What follows here is a broad, but not exhaustive, list of questions that can yield helpful information.

Etiology

How long has this been happening?

How did it begin?

How has it evolved over time?

What else was going on in your life at the time this began?

Have any other symptoms, complaints, problems accompanied this complaint?

What is your health history?

Any previous episodes? Related pathologies? Family members with this complaint?

What is the predominant emotion associated with this complaint? Did the complaint begin during a period of grief? Anxiety? Ambition?

Did the complaint begin following hard labor or an injury, stress, eating a new food, traveling out of the country, starting a new medication (and so on)?

Quality and Occurrence of the Complaint: "PQRST"

Provocation. Does anything seem to bring on the complaint? Does anything alleviate it? What makes the complaint better or worse?

Quality. What is the character of the complaint: burning, throbbing, dull, sharp, shooting, aching, and so on?

Radiation. Does the pain travel? Does this symptom affect any other organ? Is this complaint associated with any other symptom?

Severity. On a scale of 1 to 10, how severe is this? Does it interfere with sleep, activities, work, sex, relationships, child rearing, creativity, and hobbies?

Time. Timing throughout the day, throughout the month (hormonal fluctuations?), throughout the year (seasonal allergies? seasonal affective disorder? Oriental or Native American concepts of seasons?). Timing may involve an association with eating food or going without eating, an association with anxiety/relaxation, an association with menstrual cycle, or an association to a certain environment or allergen exposure. Is the complaint any better or worse with sleep?

Concomitant Symptoms

Do you feel hot or cold when this occurs?

Is there lethargy or anxiety, heart palpitations, or weakness?

Does it occur during times of stress and activity, during sleep, or after prolonged sitting?

Is there fear or fatigue, mania or depression, weakness or restlessness (and so on)?

Is there a desire to be consoled or to be left alone?

Constitution and Energetic Considerations

Note the constitution by asking the right questions and observe the person to get additional clues as follows:

Complexion. Pale? Yellow? Cyanotic? Flushed? Haggard? Dry? Damp? Inflamed? Quick or slow to perspire? Oily or dry skin?

Pulse. Strong or weak? Fast or slow? The quality? The variability?

Tongue. Large or small?

Muscles. Well developed, overdeveloped, underdeveloped? Spastic, atrophic? Soothed by pressure and massage or aggravated by pressure or massage?

Senses. Hyperacute includes sensitivity to odors, noises, bright light; pain; racing thoughts. Hypoacute includes loss of taste, hearing, sensation; poor ability to concentrate, remember, respond.

Diet and Appetite. Hungry, anorexic? Can eat large quantities or tolerate only small amounts? Hypoglycemic symptoms? Unusual cravings for a particular flavor or food? Unusual aversion or aggravation by particular foods or flavors? Are the symptoms better following meals, or between meals when the stomach is empty? What is the general diet, nutrient intake, bowel habits, and quality of the stool?

Thirst. Large thirst versus small or no thirst? Thirst due to dry mouth? Thirst due to compulsion? Thirst for large gulps or thirst for small sips? Thirst for warm fluids or thirst for cold fluids?

Sex drive. High, low? Markedly cyclical?

Sleep. Requires more than 8 hours of sleep to feel rested, able to function with little sleep? Sleeps soundly all night, or wakes many times? Takes a long time to fall asleep but then sleeps well, or falls asleep readily but wakes at a particular time unable to sleep further? Eating before bed disturbs or improves sleep? Restless during sleep or wakes with a jolt? Has nightmares or difficult dreams?

The "Triangle" Exercise

Once you have listened thoughtfully and well to a description of symptoms, I recommend starting with a traditional and simple method of thinking through the choice of the components of a formula: *the triangle*. Visualize a triangle with a horizontal base and two slanting sides that meet at a point at the top of the triangle. Like the triangle, your formula needs a *base* on which to *rest*—that's where you'll start—and it also needs two axes that "point" the formula in the right direction.

In general, the herb or herbs chosen as the base should be nourishing and nontoxic—something tonifying, restorative, alterative, adaptogenic, or nutritive to the main organ system, tissue, or issue of concern. For example, cardiovascular formulas might have *Crataegus* as a base herb on which to rest. A formula for insomnia with exhaustion might rest on *Withania*, and a skin condition formula might rest on *Calendula* or *Centella*. There is an old Wise Woman saying that all healing begins with nourishing. Everyone needs nourishment and tonification, so such herbs are always appropriate. Simple, right?

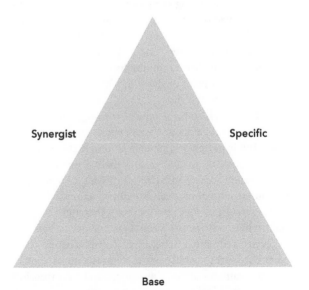

Figure 1.1. The Triangle—A Pragmatic Tool for Crafting Herbal Formulas

From there, herbs for the other two axes of the triangle are selected to point the formula in a more specific direction. The base herbs tend to be nonspecific and indicated in a very wide number of clinical situations, so to make your formula more specific for an individual patient, you will need to drive it in the right direction. Through the ages, herbalists have used varied vernacular to refer to such herbs as *synergists*, *specifics*, *energetics*, *kickers*, *directors*, and so on. The choice of names does not matter, but for the sake of discussion, I use the words *synergist* and *specific* to apply to the other two axes of the herbal formula triangle. In Traditional Chinese Medicine (TCM), royal terms such Emperor, Minister, and Assistants may be used.

The choice of a *specific* requires a detailed understanding of *materia medica*, such that one or two energetically and symptomatically precise herbs can be selected. The lists of herbs offered in "Specific Indications" at the end of each chapter help provide exacting details of symptoms for which individual herbs are most indicated. If there is more than one specifically indicated herb, use them both! If one tastes good and one tastes bad, put the tasty one in a tea and put the less palatable option in tincture or pill. The point is to identify one or two specifics that best address all symptoms possible as well as the constitution, energy, and the entire body's strengths and weaknesses.

As an example, if your case mainly involves hypertension, you may choose to rest the formula on *Crataegus*. The unique identifying symptoms are that the woman's blood pressure began going up during menopause and she is also experiencing episodes of tachycardia and heart palpitations. After reviewing the specific indications for various herbs, you might choose *Leonurus* as your specific to point or drive the formula in the right direction and to the right place. If, on the other hand, the patient is a long-term smoker with high cholesterol and vascular inflammation, your specific may be *Allium*. And if the hypertension formula is for a highly stressed Type A personality who also suffers from alarming hypertension accompanied by a throbbing headache, an appropriate choice of specific might be *Rauvolfia*.

A *synergist* is less specific but takes into account underlying contributors such as other organ systems involved and other energetic considerations that contribute to the overall case. To follow up on the hypertension examples above, the menopausal hypertension formula might include *Actaea* as a synergist to also offer hormonal and nervine support. The long-term smoker may benefit from *Ginkgo* as a synergist to combat circulatory stress, and the Type A severe hypertensive might benefit from *Piscidia* as a vasodilating synergist.

The "Specific Indications" in chapters 2 and 3 offer details about individual *materia medica* options to help choose specifics and synergists to best address the sum total of the symptoms. A synergist may be very "specific"—it's all semantics. The point is to take into consideration various individual factors and unique presentations that contribute to the case.

Returning to the example of the menopausal woman with hypertension and episodic heart palpitations, if she also was suffering from stress and poor sleep as underlying contributors, you might select *Withania* as a synergist. If, on the other hand, she was sleeping fine and did not feel particularly stressed but was having some difficulty with constipation and hemorrhoids, you might select *Aesculus* as the synergist. Or if she was having constipation, hemorrhoids, and episodes of heavy menses as part of the menopausal transition, you might select *Hamamelis* as the synergist. Review "Specific Indications" at the end of each chapter to start developing a familiarity with each individual herb and its "personality" or specific symptoms that it best addresses.

Whatever the details you are presented with, you will use the model of the triangle and select at least three herbs: a nourishing base (tonic), a synergist, and a specific. There is no reason not to select more than one nourishing base or more than one specific or synergist if something very appropriate pops out. If, for example, the menopausal woman with hyptertension had a long-standing history of anxiety, stress, mood swings, and emotional lability well before the onset of the recent hypertension, then her nervous system may need nourishment and tonification as much as, or more so than, her cardiovascular system. In that case, you might choose two nourishing herbs upon which to rest the formula—one such as *Crataegus* for the vascular system and one such as *Avena* for the nervous system. Or if she had a history of these stress-related symptoms and they also tended to cause irritable bowel and diarrhea on many occasions, then you might choose *Crataegus* and *Matricaria*, which provide a nourishing base for the vascular, nervous, and digestive systems.

Therefore, when you reflect on all the details of a case, other symptoms often emerge that are not the chief complaint, but are highly important accompanying considerations that should not be overlooked. A simple presentation of perimenopausal hypertension

More about Formula Components

Base. Also called the lead herb or the Director, the base should be a tonic that has a nourishing and restorative effect on the main organ system affected. This herb does not require a great amount of skill to choose as it is often among the herbs best known for having an affinity to a particular organ system.

Synergist. Also called the Adjuvant, the Balancer, or the Assistant, this component helps correct or complement the action of the lead herb and helps drive it to the desired tissues. The selection of these herbs requires an in-depth understanding of the case and person being treated in order to address underlying causes and give the formula the needed energetic specificity.

Specific. Also called the Kicker or the Energetic Specific, this component is selected not just for a specific condition or diagnosis, but for a specific quality, essence, or expression of any given disease or disorder. Such qualities include the pulse, the tongue, the affect, the pathology, the etiology, and the person themselves to guide the selection of the most specific medicine. The practice of learning and using specific medications is gleaned from careful study, observation, and clinical practice.

with episodic heart palpitations and no other symptom details may lead to the formula as described above: *Crataegus*, *Leonurus*, and *Actaea*. However, if you learn that this woman has had gallstones, a history of fat intolerance, and many digestive symptoms and that the heart palpitations are worse after meals, you might choose *Curcuma* or *Silybum* as synergists because biliary congestion may contribute to hypertension and, for that matter, may deter the processing of hormones and postprandial lipids. This is another example of an important underlying factor that should not be overlooked. Or, if

you find that she is a highly allergic person who sometimes has wheezing with chemical exposure, occasional hives, and chronic low-grade eczema on her hands that flares up after a day of heavy cleaning with exposure to a lot of water and cleaning products, you might choose *Angelica* as a second specific because it is specific for hypertension, asthma, eczema, and hives.

Mastering the Actions of Herbs

Herbal clinicians should have an excellent grasp of primary actions while gaining a solid knowledge of *materia medica* and specific indications. Clinicians should know which herbs are the best antispasmodics for a variety of situations, the best anti-inflammatories, the best vulneraries, the best nervines, the best antimicrobials, and so on. Such actions of herbs are also foundational considerations when creating an herbal formula or when considering which herbs to select as a tonic base, synergist, and specific of a formula triangle.

Another exercise that I often encourage my students to undertake is learning basic categories of actions of herbs. Actions include antispasmodics, antimicrobials, carminitives, alteratives, adaptogens, demulcents, vulneraries, and so on. I encourage my students to type up pages or create a "little black book" that helps to remind them of the hundreds of herbs they are learning, organized as to their categories of action. And from there, they can go deeper. For example, individual antispasmodic herbs might be categorized as having an affinity for a certain organ system or being best suited for a particular quality of spasm. Consider the following antispasmodics: *Lobelia* is especially indicated for respiratory smooth muscle and cardiac muscle spasms, *Dioscorea* is specific for twisting and boring muscle spasms about the umbilicus, *Piper methysticum* can allay acute musculoskeletal pain and urinary spasms, and *Viburnum opulus* or *prunifolium* is especially effective for spasms of the uterine muscle. Sidebars throughout this book offer quick reference lists that summarize various actions.

It is best to avoid a "What herbs are good for bronchitis?" or "What herbs are good for hypertension?" style of creating herbal formulas. When treating bronchitis, for example, be it a dry hacking cough or a cough with abundant free expectoration, rather than asking what herbs are "good for bronchitis," I encourage clinicians to ask themselves, "What *actions* do I need this formula to perform?" In this case, we need to tailor the formula to the individual and select herbs with specific desired actions—expectorant *action* for a wet cough or

Supporting Vitality Instead of Opposing Disease

Western medicine has its "differential diagnosis" where the presenting symptoms can generate a list of possible (differential) causes, ultimately leading to the diagnosis. Although this approach has some value, herbal medicine is less concerned with the formal diagnosis. Instead, herbal medicine aims to carefully consider organ system strengths and weaknesses, underlying causes (stress, toxicity, poor nutrition, poor sleep, circulatory weakness, inflammatory process, allergic hypersensitivity, and so on), and how to support and nourish basic organ function and systemic vitality. Medicines employed by herbalists are typically aimed at restoring function and supporting the innate recuperative powers of the body. The intelligent wisdom of the body to heal itself is sometimes referred to as the "vital force," and herbalists aim to support the vital force more so than to oppose the symptoms of disease. Almost all herbs offer at least some nutrition, being more like foods than drugs, and by nourishing the body and stimulating the vital force, the body is supported in healing itself.

list several herbs that perform this action. The next step is to consider any specific indications for the listed herbs to help narrow in on the best choices. And finally, the most nourishing herbs may be chosen as supporting the foundation of the triangle, a base on which the formula may rest. Other herbs from the action lists may then be chosen as specifics or synergists to best offer all the needed actions, while creating energetically specific, finely tuned formulas.

Energetic Fine-Tuning

Using the triangle method and an awareness of the actions of herbs, as detailed above, will assist you in selecting three or more herbs that address a case in its entirety, and as such, the formulas are likely to be effective and successful. To fine-tune your formulas even further, an added tier of specificity is the energetic state of your patient. TCM philosophy often depicts the energetic state as a mixture of polar opposites in keeping with the Taoist philosophy of yin and yang polarities. For example, is your patient hot or cold? Tight and constricted or loose and atrophic? Excessively damp in the tissues or excessively dry? Tired and lethargic or energized and manic—and so on. Ayurveda, the traditional medical system of ancient India, sets up a three-pronged system of doshas—vata, pitta, and kapha—rather than the two-pronged polar opposites of Taoism but is similarly aimed at addressing differing constitutions and energetic presentations. The four-elements theory of ancient Western herbalism looks for symptoms or presentations categorized into earth-, air-, fire-, and water-related symptoms, with herbal therapies being chosen accordingly. Again, the precise system, vernacular, and approach do not matter, as long as you are aware of some sort of energetic presentation. Whether or not you take it upon yourself to learn the doshas, TCM, or four-elements thinking and prescribing, you can still begin to notice whether a patient is hot or cold, damp or dry, for example, and choose herbs based on the specific clues or symptoms that you discern through thorough questioning and from simple and obvious physical exam findings.

a demulcent *action* for a dry cough. We may want herbs with an antispasmodic *action* on respiratory smooth muscle, and we may want herbs with strong antimicrobial *actions* for bronchitis due to respiratory viruses and other microbes. Throughout this text, the comments that accompany each formula mention actions we are attempting to accomplish with the recipe. For example, "*Lobelia* is included here as a reliable bronchodilator and respiratory antispasmodic," or "*Lomatium* is one of the most well-known respiratory antiviral herbs."

When thinking through an herbal formula, especially for difficult or complex cases, it is useful to write down what actions you wish the formula to perform and then

For example, a hypertension patient may present with episodes of elevated blood pressure related to ongoing job and family stress. The person also suffers from insomnia, frequent muscle tension headaches, and mild digestive upset, also stress related. A hypertension formula for this case might be based on the nervine vasodilators, such as *Passiflora*, *Tilia*, or *Rauvolfia*, resting on a base of *Avena* as a nervous system trophorestorative. The actions desired

Pharmacologic versus Physiologic Therapy

Pharmacologic Therapy. Pharmacologic prescribing is the use of a potent medicine to force a rapid pharmacological response. The energy to catalyze changes, movement, and homeodynamic balance seems to come from the medicine itself. The chemical constituents have strong actions in the body and act as cardiosedatives, emetics, antibiotics, vasoconstrictors, antispasmodics, diuretics, and so on. Pharmacologic medicine can be used heroically and can save lives, but it doesn't build the vital force and restore organ function, plus its use may be needed repeatedly. Most pharmaceutical drugs are pharmacologic in nature and are often suppressive to the body's vital force—for example, acetaminophen to suppress a fever, antibiotics to kill pathogens, or bisphosphonate drugs to halt bone cell turnover rates.

Some of the more toxic herbs can have pharmacologic activity, but in general, herbs are more like foods, offering nutrition and physiologic support. Pharmacologic medications do very little to nourish, tone, or deeply "cure" anything; the symptoms return as soon as the medication is removed. For example, steroids may suppress wheezing in the lungs or suppress eczematous skin lesions, but because they do not alter the underlying condition, wheezing and itching will recur as soon as the medications are stopped. At times, for serious and acute situations, an herb with a pharmacologic action may be chosen as a synergist or a specific in an herbal formula, but never as a lead herb upon which to base a formula.

Physiologic Therapy. Physiologic prescribing involves the use of gentle medicines over an extended period of time to nourish organs and restore normal tone and function. Physiologic medications do not have rapid or strong pharmacological actions and even if prescribed inappropriately would be unlikely to push the limits of homeostasis or cause undesirable side effects. Physiological medications balance, nourish, and tone, and they help restore optimal physiology through gentle support of assimilation, elimination, detoxification, metabolism, perfusion, and nerve function. Most herbs are physiologic in nature, capable of restoring normal functioning, organ tone, and homeostatic balance. In contrast to the above pharmacologic examples, herbs may be used to support a fever when needed to allow the body to fight infections, or to encourage the body to build new bone cells rather than halt all bone cell turnover. Because physiologic medications repair and restore tissue and organ function, they can usually eventually be stopped as the body becomes capable of maintaining balance without them. Herbs with nutritive physiologic actions should be those tonics upon which all formulas are based.

for stress-related hypertension are vasodilators with anxiolytic and calmative effects. If, on the other hand, hypertension is unrelated to stress, but rather has slowly progressed over a decade related to worsening diabetes and long-term smoking, improving blood vessel tone and elasticity would be desired actions, as well as offering antioxidant and hypoglycemic effects to help protect the blood vessels from high blood sugar and the toxic effects of smoking. In this case, *Ginkgo* and *Angelica* might offer mild vasodilating effects while helping to protect blood vessel walls. *Salvia miltiorrhiza* might be capable of improving circulation damaged by smoking and diabetes, and the formula might rest on a nourishing base of *Crataegus*, whose procyanidins might improve the integrity of vein and artery walls. With these examples, it is easy to see how herbal formulas for hypertension might be very different when the underlying causes are considered and herbs with specific actions are selected.

When I teach, I often present sample cases and lead discussion with my students to explore how to think through the choices of herbs for the base, synergists, and specifics, based on details of the particular patient. It can be helpful to work through two examples of people with the same "diagnosis," such as insomnia or acne, and discover how differently a formula evolves based on the patient's unique constitution, energetics, and symptoms. In the first sample case, I offer an extensive discussion that arrives at a specific formula. Following that, I present a more condensed sample case, with several options for base, synergist, and specific herbs. See how you do at finishing the selection process for the sample cases described below. There is no single right or wrong answer.

SAMPLE CASE: VARIATIONS OF INSOMNIA

Patient 1: This insomnia patient has been exhausted for years, with an accompanying history of chronic hay fever and occasional respiratory infections requiring medical attention. The insomnia is of recent onset and involves not being able to fall asleep for many hours, lying in bed very tired and exhausted, but awake. The person will finally fall asleep and wake in the morning groggy and still exhausted. She is cool with cold hands and feet in bed at night and even during the day. She experiences gas and bloating if she eats raw broccoli and onions. For such a patient, you might consider basing a formula on "energy or chi" tonics, such as *Panax, Eleutherococcus, Astragalus, Rhodiola, Ganoderma, Cordyceps,* or *Oplopanax*. You might settle on a base of *Panax* because it is more warming than some of the others. For the specific, you might select *Astragalus* because it is also a chi and immune tonic, and it is also specific for allergies and respiratory infections that linger and exhaust. The

synergist might be *Zingiber* because it enhances circulation in the cold extremities and improves digestion by warming and stimulating core organs. It could help drive the other ingredients in the formula where they need to go by being a heating, stimulating, and moving herb.

Patient 2: This second patient has suffered from restless sleep for many years, but it is presently worse. The pattern has been to fall asleep readily but to wake after a few hours and then fall back to restless sleep for the remainder of the night, waking briefly every half hour or so, tossing and turning with stiff muscles, and going back to sleep for short stretches only. She has occasional episodes of feeling too hot in bed and rolling away from her partner feeling uncomfortably warm, and taking a while to fall back asleep. The person has come for a consultation because recently, instead of frequently waking and rolling over and going back to sleep, she now lies awake for half an hour, sometimes several hours, before falling back to sleep. She also experiences back and joint stiffness at night in bed and for a short while in the morning upon waking. She reports having only a few minor colds per year, recovers quickly without needing treatment, and doesn't even feel terribly ill with them. She reports having oily skin, still prone to breakouts on the central face, especially premenstrually, and having some minor PMS symptoms, primarily emotional lability and crankiness.

There are many relevant details here. For this restless, stiff insomnia patient, who is excessive in movements, muscle tension, nervous tension, and heat, you want to rest a formula on something cooling and relaxing, particularly to the nervous and musculoskeletal systems, such as *Avena, Passiflora,* or *Scutellaria*. While *Valeriana* and *Piper methysticum* are two of the most powerful herbs specific for both sleep and muscle relaxation, you might choose not to "rest" the formula on them because they are rather hot and could be a problem long term or if put in the formula in too great a proportion. Not only are they heating, but they are not particularly nourishing and never listed in the folkloric literature as nervous system restoratives or daily long-term use herbs. *Valeriana, Actaea,* and *Piper methysticum* could certainly be used in the formula, as long as care is taken to use them as synergists or specifics in smaller proportion than the other ingredients in the formula, while basing the formula on something more restorative to the nervous and musculoskeletal system and cooling to the body, such as *Avena* or *Scutellaria*. Another approach might be to consider

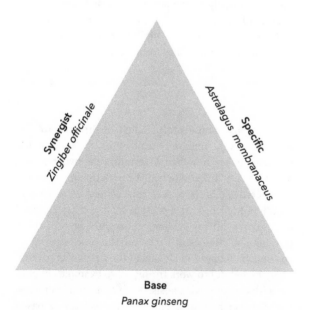

Base
Panax ginseng

Figure 1.2. Formula for Insomnia Patient with Exhaustion

adding *Valeriana* or *Piper methysticum* to a formula for short-term use or to be taken only before bed to reap the benefits of the more powerful muscle-relaxing effects of these herbs. In addition, you would create a separate formula to be used during the day and over the long term, attempting to restore the nervous and muscular tone so that the more powerful muscle relaxers and before-bed

formula are no longer needed. *Scutellaria, Matricaria, Eschscholzia, Hypericum, Passiflora, Avena,* and *Melissa* are cooling, soothing, nourishing, and restorative to the nervous system, with *Scutellaria* and *Passiflora* having the greatest effects on muscle tension as well. Many people with insomnia, who are not particularly exhausted like the first insomnia patient, benefit from sedating relaxing herbs like *Valeriana*, while patients with insomnia and long-term exhaustion might only become further tired or even lethargic and depressed with strong sedatives like *Valeriana* or *Piper methysticum*.

Other considerations or clues for the second insomnia patient are her oily skin, skin eruptions, aggravation by elevated premenstrual hormones, and hormone-related emotional tension. Although these are all very minor and very common in the general population, they are all heat symptoms. These symptoms also suggest that the liver and hormonal metabolism may be contributing underlying factors in this person's overall constitution and balance. The liver not only processes hormones and removes waste products from the blood that can otherwise contribute to acne and skin eruptions, but, in TCM, the liver is said to "rule" the joints and tendons. In many traditions, including TCM, liver herbs are often said to be specific for vague muscle stiffness that does not represent tendonitis, arthritis, fibromyalgia, or other condition. Therefore, a good synergist for this case might be a cooling liver herb noted to improve skin and hormonal balance. Some choices here might be a simple alterative such as *Taraxacum, Arctium, Silybum,* or *Curcuma*. Thus, for the second insomnia patient you might end up with a formula such as *Hypericum, Passiflora,* and *Curcuma*, to be taken multiple times per day for many months, with a before-bed formula of *Avena, Piper,* and *Valeriana*. Yes, there is some arbitrariness in the selection of herbs, but only to a degree. The arbitrariness stems from a somewhat capricious choice between herbs that have very similar energies and specific indications, or mixing and matching formulas in such a way that the overall base and specific energy is the same. It is possible to create several variations of a formula having nearly identical action. For example, both *Taraxacum* and *Arctium* roots may be somewhat interchangeable as an alterative base in a formula. Either fennel or caraway seeds may offer interchangeable carminative effects in an IBS formula. *Valeriana* or *Piper methysticum* might both be effective in the above insomnia case. Or a synergist of *Curcuma* or *Silybum* might both be logical to help the liver process hormones.

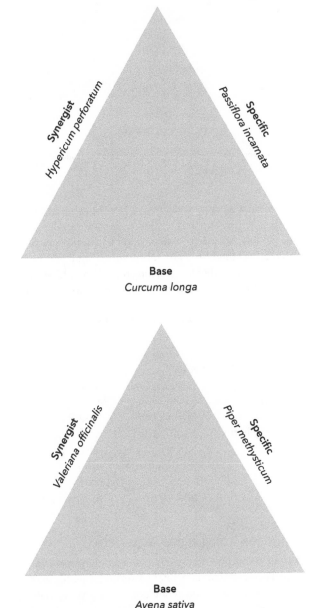

Synergist
Hypericum perforatum

Specific
Passiflora incarnata

Base
Curcuma longa

Synergist
Valeriana officinalis

Specific
Piper methysticum

Base
Avena sativa

Figure 1.3. Two Formulas for Restless Insomnia

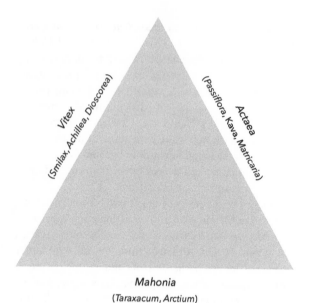

Mahonia
(*Taraxacum, Arctium*)

Figure 1.4. Formula Possibilities for Acne with Fiery Symptoms

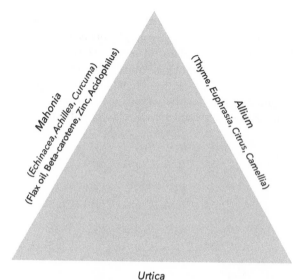

Urtica
(*Taraxacum, Calendula*)

Figure 1.5. Formula Possibilities for Acne with Dampness and Infections

SAMPLE CASE: VARIATIONS OF ACNE

Patient 1: The first case is a 25-year-old woman with acne. History reveals frequent heavy menses, PMS including menstrual headaches, breast tenderness, and significant anxiety, along with chronic constipation as concomitant complaints. She rarely suffers from infections but when she does, they will come on quickly with painful sore throats, high fevers, and acute illnesses. She will recover quickly in a day or two. She also has frequent episodes of insomnia and muscle pain in the neck and upper back, most often premenstrually.

Figure 1.4 shows possible herbal combinations based on these considerations of the patient's constitution, action of herbs, and energetics.

4 Elements	Actions of Herbs	Energetics
Seek water to balance fiery symptoms	Seek alteratives, cholagogues	Seek cooling
Seek earthy therapies and herbs to ground the volatile tendencies	Seek hormonal-balancing agents	Seek sedating
	Seek nervines	Seek yin tonics
	Seek detox	

Patient 2: The second case is also a 25-year-old woman with acne, but her history reveals minor allergies, hay

fever, frequent upper-respiratory infections, occasional yeast vaginitis, and tetracycline use for several years. She is chilly, has a damp constitution with much mucous production, tends toward loose stools, and has low-grade infections that linger a long time. Figure 1.5 shows possible formula combinations appropriate for this case.

4 Elements	Actions of Herbs	Energetics
Seek warming herbs to balance cold	Seek immune stimulants	Seek chi tonics
Seek drying herbs to balance damp	Seek antimicrobials	Seek yang tonic
Seek fiery herbs	Seek intestinal, vaginal, and respiratory astringents	Seek warm and drying tonics
	Seek antiallergy herbs	Seek to purge fluid and dampness
		Seek to move medicine to head, skin, upper respiratory tract

SAMPLE CASE: VARIATIONS OF RHEUMATOID ARTHRITIS

Patient 1: Our first example is a 50-year-old woman with a chief complaint of rheumatoid arthritis, primarily in the hands, wrists, neck, and shoulders. Onset was associated with conflicts and issues with children, loss of control in influencing children's lives, and disappointments. She reports minor anxiety and frequent bouts of insomnia. She also suffers from frequent constipation, a chronic cough due to a dry scratchy throat, and occasional brief episodes of cystitis. Her hands become red and swollen in acute episodes that are experienced as aching, burning, sore, and tender and then settle down over a month's time. Her neck becomes tight and spastic, with a throbbing headache and burning sensation on the skin. Symptoms wax and wane independently of diet or activity, but perhaps correlate to stress. She has a trim build, is often warm, is often thirsty, has a big appetite and a fast metabolism, and is very active. See Figure 1.6 for formula possibilities for this case.

4 Elements	Actions of Herbs	Energetics
Seek to cool fire	Seek nervines	Seek to cool and moisten
Improve dryness, heat with watery, cooling herbs	Seek anti-inflammatories	Aim to soften, lubricate, and smoothe
Improve insomnia, restlessness, and worry with grounding, earthy herbs	Seek antispasmodics	Seek to quiet and calm energy
Aim to ground with moist earthy herbs	Seek tissue lubricants	Aim to move medicines to nerves
	Seek nerve and connective tissue tonics	

Patient 2: The second example is a 50-year-old woman with a chief complaint of rheumatoid arthritis—the arthritic pain is in multiple joints including the low back, hands, wrists, shoulders, and hips. There is a family history of rheumatoid arthritis and osteoarthritis. Constant mild to moderate stiffness is worst after sleeping or prolonged inactivity and also is aggravated with exertion. She has some minor arthritic and degenerative changes in the low back and some bony deformity beginning in the finger joints. The sensation is heavy, stiff, and aching and is better when resting the affected limb. She tends to be chilly, has chronic postnasal drip, is on hormone replacement therapy, retains fluid, has mild constipation, is slightly overweight, and reports low energy. Figure 1.7 shows formula possibilities for this case.

4 Elements	Actions of Herbs	Energetics
Seek to dry out excess water	Seek anti-inflammatories, anodynes	Seek warming and drying
Seek to warm, fire	Seek to move fluid, diurese	Seek stimulating, moving therapies
Seek to lift, lighten with airy or volatile compounds	Seek connective tissue tonic	
	Seek alterative detox therapies	

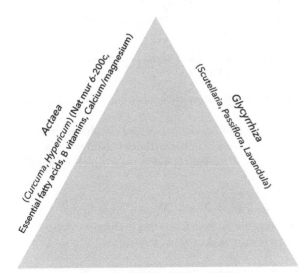

Piper methysiticum

(*Avena*, Gotu kola, *Althaea*, *Symphytum*)

Figure 1.6. Formula Possibilities for Rheumatoid Arthritis with Stress

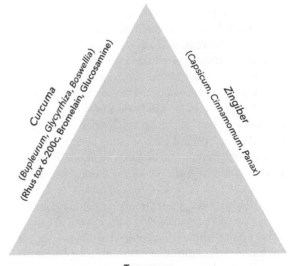

Taraxacum

(Gotu kola, Nettles, *Equisetum*)

Figure 1.7. Formula Possibilities for Rheumatoid Arthritis with Cold Constitution

The Use of "Toxic" Herbs in Formulas

In the sample case of insomnia with exhaustion where *Zingiber* is used as a synergist because of its heating and moving energetic qualities, *Zingiber* might be used in a lesser proportion than the other two herbs, in order to ensure the formula is overall nourishing, restorative, and tonifying. *Zingiber* is not particularly toxic, but due to its strong energy, there is a minor concern over its proportion in a formula. This is especially true for caustic, irritating, or outright toxic botanicals. Some such toxic or otherwise powerful herbs have strong moving or driving action: *Phytolacca* is a lymph mover, *Gentiana* is a bile mover, *Rhamnus* is an intestinal smooth-muscle mover, *Sanguinaria* is a tissue mover, and *Pilocarpus* is a secretory stimulant. All are used in formulas in a small or lesser quantity than the lead herbs, and yet their presence is equally contributory and as powerful as the other primary ingredients.

The case is the same when energy or motion is excessive, and a goal of the formula is to calm and soothe or to quiet down an excessively hot or hyperfunctioning action in a tissue or organ. *Veratrum*, *Aconitum*, *Atropa belladonna*, *Digitalis*, *Lobelia*, and *Conium* all have powerful relaxing, sedating, diminishing effects on nerve, cardiac, respiratory, and musculoskeletal tissues, appropriate only in small amounts in the overall formula.

In many cases, herbs with such extremely strong energies are added to formulas only when extremely strong imbalances are occurring. In general, the stronger and potentially toxic or harsh the herb, the smaller the dose in the formula. And yet even that milliliter, or as little as five drops, drives the formula and contributes equally to the other herbs occurring at a dose of 10 or even 100 times the amount. The smaller the degree of atrophy/hypertropy or the more minor the degree of imbalance of hot/cold or excess/deficient status of the body, the less the need for any strong energy herbs, and the more the formula can be based on purely nourishing and restorative ingredients for the base, synergist, and specific alike.

In the previous example, the insomnia patient is so exhausted and so cold that a little bit of *Zingiber* to warm up our formula is appropriate. *Zingiber* is quite hot but overall nontoxic and safe, even long term. You would not include *Podophyllum* or *Iris* right off the bat to warm the bowels, and you would not choose *Thymus* or *Allium* (she is sensitive to onions and might react) to warm the lungs and reduce infections. None of those herbs are as indicated or specific as *Zingiber* to drive the formula into the proper locations. Thus, for this insomnia patient, you might end up mixing a formula of 30 milliliters *Panax*, 20 milliliters *Astragalus*, and 10 milliliters *Zingiber* to fill a 60-milliliter, or 2-ounce, dropper bottle of tincture.

Further Guidance on Creating Warming and Cooling Formulas

The actions of herbs—both traditional folkloric and mechanistic—can be organized into simple categories of warming or cooling. Vasodilators, for example, and the hot spicy "blood movers" are generally warming, and demulcent herbs are generally cooling.

Warming stimulants. Here are some general guidelines for the use of warming stimulants in treating acute conditions.

- Use for colds, fever, and chills with acute onset
- Use for abundant mucous, phlegm in the throat and lungs
- Use as diaphoretic for those who fail to mount a useful fever in acute illness
- Use for many common infections of childhood in small doses
- Discontinue therapy when improvement is achieved
- Contraindicated in yang constitutions, for those who feel uncomfortably hot
- Contraindicated in hemorrhage, free sweating, or night sweats

In cases of chronic disease, observe these guidelines for warming stimulants.

- Use for circulatory enhancement, excessive clotting, blood stasis
- Use for fluid accumulation, poor circulation to organs
- Use for chronic inflammation with stiff and swollen character that feels better in hot weather or after a hot bath
- Use for cold hands and feet
- Use for atonic, feeble, sluggish constitutions
- Don't use for high fevers due to chronic debility, rather for weakness and sense of chill

Cooling remedies. These are guidelines for the use of cooling remedies in treating acute conditions.

- Use for acute mucosal inflammations, cool tissues with demulcent anti-inflammatories

Warming Stimulants

ACTION	HERBS
Peripheral vasodilators	*Achillea millefolium* *Allium sativum* *Cinnamomum* spp. *Ginkgo biloba* *Zingiber officinale*
Secretory stimulants	*Armoracia rusticana* *Capsicum* spp. *Iris versicolor* *Pilocarpus jaborandi*
Diaphoretics	*Achillea millefolium* *Capsicum annuum* *Zingiber officinale*
Warm antimicrobials	*Allium sativum* *Curcuma longa* *Origanum* spp. *Thymus vulgaris*
Chi tonics	*Eleutherococcus senticosus* *Panax* spp. *Withania somnifera*

Cooling Remedies

ACTION	HERBS
Demulcents	*Aloe vera, A. barbadensis* *Althaea officinalis* *Ulmus fulva* *Verbascum thapsus*
Bitters/alteratives	*Arctium lappa* *Berberis aquifolium* *Rumex crispus* *Taraxacum officinale*
Diuretics	*Apium graveolens* *Equisetum arvense* *Galium aparine* *Petroselinum* spp.
Cooling antimicrobials	*Echinacea purpurea* *Hydrastis canadensis* *Lomatium* spp. *Mentha piperita*
Astringents	*Geranium maculatum* *Hamamelis virginiana* *Rubus idaeus*
Energy dispersants	*Achillea millefolium* *Galium aparine* *Iris tenax, I. versicolor* *Mentha piperita*

- Use for burning sensations in the throat, intestines, or skin
- Use for tight, dry sensations in the mucous membranes or joints
- Use for acute infections with fever or sensation of heat; cool with antimicrobials and alteratives
- Use for acute toxicity states, joint pain, or headache; cool with bitters, alteratives, diuretics
- Contraindicated for acute disease with chills, abundant watery mucous, consolidation in the lungs

In cases of chronic disease, follow these guidelines when using cooling remedies.

- Use for dry, hot constitutions; use bitters, demulcents, alteratives, and diuretics
- Use for general yang states with warmth, redness, and heat in the body
- Use for tendency to acute fevers, infections, inflammation, and joint pain
- Contraindicated for chronic diseases associated with excessive dampness, coldness, deficiency, and fluid stasis

Types of Herbal Medicines

Herbs are available as teas, tinctures, powders, encapsulations, syrups, solid extracts, and other forms. Following are pros, cons, and indications for the most common types of these herbal preparations.

Herbal Teas

Teas are especially indicated for individuals with digestive and bowel weakness to avoid irritation by pills or sometimes tinctures and to improve absorption of desired substances. Teas are also indicated for urinary ailments where substances in water quickly reach the urinary passages. Demulcents to soothe inflamed mucosal surfaces such as the esophagus, stomach, intestines, and bladder are best delivered via teas. For respiratory complaints, teas are also indicated to help soothe coughs and sore throats, and simply for the heat and steam they offer to help loosen chest congestion.

Teas are also relatively inexpensive and, when of good quality, can be a very effective method of getting nourishing herbs into the body, such as *Crataegus* as a daily circulatory tonic, or *Centella asiatica* to support the connective tissue in the lungs in cases of emphysema or chronic obstructive pulmonary disease. When the desired herbs are particularly unpalatable, tinctures or pills would be friendlier. Or when a high dosage of an herb, action, or chemical is desired, it may be easier to accomplish that with pills or tinctures. Because many respiratory and vascular conditions may be chronic and lifelong, teas can be a valuable part of making an aggressive protocol, where teas, tinctures, pills, syrups, topical applications, and medicinal foods are all used in tandem, such as to help prolong life in cases of heart failure.

Tinctures

Tinctures are especially handy when fresh plant juices are desired for preservation. For the home herbalist, tinctures are a practical way to stock the medicine chest with inexpensive, valuable medicines with a long shelf life. Tinctures will store for many years, whereas dried herbs age as the months go by and should be replaced about every two years to be of medicinal quality.

To any clinical herbalist, another important virtue of tinctures is that they can be combined into formulas as precisely indicated for a given individual, where commercially available pills are a one-size-fits-all formula. Tincture prescriptions can be formulated in an exacting manner to reduce the need to take many different medications to address many different organs, imbalances, and pathologies. Tinctures, when well thought out, can address many different pathologies, levels, and energies all at once.

Pills

Tablets and capsules can be convenient and practical ways to combine herbs with vitamins, minerals, amino acids, protomorphogens, bile, and numerous other diverse substances. Formulas designed for treating everything from prenatal nutrition to congestive heart failure or hypothyroidism are available, and when well indicated they are very helpful. However, unless you are producing your own encapsulations, you are usually limited to prescribe the pill combinations as the proprietor has formulated them. Because they might not be as specific for particular presentations as your own formulas, you might complement proprietary encapsulations with more specific herbs in tea or tincture form as needed to round out the herbal prescription.

Pills are also useful when high dosages are being pushed for urgent health problems or when numerous different medications are being employed at once. Herbal pills may be simply dried and powdered plant material, or they may contain purified and concentrated plant constituents, such as curcumin, manipulated to boost its absorption and assimilation.

Knowing When to Use What Form of Medicine

When there are urgent circumstances, standardized or concentrated medicines might be desirable for their concentration and known potency, such as using standardized *Ginkgo* for serious ischemic disease or *Convallaria* for heart failure or silymarin concentrate for acute *Amanita* toxicity. On the other hand, there is no reason to take expensive standardized *Matricaria* when a dollar's worth of good quality chamomile tea would likely do the trick for a flare-up of indigestion or irritable bowel.

When treating complex and chronic disorders, it is often appropriate to use many different herbs or supplements at once. For example, a protocol might call for ginkgo, garlic, milk thistle, passionflower, and echinacea, along with a multivitamin, flaxseed oil, and some additional beta-carotene, zinc, magnesium, and calcium. A person on a budget (which is almost everyone) would not be able to keep up with such a program for very long if each item had to be purchased as a separate bottle of pills. Not to mention how difficult it is for many people to take handfuls of pills day in and day out. However, if the herbs could be combined in a tincture and a tasty tea and supported with the perfect vitamin and herbal/nutritional combination, the cost and convenience could both be improved.

Preparing Formulas for Use

Teas and tinctures are used throughout this formulary. With regard to herbal teas, as a general rule, delicate plant parts such as flowers are steeped rather than boiled, because the aromas and flavors can be lost or destroyed with vigorous boiling. Green leafy herbs such as nettles, alfalfa, and mint are also gently steeped, but harder, denser plant parts such as roots, barks, dried rose hips or hawthorn berries, and seeds are best simmered to extract the medicinal components. Simply steeping herbs is referred to as preparing an infusion, while gently simmering herbs is referred to as a decoction. Unless otherwise specified, it is ideal to infuse teas for

10 minutes. For herbs that are decocted, they simmer gently for 10 minutes and then are removed from the heat to stand 10 to 15 minutes more. In a few cases in this text, I recommend steeping or simmering for a longer period, when the intention is to liberate minerals or other compounds that are not readily released with simple infusions or decoctions. In some cases, it is recommended that vinegar or lemon juice be added to the water to best extract minerals. In other cases, it is recommended that mucilaginous herbs be macerated in cold water for many hours or overnight before bringing to a brief simmer to best extract mucilage.

A general dosage for a tea is a minimum of 3 cups a day. For an acute or urgent situation, however, such as a chest cold threatening to turn into pneumonia, a more aggressive approach would be recommended, such as preparing an 8- or 10-cup pot of tea with a goal of drinking 1 cup every hour for several days.

Herbal tinctures are also used throughout this formulary with recipes given for the proportions and ratios. Because commercial tinctures are readily available, I do not go into detail discussing the making of tinctures. However, in some cases, herbal oils or vinegars that are not readily available are used in formulas, and in these cases, I offer brief details on how to prepare them. I also provide a few unique methods of preparing formulas, such as using solid extracts to thicken formulas to help them cling to the esophagus, or placing the formula in a spray bottle to use as a throat spray.

A general dosage strategy for tinctures is to use 1 dropperful (½ to 1 teaspoon) three times a day when treating chronic conditions. Most people have difficulty taking medicines any more than this, unless they are acutely ill or uncomfortable and motivation is high. In acute situations, tinctures may be taken as often as hourly for acute pain or infectious illnesses, or even as often as every 10 minutes. When tinctures or other herbal medicines are taken at this aggressive dosage, it is always for a very limited length of time, such as every 10 minutes for an hour, or every hour for a day, with instructions to reduce the frequency as symptoms improve.

Although the focus of the book is to create effective herbal formulas, in some cases, protocols are offered to create comprehensive therapies for chronic conditions such as congestive heart failure or emphysema. In such cases, I may suggest the use of a tea, a tincture, medicinal foods, and herbal encapsulations to treat such challenging conditions; additional examples are circulatory insufficiency related to diabetes or long-term smoking or autoimmune disorders or asbestos exposure causing irreparable lung damage. When herbal capsules are employed, they are often dosed as little as two per day or as much as two or three at a time, three times a day. While teas and tinctures can be taken with or without food, herbal capsules are most often taken with meals to enhance their digestion and to prevent them from being nauseating on an empty stomach.

Extending the Triangle Philosophy

As explained, herbal formulas exemplified throughout this book aim to use nourishing herbs as base ingredients in all formulas, complementing them with specifics and synergists. All the formulas include an introductory sentence or two explaining why the herbs are chosen, cite any research to support its use in specific conditions and situations, or comment on why the formula is appropriate for a specific presentation of the condition being discussed. I encourage readers to refer to "Specific Indications" at the end of each chapter to help master *materia medica* knowledge. Staying focused on the desired actions can help you avoid creating a formula that is too broad. With so many available herbs to choose from, knowledge of niche indications for various herbs helps prevent a "kitchen sink" approach. Instead of choosing every herb you know of that may be effective for hypertension or respiratory viruses, you can select the best choices with a solid knowledge of *materia medica*. It is a common mistake to place too many herbs with the exact same function in a formula and miss including herbs that address the underlying cause or other special considerations. For example, bronchitis could occur when stress, poor sleep, and inadequate nutrition have simply weakened the immune system, allowing the common cold to linger and progress to the lungs. For such a "rundown" individual, lifestyle changes are needed and may be complemented by the immune modulators *Astragalus*, *Ganoderma lucidum*, or *Andrographis*. In other cases, underlying allergic reactivity in the airways may lead to abundant secretions that then become a welcoming breeding ground for secondary opportunistic infections. While antimicrobial herbs may be needed in the acute formula, those with a pattern of such infections may be improved with herbs that address allergic hyperreactivity such as *Petasites*, *Euphrasia*, or *Tanacetum*. Bronchitis may also occur in the elderly during the chilly months of the year where winter colds progress to bronchitis, and warming drying herbs such as *Thymus*, *Eucalyptus*, and *Zingiber* might be useful synergists in teas or tincture formulas. In other cases still, bronchitis may not be infectious at all, but rather due to smoking, exposure to a brush fire, or chemical irritation. In such cases, soothing anti-inflammatory herbs such as *Tussilago* or *Verbascum* may allay symptoms, or powerful antioxidants such as *Curcuma longa* or encapsulated supplements such as resveratrol may speed recovery. In all cases, herbs are chosen NOT for the diagnosis of "bronchitis" but rather for the underlying causes and for the nature of the presenting symptoms. Harsh and spastic coughing requires bronchial antispasmodics, colloquially known as antitussives, such as *Lobelia*, *Prunus*, or *Tussilago*. A tiny cough driven by a dry or tickling sensation in the throat does not require such big gun antispasmodics, but rather soothing demulcent herbs such as slippery elm lozenges or the inclusion of *Verbascum* or *Glycyrrhiza* in a tea or tincture. Over-abundant mucous production can be dried with sage or thyme, while insufficient secretions or scant rubbery tenacious mucous can be thinned and expectorated more easily with *Grindelia*, *Eriodictyon*, or *Armoracia*. Underlying viruses can be addressed with *Lomatium* or *Ligusticum*, and chronic bacterial infections may be treated without antibiotics with the use of oregano oil, garlic, or steam inhalations using essential oils. Once one has a solid knowledge of the *materia medica*, the specific indications of the herbs, prescribing and formulating become somewhat intuitive. This becomes easier, and even second nature, the longer one focuses on making formulas and the more experience one obtains.

Many herbalists believe that the plants are their teachers and that they learn from the plants themselves over time, as the plants teach us what they are capable of. Many herbalists are deeply aligned with nature and often have cultivated nature-based spirituality. Herbalists may embrace the notion that clinical intuition is evidence of a connection with the plants, and that this is the real root of herbal medicine as a healing discipline.

— CHAPTER TWO —

Creating Herbal Formulas for Cardiovascular, Peripheral Vascular, and Pulmonary Conditions

Formulas for All-Purpose Vascular Support	27
Formulas for Hyperlipidemia and Atherosclerosis	32
Formulas for Angina and Coronary Artery Disease	38
Formulas for Hypertension	41
Formulas for Congestive Heart Failure and Cardiomyopathy	47
Formulas for Heart Stress at High Altitudes	52
Formulas for Arrhythmias	52
Formulas for Peripheral Vascular Insufficiency	57
Formulas for Cerebral Vascular Insufficiency	61
Formulas for Capillary Fragility and Telangiectasias	65
Formula for Raynaud's Syndrome	67
Formulas for Anemia	68
Treating Hemochromatosis	70
Formulas for Vascular Infections and Endocarditis	72
Formulas for Venous Congestion and Varicosities	73
Formulas for Vasculitis	82
Formula for Lymphedema	82
Formulas for Cardiopulmonary Disease	83
Formulas for Hypotension	85
Formulas for Impotence	87
Treating Sleep Apnea	88
Specific Indications: Herbs for Cardiovascular, Peripheral Vascular, and Pulmonary Conditions	89

Cardiovascular disease (CVD) has been the number one cause of death for generations,[1] and as many as 30 percent of all deaths worldwide are due to underlying cardiovascular disease. Cardiovascular disease continues to increase in prevalence along with the global diabetes epidemic. CVD results from continuous deterioration of vascular muscle and endothelial cells, deterioration of vascular muscle function and accumulation of lipid-rich atherosclerotic plaque associated with oxidative stress, and chronic inflammation.[2] The major risk factors for developing cardiovascular disease are hypertension, hyperlipidemia, diabetes, smoking, heavy alcohol consumption, lack of exercise, hyperlipidemia, and high blood pressure. Hypertension leads to tissue changes associated with peripheral artery disease, stroke, atherosclerosis, heart failure, and coronary artery disease, and may lead to kidney damage, cerebrovascular insufficiency, and associated dementia and blindness.

Therefore, quitting smoking, optimizing nutrition, exercising, and losing weight where necessary are key approaches to managing cardiovascular disease, but these actions require significant personal effort, self-discipline, and personal responsibility. Hypertension leads to tissue changes associated with peripheral artery

Herbal Flavonoids for Protecting the Vasculature

For the most part, cardiovascular diseases develop gradually over many years, if not decades. Hypertension, cardiac hypertrophy, and atherosclerosis are all long-term processes that can culminate in myocardial infarction and heart failure. In other cases, atherosclerosis in the small blood vessels can lead to peripheral vascular disease including intermittent claudication, circulatory and venous insufficiency, and cerebrovascular insufficiency. While modern medicine has numerous heroic measures for severe acute events and diseases such as coronary bypass surgery, placement of artificial valves, and even heart transplants, dietary and herbal measures are invaluable as preventive measures and to slow the progression of early disease.

Ginkgo contains valuable flavonoids, called the ginkgolides, along with the ginkgo heterosides, a group of terpenes; both are useful to protect the vasculature. *Crataegus* berries contain proanthocyanidins (PCOs), which are bright orange-red flavonoid pigments, considered antioxidant, anti-inflammatory, and nourishing. The Eclectic physicians of the late 1800s and early 1900s purported *Crataegus* to "improve nutrition" to the heart muscle itself, fitting the definition of a cardiovascular trophorestorative. The fruits called "hips" produced by *Rosa canina*, like *Crataegus* berries, contain heart- and blood vessel–strengthening flavonoids. Hawthorn berries and rose hips are both available as solid extracts to prepare medicinal foods, and the dried berries and berry powders are useful in herbal tea recipes. Tinctures and other concentrates are also available to include in a broad variety of vascular protection formulas for use long term to slow the progression of cardiovascular and peripheral vascular diseases.[3] Citrus fruits are also high in flavonoids. Hesperidin occurs in citrus fruits; it may attenuate hyperglycemia and help control neuropathic pain in diabetic neuropathy via reduction in inflammatory cytokines, as well as help to control blood pressure and improve inotropic function in the heart.

This chapter also details herbal therapies to address the additional risks factors of hypertension and endothelial inflammation on the development of cardiovascular disease. You'll also find herbal treatment ideas for treating frank cardiovascular disease, arrhythmias, as well as other issues of the blood and vasculature such as anemia, hypotension, sleep apnea, varicosities, phlebitis and blood clots, endocarditis, and pulmonary vascular disorders. Because some of the main contributors to cardiovascular disease are hyperglycemia and hyperlipidemia, herbal medicines that improve blood glucose and lipid levels are also valuable in a variety of vascular formulas.

Crataegus oxyacantha, hawthorn

disease, stroke, atherosclerosis, heart failure, and coronary artery disease. In addition, it may lead to kidney damage, cerebrovascular insufficiency, and associated dementia and blindness. The pathogenesis of CVDs is multifactorial but involves the accumulation of atherosclerotic plaques within the walls of the arteries. Herbal medicines and medicinal foods may offer significant protection from CVD over a lifetime. Numerous herbs have effects on vascular smooth muscle cells, improving blood vessel tone via effects on the contractile proteins actin and myosin. The main mechanisms of action of the leading hypertension drugs are calcium channel blockade, β-adrenergic blockade, and angiotensin-converting enzyme inhibition. Herbs have been shown to act by all of these mechanisms as well. The herbs used to treat heart disease are the same herbs used to treat hypertension and the same herbs used to treat atherosclerosis. Thus the formulas provided in this chapter become rather repetitive. You'll notice that *Allium sativum* (garlic), *Crataegus* spp. (hawthorn), *Salvia miltiorrhiza* (red sage), and several other herbs are listed again and again, and a formula for treating heart disease may look very similar to a formula for hypertension or atherosclerosis or hyperlipidemia.

Formulas for All-Purpose Vascular Support

Because hypertension, atherosclerosis, hyperlipidemia, and hyperglycemia frequently occur in tandem with one another, it is often useful in clinical practice to have broad-acting therapies that address all of these issues simultaneously. Many of the foundational cardiovascular herbs serve multiple purposes, and high-flavonoid herbs, alteratives, metabolic support, and vascular anti-inflammatories can be combined to create a variety of all-purpose formulas.

Pueraria montana var. *lobata* (kudzu or gegen) is a rich source of polyphenolic compounds, including isoflavones, isoflavonoid glycosides, coumarins, and puerarols, all contributing to hypotensive and vascular protectant effects. Many mechanisms of action have been identified, including blockade of β-adrenergic receptors and angiotensin-converting enzyme inhibition, thereby reducing blood pressure and helping to slow the progression of atherosclerosis. *Salvia miltiorrhiza* (red sage) is long-standing in Traditional Chinese Medicine (TCM) and is now becoming popularized in the West due to the sheer volume of impressive research related to protecting the vasculature. *Salvia miltiorrhiza* has hypotensive and endothelial protective properties and numerous antioxidant, anti-inflammatory, and antiproliferative properties. It is widely used in TCM for treating coronary artery disease, hypertension, diabetes, atherosclerosis, and chronic heart failure. Tanshinone is credited as one active constituent against stroke and coronary artery disease. The salvianolic acids increase cerebral blood flow after ischemic stroke and inhibit thrombosis, via modulating effects on thromboxane B_2 formation and platelet aggregation. *Ginkgo biloba* and the high-flavonoid berries can be included in formulas to protect the vasculature. *Angelica sinensis* (dong quai) has many anti-inflammatory mechanisms in the vasculature and reduces inappropriate platelet activation and aggregation. It also has antiallergy and antihistamine effects. Because these herbs are safe and gentle and work via multiple mechanisms, they can be the foundation of numerous long-term formulas for chronic and peripheral vascular disease.

All-Purpose Cardiovascular Support Tea

There is often the need for herbs tasty enough to use long term to help protect the vasculature for those with risk factors for developing heart and blood vessel disease due to smoking, obesity, hypertension, and elevated lipids or glucose.

Salvia miltiorrhiza
Pueraria montana var. *lobata*
Angelica sinensis
Berberis aquifolium
Arctium lappa
Cinnamomum verum
Zingiber officinale
Glycyrrhiza glabra
Hibiscus sabdariffa calyces
Rosa canina chopped hips

Combine several ounces of each ingredient and store the mixture in an airtight container. Simmer 2 tablespoons mixed herbs in 6 cups of water for 10 minutes. Let stand covered for 10 minutes more and then strain. Drink throughout the day.

Herbs for Vascular Headaches

The underlying causes of headaches range from muscle tension to toxicity states to drug reaction, hypertension, and vascular reactivity. The following herbs may help when headaches are due to hypertension or in allergic people with vascular hyperreactivity.

Angelica sinensis	*Rauvolfia serpentina*
Matricaria chamomilla	*Tanacetum parthenium*
Petasites hybridus	

Ganoderma lucidum for Vascular Support

Reishi mushrooms (*Ganoderma lucidum*) are woody inedible mushrooms with a long history of use in China. They are greatly esteemed as a chi tonic, an agent capable of improving vitality and promoting health and longevity. Reishi is primarily used for immune modulation, but significant research also shows that it reduces toxic and inflammatory reactions in the body.[4] Reishi mushrooms have been shown to lower the risk of cancer and heart disease and to boost the immune system. Reishi may reduce inflammation in the liver, kidney, and blood and thereby help protect the blood and blood vessels.[5] Reishi mushrooms may protect the heart via antioxidant mechanisms,[6] as well as protective effects on individual mitochondria,[7] and support of mitochondrial energy production.[8] Reishi is also shown to protect the heart in animal models of ischemic injury[9] and ethanol toxicity.[10] Many of reishi's medicinal virtues are credited to polysaccharides, which are shown to lower blood glucose and have vascular protective effects in animal models of diabetes.[11]

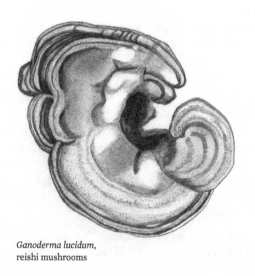

Ganoderma lucidum,
reishi mushrooms

Tincture for Poor Circulation with Poor Metabolism

This tincture can be helpful for those with poor circulation due to hypothyroidism or metabolic insufficiency. The herbs in this formula may promote metabolism and improve lipid profiles via supporting basal metabolic rate. *Commiphora mukul* promotes thyroid function and basal metabolic rate while *Fucus* supplies organic iodides for the thyroid to transform into thyroxine. *Opuntia*, the prickly pear cactus, is a possible complement to this tincture and best consumed as a juice using a tablespoon each day in a medicinal tea or sparking mineral water. Lipid and thyroid profiles may be rechecked in 3 months.

For a more aggressive protocol, complement the tincture with encapsulations containing copper, selenium, iodine, and tyrosine. *Iris* has not yet been robustly researched, but it is traditionally used for hypothyroidism with fluid stagnation.

Commiphora mukul	30 ml
Fucus vesiculosus	15 ml
Zingiber officinale	10 ml
Iris versicolor	5 ml

Take 1 or 2 dropperfuls of the combined tincture 3 or more times a day.

Green Tea Simple for Heart Protection

As little as 1 cup per day of *Camellia sinensis* may reduce the risk of coronary artery disease by providing polyphenols.

Camellia sinensis

Steep 1 tablespoon per cup of hot water and then strain. Consume as often as possible.

Green Tea Chai for the Heart

Chai is traditionally prepared with black tea and spices, but since green tea appears to have greater health benefits, here is an alternative using green tea as the base. Spices such as ginger, cardamom, cinnamon, and pepper create a palatable brew, plus offer many health benefits. Ginger is useful in the treatment of cardiovascular disease, and numerous anti-inflammatory effects are demonstrated, including inhibition of the production of nitric oxide, inflammatory cytokines, cyclooxygenase, and lipoxygenase. Benefits in relation to heart disease include antioxidant, antiplatelet, positive inotropic, hypotensive, hypoglycemic, and hypolipidemic effects. Black pepper (*Piper nigrum*) and cinnamon both have hypolipidemic and cardioprotective effects. *Cinnamomum verum* helps lower blood pressure, improve blood glucose control, and support endothelium nitric oxide and potassium channels.

Camellia sinensis	6 ounces (180 g)
Zingiber officinale	1 ounce (30 g)
Citrus paradisi (grapefruit) peel	1 ounce (30 g)
Elettaria cardamomum pods, chopped	1 ounce (30 g)
Cinnamomum verum	1 ounce (30 g)
Piper nigrum	½ ounce (15 g)

Gently simmer 1 teaspoon mixed herbs per cup of hot water for just 5 minutes. Let stand in a covered pan for 10 minutes more and then strain. Drink as desired.

Citrus Zest to Protect the Heart

Drinking large quantities of fruit juice is not recommended due to the quantity of sugar contained and possible impact on blood glucose and weight gain, but eating oranges, tangerines, and grapefruit is beneficial to health. Naringenin is a primary flavonoid in citrus fruits, especially grapefruit, and is credited with potent antioxidant effects, which is an explanation of why regular consumption of citrus may reduce the risk of cardiovascular disease. Naringenin also has insulomimetic activity, helping to improve hyperglycemia and protect against the development of metabolic syndrome, a condition sometimes referred to as the "deadly quartet" where hypertension, elevated lipids, insulin resistance, and obesity overlap. Naringenin may delay the absorption of glucose into the blood and increase the uptake and utilization of glucose by the muscle cells. Naringenin may also protect against the development of metabolic syndrome, improving glucose tolerance and optimizing insulin sensitivity. Citrus peels also contain vitamin C, calcium, and potassium, which support healthy blood pressure, and limonene, an aromatic terpene credited with anticancer effects. Citrus zest contains naringenin and can be collected each morning and added to oatmeal, the day's drinking water, or salads. Be sure to choose organic citrus to avoid exposure to pesticides. Try adding 1 to 2 tablespoons of orange, grapefruit, or tangerine zest to 1 quart (1 liter) water or prepared herbal tea. Add 1 tablespoon zest to a bowl of prepared oatmeal. Add 2 or 3 tablespoons zest to oil and vinegar or any salad dressing recipe, as much or as little as desired.

Vascular Protection Tea from Traditional Chinese Medicine

The combination of *Angelica sinensis* and *Astragalus* roots have been used at least since the thirteenth century in TCM to treat vascular inflammation such as diabetic nephropathy, pulmonary fibrosis, liver fibrosis, and heart disease.

Angelica sinensis root	½ pound (225 g)
Astragalus membranaceus root	½ pound (225 g)

Combine the two herbs, store in a cool dark place, and decoct 1 teaspoon per cup of hot water. Drink 3 or more cups per day.

Tincture for General Vascular "Heat" and Inflammation

Because platelets harbor histamine and white blood cells harbor cytokines and inflammatory mediators, using herbs that act directly on blood cells may help treat blood vessel inflammation, clotting disorders, and atherosclerosis. Those with atopic tendencies, hives, or allergies occurring in tandem with cardiovascular complaints would especially benefit from herbs such as *Petasites* and *Tanacetum*, both known to reduce the release of inflammatory products from blood cells and have general anti-inflammatory effects. Because *Petasites* contains pyrrolizidine alkaloids, the use of this formula should be limited to a few months' time, which should be sufficient to treat an acute blood vessel issue or hives.

Tanacetum parthenium	30 ml
Petasites hybridus	30 ml
Crataegus spp.	30 ml
Ginkgo biloba	30 ml

Take 1 or 2 dropperfuls of the combined tincture 3 or more times daily.

Umbel Seeds for the Heart and Kidneys

The umbel family, known botanically as the Apiaceae family, is high in coumarins, which have many benefits for the cardiovascular system. The seeds of this family are also high in organic mineral complexes that offer nutritional benefits. Add umbel seeds to foods, herbal teas, salad dressings, and other medicinal preparations.

Ammi visnaga	*Coriandrum sativum*
Anethum graveolens	*Daucus carota*
Angelica spp.	*Foeniculum vulgare*
Apium graveolens	*Petroselinum crispum*
Cnidium monnieri	*Pimpinella anisum*

All-Purpose Vascular Support

General Tincture for "Cold" Vascular Stasis

Those with cold hands and feet, vascular congestion, and a tendency to clots might be said to have "coldness" in the blood in the terminology of various energetic traditions. The use of warming and "blood-moving" herbs is appropriate in such situations. Warming herbs might include *Zingiber* or *Capsicum*, and blood-moving herbs might include *Achillea*, *Ginkgo*, *Angelica*, or *Salvia miltiorrhiza*.

Angelica sinensis	25 ml
Salvia miltiorrhiza	25 ml
Zingiber officinale	10 ml

Take a dropperful of the combined tincture 3 or more times a day.

Tincture for Blood Cell Reactivity

Platelets harbor histamine, serotonin, and other inflammatory substances and may be involved in initiating hives, eczema, and atopic phenomena. Modern research has demonstrated that many herbs used folklorically for hives, asthma, and allergies reduce inappropriate platelet activation. *Ephedra* tincture is no longer on the market and needs to be made for one's self. The formula can also be altered to use 20 milliliters each of the herbs without *Ephedra*. Because *Petasites* contains pyrrolizidine alkaloids, the use of this formula should be limited to a few months' time, which should be sufficient to treat an acute blood vessel issue or hives.

Angelica sinensis	15 ml
Ephedra sinica	15 ml
Petasites hybridus	15 ml
Tanacetum parthenium	15 ml

Take as much as 1 teaspoon of the combined tincture every half hour for acute hives and atopic reactivity, reducing as symptoms subside.

Cocoa Powder for Polyphenols

Unsweetened chocolate, cocoa, and cocoa products provide a substantial quantity of dietary polyphenols, including resveratrol, and modulate cardiovascular risk factors, improving blood flow, platelet activation, and insulin resistance. Chocolate candy does not have this effect due to the added sugar, but there are some quality stevia-sweetened dark chocolate options available. Find ways to use chocolate in the daily diet, such as adding 1 to 2 tablespoons to a bowl of oatmeal, a smoothie, or preparing medicinal truffles by blending dark cocoa powder and a non-caloric sweetener such as stevia, or with honey or maple syrup, into nut butter and rolling the "dough" into balls. The resulting truffles can also be rolled in coconut or sesame seeds. These can help wean those with a sweet tooth on to something healthier than commercial chocolates and candies.

Healthier Mocktails and Spritzers

While drinking the occasional glass of red wine offers cardiovascular benefits, the regular consumption of alcohol has significant detrimental effects on the heart and blood pressure. Having an array of appealing beverages available can help patients eschew alcohol, soda pop, and fruit juice. Here is an example and some alternatives for inspiration. *Punica granatum* (pomegranate) is high in anthocyanins, proanthocyanidins, and other flavonoids with strong antioxidant and endothelial protective effects, and pomegranate juice makes a good base for drinks.

Pomegranate juice
Citrus zest
Sparkling water

Dilute the pomegranate juice with the sparkling water to taste, using anywhere from 1 tablespoon to 1 cup of juice in a large glass. Add citrus zest or other embellishment. Substitute another high flavonoid juice, such as blueberry, cherry, grape, or prickly pear, for pomegranate juice as desired.

Opuntia Mocktail for Metabolic Syndrome

Here is another medicinal beverage based on the insulin-enhancing activity of the nutrients vitamin D and inositol, both shown to improve signal transduction of insulin receptors. *Opuntia* (prickly pear) fruits, hibiscus, and *Stevia* are all tasty herbs that also improve insulin responses, along with antioxidant and vascular protective effects.

1 cup (240 ml) *Hibiscus* and *Stevia* tea, chilled
1 teaspoon inositol powder
1 tablespoon *Opuntia ficus-indica* juice
1,000 IU liquid vitamin D
2 cups (480 ml) sparkling mineral water, chilled

Brew the *Hibiscus* and *Stevia* tea by steeping ½ tablespoon of each herb in a cup of hot water for 10 minutes. Strain and chill in the refrigerator. Dissolve the inositol powder in the *Opuntia* juice in a large glass. Work in a drop of vitamin D using a product where 1 drop provides 1,000 IUs. Vitamin D is a fatty substance and needs to be stirred vigorously to blend in. Add the chilled herbal

Resveratrol and the French Paradox

The so-called French Paradox is the phenomenon whereby the French population eats a high-saturated-fat, rich cuisine, and yet the French have relatively low incidence of cardiovascular disease. The fact that they also consume substantial amounts of red wine is, in part, what allows this. Grapes are high in the polyphenol resveratrol, which is shown to exert potent antidiabetic, antioxidant, and anti-inflammatory actions, all serving to protect against heart disease. Resveratrol (a fusion of the words *resorcinol* and *veratrine*) is a polyphenolic compound also found in vine and bramble fruits, peanuts, chocolate, and *Pinus* species, such as maritime pine bark. In plants, polyphenols are made from coumaroyl-CoA and malonyl-CoA in response to stress, injury, and simple exposure to UV light. Resveratrol gives plants disease resistance. Resveratrol is reported to have antiatherogenic effects and to reduce thromboxane synthesis.[12] In addition, these polyphenols are noted to have antioxidant and lipotropic actions. Resveratrol has also been noted to inhibit platelet aggregation.[13]

Vitis vinifera, grapes; *Rubus* spp., blackberries and raspberries; *Arachis hypogaea*, peanuts

tea and sparkling water. Drink once or twice a day for 3 months or long term.

Breakfast Vegetable Juice for Heart Health

This recipe requires a Vitamix, a centrifugal juicer, or other high-powered juice extractor. *Apium graveolens* (celery) seed medicines and juice from celery stalks may be safe and nourishing for treating hypertension and hyperlipidemia. *Apium* can reduce cholesterol without interfering with general metabolism of hepatic sterols[14] via enhanced biliary excretion of cholesterol and other mechanisms.[15] *Apium* is also noted to have anti-inflammatory effects in the blood,[16] and may help treat hypertension by promoting the excretion of sodium. Beets contain betaine, which can support liver detoxification enzymes and cholesterol metabolism.

1 cup (60 g) coarsely chopped pineapple
½ cup (30 g) coarsely chopped celery stalks
½ cup (30 g) coarsely chopped beets
¼ cup (15 g) chopped onion
1 tablespoon (15 ml) prickly pear juice
2 teaspoons (10 ml) flaxseed oil
1 teaspoon (15 ml) minced ginger root
½ teaspoon (3 g) minced fresh tumeric root
½ teaspoon (3 g) chopped fresh garlic
½ teaspoon (3 g) coconut oil
½ teaspoon (3 g) psyllium seeds

Place the coarsely chopped fruit and vegetables in the Vitamix or juicer and liquify. Add the oils, spices, and psyllium seeds and blend again on high speed. Pour into a glass and drink promptly.

Hyperlipidemia and Atherosclerosis

Formulas for Hyperlipidemia and Atherosclerosis

Dyslipidemia is a risk factor for cardiovascular disease, when elevated low-density lipoprotein (LDL) and decreased high-density lipoprotein (HDL) cholesterol levels are associated with an increased risk of damaging blood vessels, raising blood pressure and contributing to the development of coronary artery disease and the risk of suffering an acute myocardial infarction (MI) or other cardiovascular event. Hyperlipidemia and atherosclerosis are grouped together in this chapter because the two conditions are intimately linked and can be approached therapeutically in identical ways.

Atherosclerosis involves narrowing of blood vessels due to lipid deposition and vascular smooth muscle cell proliferation. Atherosclerosis is considered a chronic inflammatory disease where LDL is deposited on blood vessel walls, contributing to the formation of foam cells and plaque. Research into herbal therapies typically involves investigating traditional heart disease plants for specific mechanisms such as lowering lipids and protecting endothelial cells and antiproliferative effects on vascular smooth muscle. When numerous mechanisms of action are demonstrated for a select herb or, in China, for a traditional formula, the research may mature to clinical trials.

Many plants can help reduce the risk of cardiovascular disease. Traditional blood movers such as ginger, cayenne, cinnamon, and garlic can be used in the diet or consumed in capsules, tinctures, and teas. *Allium* species (garlic and onions) are among the most studied and widely utilized herbs to help treat elevated lipids that contribute to atherosclerosis. *Allium* species contain allicin and related organic sulfur compounds that can help reduce total cholesterol and lipid ratios. *Zingiber officinale* (ginger) has many medicinal virtues, including beneficial effects on lipids, vascular inflammation, and blood sugar regulation. Ginger can be used in cooking, in prepared medicinal teas, and in tincture formulas. In animal studies, blood movers also include *Angelica* and *Apiaceae* species, *Salvia miltiorrhiza*, and *Viscum*. *Viscum* species (mistletoe) are traditionally used for heart disease, and research has shown viscolin to have antiproliferative effects on aortic smooth muscle cells,[17] as have related species in the *Phoradendron* genera, also known as mistletoe. *Angelica sinensis* is widely studied for vascular protective, anti-inflammatory and antiallergy effects, with many of the medicinal effects credited to ferulic acid.[18] *Salvia miltiorrhiza* is used widely in China to treat cardiovascular disease, and 50 different diterpenes known as tanshinones are shown to reduce atherosclerotic plaque formation and have many vascular protective effects.[19]

Flavonoids in many colorful fruits and berries are among the foodlike herbs shown to improve lipids and help protect the vascular system from oxidative stress and tissue changes induced by elevated blood pressure, lipids, and glucose. For example, nourishing *Crataegus* (hawthorn berry) is a high-flavonoid berry and one of the Western herbs most widely used in a variety of cardiovascular formulas. The flavonoids are credited with hypolipidemic effects. *Crataegus* berries, as well as the leaves and flowers, may help protect the tunica intima from atherosclerotic changes, promote oxidation enzymes in the liver (helping to metabolize lipids), and upregulate the influx of cholesterol from the plasma into the liver for processing. *Crataegus* is in the Rosaceae (rose) family, and rose hips also have cardioprotective effects. *Rosa canina* hips can be added to teas and taken as a solid extract. Rose hips have been shown to promote brown fat, supporting lipolysis and reducing fat storage.[20]

Agents that optimize blood lipids are often important aspects of protocols that reduce cardiovascular disease risk or progression. The regular consumption of omega-3 fatty acids can help reduce the risk of cardiovascular disease and several meta-analyses have shown the omega oils to lower triglyceride without significant effects on other serum lipids.[21] *Commiphora mukul* supports thyroid function and, via this and other mechanisms, can help reduce hyperlipidemia. *Commiphora* compounds known as guggulsterones inhibit the development and maturation of fat-storing cells, called adipocytes. Guggulsterones exert direct inhibitory effect on adipocytes, inducing decreased synthesis of new cells, decreased fat accumulation in existing cells, and increased destruction (apoptosis) of fat cells.[22] *Astragalus membranaceus* polysaccharides improve lipids, and the herb is a traditional medicine for numerous inflammatory diseases. The berberine in *Coptis chinensis* is credited with lipid-lowering effects in both animal and human studies.

Milk Thistle Capsules for Hyperlipidemia

Thistle family plants are helpful to the liver's processing of lipids and carbohydrates and can improve portal circulation and thereby reduce hypertension. *Silybum marianum* and *Cirsium japonicum* contain silibinin, a polyphenol that inhibits the release of inflammatory cytokines and antagonizes angiotensin receptors, contributing to the hypotensive effects. *Cynara scolymus* (artichoke) offer similar benefits and can be consumed as a medicinal food. Add artichokes to salads, side dishes, and snack trays.

Silybum marianum capsules

Take 2 or 3 capsules at a time, 2 or 3 times a day.

Seedy Sea Salt

This salt, seed, and herb blend offers medicine, while helping salt lovers to rein it in.

2 tablespoons raw *Silybum marianum* (milk thistle) seeds
2 tablespoons toasted sesame seeds
2 tablespoons raw walnut pieces
1 tablespoon dried herbs (such as cumin, caraway, and fennel seeds, smoked paprika, or basil)
1 to 2 tablespoons sea salt

Put all of the dry seeds and herbs in a coffee grinder, nut mill, or mortar and grind into a fine to slightly coarse powder as desired. Transfer to a small jar or shaker and add the salt. Coarse salt may also need to be ground to be able to flow through standard shakers. The blend may also be stored in a honey pot or pottery to use by the spoonful in making sauces, dressings, and garnishes.

Outrageous Oats

Avena sativa (oats) may help lower blood pressure and improve cholesterol. Including oatmeal in the daily diet may have beneficial effects on lipid profile and may help decrease the need for blood pressure medications. The fiber in oats is credited with much of the medicinal benefits on cholesterol and glucose regulation. Oats help prevent endothelial dysfunction, improve insulin sensitivity, and promote arterial relaxation. This recipe turns a bowl of oatmeal into a superfood, adding other ingredients of benefit to cardiovascular health.

⅓ cup (50 g) rolled oats
⅓ cup (50 g) oat bran
2 tablespoons coconut flakes
1 heaping tablespoon *Astragalus* powder
1 to 2 teaspoons cocoa powder (to taste)
½ teaspoon cinnamon (or to taste)
1½ cups (360 ml) water
½ cup (100 g) fresh or frozen berries
½ cup (120 ml) oat or other milk
1 tablespoon flaxseed oil

Continued on page 35

Easy Garlic Cloves

It can be tempting not to bother with garlic when you have to stop to break apart the bulbs and peel all the cloves while trying to get dinner on the table. Encourage patients to spend half an hour now and then peeling the papery covering from the individual cloves of five or more whole bulbs of garlic. These peeled cloves will keep well for several months when stored covered with olive oil in a small glass jar in the refrigerator. Either spoon out individual cloves as needed, or use a spoonful of the cloves with some of the oil in cooking. Other options are to put all of the peeled cloves in the blender with olive or quality nut oils or vinegar and puree. Put the puree in a small jar and use by the tablespoon in sauces, stir-fries, salad dressings, and vegetable dishes over the next few days.

Flavonoids for Atherosclerosis with Hyperlipidemia

One of the main problems with elevated lipids is that they become deposited on blood vessel walls, leading to loss of elasticity and fibrotic changes. Flavonoids, such as berry pigments, and high-flavonoid herbs can help protect the blood vessels from the damaging effects of high cholesterol and triglycerides. The following herbs and other substances are high in flavonoids and can be used in teas, tinctures, capsules, and medicinal beverages and foods for patients with hyperlipidemia.

Angelica sinensis	*Hypericum perforatum*
Berries	*Opuntia ficus-indica*
Centella asiatica	*Punica granatum*
Crataegus spp.	Resveratrol
Ginkgo biloba	*Rosa canina*
Hibiscus rosa-sinensis,	*Vaccinium myrtillus*
H. sabdariffa*	

Hypolipidemic Agents

The following hypolipidemic agents include the most studied herbs for lowering cholesterol. Many hypolipidemic agents are foodlike and are able to be used in the diet as well as in herbal medicines. (Keep in mind that a healthy overall diet with plenty of fiber will also help lower cholesterol.)

Allium sativum.[23] Garlic lowers cholesterol and triglycerides by inhibiting biosynthesis in the liver and inhibiting oxidation of LDL.

Ammi visnaga. Preliminary research suggests that khellin, a flavonoid in the plant, increases HDL and improves the HCL/LDL ratios.

Apiaceae family plants.[24] These include *Apium, Foeniculum,* and *Angelica,* and they are beneficial for the osthol[25] they contain.

Apium graveolens.[26] Celery seed medications are a traditional remedy for hypertension, and the coumarins in the plant also significantly reduce serum levels of total cholesterol and low-density lipoprotein (LDL) and enhance triglyceride processing by the liver.

Glycine max.[27] Soy may reduce plasma cholesterol via several mechanisms including inhibition of adipogenesis, appetite suppression, displacement of fat intake, and increased satiety.

Commiphora mukul. Guggulsterones may interfere with lipoprotein synthesis by inhibiting the production of cholesterol in the liver.[28]

Fish oil.[29] Fish oils contain the omega fatty acids eicosapentaenoic acid (EPA) and docosahexaenoic acid (DHA), and ingestion reduces lipogenic genes in adipose tissues affecting lipid synthesis and may also significantly reduce triglyceride levels.

Allium for High Cholesterol and Heart Health

Allium cepa (onions) and *Allium sativum* (garlic) support healthy blood glucose, cholesterol, and blood pressure, with much of the action credited to sulfur compounds, alliin and allicin. Garlic has been the subject of more research than onions, but both garlic and onions have cardiovascular benefits and should be consumed on a daily basis. Garlic improves cholesterol, triglyceride, and lipid ratios,[30] has a vasodilating action,[31] and has cardioprotective effects via a variety of molecular mechanisms including cyclooxygenase inhibition,[32] improved elasticity of the aorta,[33] nitric oxide promotion,[34] and inhibition of platelet aggregation.[35] Allium may also reduce vascular inflammation in diabetes, reducing blood glucose, improving insulin sensitivity, and reducing oxidative stress in animals fed a high-fructose diet. Human clinical trials have shown that the consumption of 400 milligrams of garlic powder can help reduce body weight and body fat in patients with nonalcoholic fatty liver disease.[36]

Allium sativum, garlic

Place the rolled oats, oat bran, coconut flakes, and astragalus, cocoa, and cinnamon powders in a small sauce pan and cover with water, and bring to a gentle simmer. Add the berries and oat milk and thoroughly warm. Add the flax oil at the time of serving because the essential fatty acids are heat labile and should not be cooked.

Breakfast Drink for High Lipids with High Glucose

This beverage features herbs and nutrients well researched for improving lipids. *Opuntia* (prickly pear) juice may require some searching but it is commercially available, and the fruits are available fresh in some southern states. Many kinds of berries are well documented to help protect blood vessels from the damaging effects of elevated lipids. Fish oils help improve the HDL/LDL ratios as does the ginger, and vitamin D may improve insulin response.

2 cups (470 ml) soy milk, chilled
½ cup (100 g) fresh or frozen berries
1 tablespoon *Opuntia* juice
1 teaspoon cod liver oil
1 teaspoon coconut oil
1 teaspoon freshly chopped ginger
1,000 IU vitamin D liquid

Place all the ingredients in a blender and blend several minutes until homogenized. Drink promptly.

Tincture for High Lipids with High Glucose

Some hypolipidemic herbs are tasty and suitable for teas, while other less-tasty herbs such as these are used in a tincture. These hypolipidemic herbs do double duty in treating diabetes or milder states of hyperglycemia as well.

Berberis aquifolium	15 ml
Allium sativum	15 ml
Commiphora mukul	15 ml
Gymnema sylvestre	15 ml

Take 1 or 2 dropperfuls of the combined tincture 2 to 4 times daily.

Spicy Tea for High Lipids with High Glucose

The palatable herbs in this tea help reduce both lipids and glucose, and quantities can be adjusted to suit the tastes of the person drinking it.

Glycyrrhiza glabra shredded roots	4 ounces (120 g)
Apium graveolens seeds	2 ounces (60 g)
Cinnamomum spp.	2 ounces (60 g)
Crataegus spp. berries	2 ounces (60 g)
Hibiscus rosa-sinensis flowers	2 ounces (60 g)
Syzygium jambos	1 ounce (30 g)
Trigonella foenum-graecum seeds	1 ounce (30 g)

Combine all dry herbs, blend well, and store in a covered jar or sealed bag in a dark place. Use 1 to 2 heaping teaspoons per cup of hot water. Bring to a gentle simmer for 10 minutes in a covered pan. Remove from the heat promptly and allow to sit, covered, for 10 to 15 minutes and then strain. Drink 3 or more cups per day to help reduce elevated blood sugar and fats. The tea may be consumed at a lesser dose or just several times per week to maintain the effects.

Liver Herbs and Alteratives for Hypertension and Hyperlipidemia

Liver herbs are appropriate to include in formulas for high cholesterol and to protect the liver from fatty degeneration. *Silybum marianum* is the most researched liver herb, but other alterative herbs may be valuable to include in formulas as well. *Arctium lappa* (burdock) is a well-known and well-used alterative agent and may be appropriate to include in formulas for diabetes, hyperlipidemia, and heart disease.

Animal studies suggest that burdock root protects against atherosclerosis in hyperlipidemic animals. Artigenin in burdock may ameliorate endothelial dysfunction and may antagonize mineralocorticoid receptors contributing to the hypotensive effect. The "do all" herb, *Achillea millefolium* (yarrow), acts as a styptic, vascular decongestant, peripheral vasodilator, antispasmodic,[37] and anti-inflammatory.[38]

Chinese Herbs with Hypolipidemic Effects

The following herbs are traditional in China for treating heart disease, vascular inflammation, high blood pressure, and elevated lipids. They have all had significant research and are commercially available in the United States to include in teas or tincture formulas.

Epimedium brevicornu, E. grandiflorum. Horny goatweed is a traditional Chinese herb that contains the flavonoid icariin, shown to improve blood lipids.

Panax ginseng. Ginseng, or renshen, has broad medicinal virtues, and the triterpenoid saponins and steroidal glycosides, known as ginsenosides or panaxosides are credited with numerous medicinal effects, including hypotensive and hypolipidemic actions.

Reynoutria japonica. Japanese knotweed is high in resveratrol, noted to help protect vein and artery walls and to help reduce cholesterol.

Reynoutria multiflora. He shou wu is used both raw and prepared, and both forms may improve serum lipid profiles.

Pueraria montana var. ***lobata.*** Kudzu reduces serum lipids and improves hepatic cholesterol levels in animal models of hyperlipidemia, improving hepatic lipid metabolism.

Rheum palmatum. Chinese rhubarb contains rhein, aloe emodin, emodin, chrysophanol, and physcion, all credited with improving triglyceride lipid levels, although high doses of rhein or emodin can cause diarrhea and intestinal cramping.

Salvia miltiorrhiza. Red sage is widely used in China to treat coronary artery disease, and animal and human clinical studies have shown hypolipidemic effects. Many of the medicinal effects are credited to the component tanshinone.

Scutellaria baicalensis. Huang qin or Chinese scute is widely used in China as an anti-inflammatory and antiallergy agent. Huang qin roots are credited with lipid-lowering effects and may promote cholesterol-processing enzymes in the liver.

Tincture for Hyperlipidemia in Diabetics

When cells do not process glucose well, some of the excess sugar may be converted and stored as lipids. Helping support the metabolism of both glucose and lipids will be helpful. Inositol, chromium, magnesium, vitamin D, and selenium would be excellent complements to this formula.

Gymnema sylvestre	15 ml
Allium sativum	15 ml
Ginkgo biloba	15 ml
Vaccinium myrtillus	15 ml

Take 1 teaspoon of the combined tincture 3 or more times daily.

Tea for Hyperlipidemia with Insulin Resistance

Many leguminous herbs such as licorice, *Astragalus*, and alfalfa are noted to improve insulin resistance. This formula, which is appropriate for hyperlipidemia with slow metabolism and insulin resistance, features *Astragalus*, *Stevia*, *Cinnamomun*, *Hibiscus*, and *Opuntia*, all of which also improve the action of insulin at insulin receptors.

Hibiscus sabdariffa flowers, cut and sifted
Cinnamomum verum bark, finely chopped
Stevia rebaudiana
Vaccinium myrtillus leaves
Astragalus membranaceus roots, finely chopped
Opuntia ficus-indica juice

Combine equal parts of the dry herbs and blend. Gently simmer 1 teaspoon per cup of hot water for just 1 minute. Cover the pan, let stand for 10 to 15 minutes, and then strain. House the tea in a covered jar or thermos. Add 1 teaspoon of *Opuntia* juice to each cup at the time of serving.

Tincture for Hyperlipidemia with Slow Metabolism

Commiphora mukul (guggul) helps promote the basal metabolic rate, and *Coleus* may help the liver and muscle cells take up substrates for basic metabolism. The garlic, cayenne, and turmeric are all warming, blood moving, and anti-inflammatory, and although they make this formula "warm," they are usually well tolerated by hypothyroid patients, who are typically cold due to slow metabolism.

Commiphora mukul	7 ml
Allium sativum	7 ml
Coleus forskohlii	7 ml
Curcuma longa	7 ml
Capsicum annuum	2 ml

Take 1 or 2 dropperfuls of the combined tincture 3 or more times daily to help reduce elevated lipids. The effects may be able to be maintained with a lesser dose and frequency once improvements are made.

Tincture for High Lipids with Excessive Clotting

Those with a history of blood clots and/or elevated platelets or C-reactive protein may benefit from herbs known as "platelet antiaggregators." Bromelain would be an excellent supplement to complement formulas to reduce the tendency to excessive or inappropriate platelet activation and clotting. Patients on an anticoagulant medication should use this only under a doctor's supervision, as lab work will be required to monitor progress. See also "Fibrinolytic Botanicals" on page 73, "Platelet Antiaggregators and PAF Inhibitors" on page 77, and the Tea for History of Blood Clots formula on page 81.

Allium sativum	30 ml
Angelica sinensis	15 ml
Salvia miltiorrhiza	15 ml

Take 1 teaspoon of the combined tincture 3 or more times a day, along with 500 to 1,000 milligrams of bromelain between meals 3 or 4 times daily.

Tincture for High Lipids with Liver Congestion

High lipids can sometimes be the result of weak liver function where fats are not emulsified, processed, or metabolized readily. Herbs that support liver function and intestinal health can be helpful in such cases. Consider a formula such as this, especially when patients have symptoms of toxemia, digestive issues, or acne or other skin issues. This tincture would be complemented by the use of a commercial lipotropic encapsulation formula.

Aesculus hippocastanum	15 ml
Silybum marianum	15 ml

Red Yeast Rice for High Cholesterol

Monascus purpureus is a type of red yeast used as early as the Tang Dynasty in China for circulatory support by fermenting rice with the yeast and eating the resulting fermented product. Modern investigation shows that monacolins in the rice have properties similar to statin drugs, in that the ferment acts as a natural HMG-CoA reductase inhibitor.[39] The statin pharmaceuticals have many serious side effects including muscle damage and depletion of coenyzme Q10 (CoQ10). Red yeast may also deplete CoQ10 in the heart muscle, [40] but may spare the skeletal muscle from damage,[41] as seen with the synthetic statins. Red yeast rice has not been shown to inflame the liver,[42] as do the statin pharmaceuticals, and it has fewer side effects, overall,[43] but it would be wise to take red yeast rice in tandem with CoQ10. Red yeast rice may improve lipids when dosed at 600 milligrams twice a day, and the higher dosage of 3,500 milligrams per day may yield the best results. Several human clinical studies report red yeast rice to moderately improve lipids[44] and especially to lower elevated LDL and total cholesterol;[45] to lower C-reactive protein, which is an indicator of vascular inflammation,[46] to help protect the vasculature from inflammatory damage;[47] and to promote the repair of renal tissue when damaged by elevated lipids.[48] A large study conducted in China reported red yeast rice to significantly reduce the occurrence of major adverse cardiovascular events and to improve survival following myocardial infarction.[49]

Allium sativum	10 ml
Curcuma longa	10 ml
Iris versicolor	10 ml

Take 1 to 3 dropperfuls of the combined tincture 3 times per day. Complement with 2 lipotropic capsules with each meal.

Formulas for Angina and Coronary Artery Disease

Angina is a symptom of coronary artery disease (CAD) and can progress to myocardial infarction, the leading cause of death. CAD progresses from atherosclerosis, hence, all of the herbal formulas proposed for high cholesterol and atherosclerosis are also appropriate for CAD. As plaque builds up in the coronary arteries, the blood vessels narrow and can cause acute anginal pain as the oxygen-poor blood vessels spasm. If the coronary arteries become entirely occluded, or if the blood flow is so blocked that blood cells pools and clot, a thrombus forms, and acute ischemia in the cardiac muscle may result and cause cell death. When extensive, an infarct may result. Acute myocardial infarction (MI) is frequently fatal, therefore recognizing CAD as early as possible, addressing the underlying contributors, and slowing the progression are important medical goals.[50] All patients suffering from angina should be evaluated by a cardiologist, and it is likely that such patients will be given pharmaceutical vasodilators such as nitroglycerin, beta blockers, or possibly blood-thinning agents. Herbal medicines may be capable of reducing vasospasms, but due to the severity of the pathology and uncertainty over their reliability, most patients and herbalists alike will opt for nitroglycerine while complementing with herbal vascular support. Mild or early cases may respond to herbal vasodilators while initiating an aggressive and comprehensive protocol to improve heart health, reduce risk factors, and address contributors.

Atherosclerosis was recognized as a prime contributor to cardiovascular disease and was described as early as 500 BC in an early Chinese medical canon. Atherosclerosis of the coronary arteries does not always present with acute episodes of angina, but may involve an insidious loss of blood vessel elasticity due to chronic inflammation, resulting in eventual sclerosis and stenosis of blood vessels. Altered endothelial function and disturbed texture of the vessel walls promotes hypercoagulation in the blood, increasing the risk of micro clots and larger thrombi. Herbs may be used in tandem with some pharmaceuticals or, in skilled hands and in certain cases, may be used in lieu of pharmaceuticals as part of a broader treatment protocol. CoQ10, arginine, and concentrated *Allium* capsules are all appropriate and can be combined to create an aggressive multipronged approach.

Tincture for Angina

This formula is a basic circulatory and platelet antiaggregating blend, while the Tincture for Angina with Acute Squeezing Pain that follows is a more powerful vasorelaxant. *Arnica* is not often ingested orally, but one of the folkloric indications is for chest pain that is aching and sore in character, as if bruised.

Ginkgo biloba	15 ml
Angelica sinensis	15 ml
Crataegus spp.	15 ml
Viscum album	10 ml
Arnica montana	5 ml

Take 1 or 2 teaspoons of the combined tincture 3 to 6 times daily.

Sini Tang Decoction for Use after Myocardial Infarction

This formula is based on a TCM formula used to support heart muscle recovery, improve cardiac ejection, reduce the chance of sepsis, and lessen the effects of shock and immune assault. Do NOT confuse the processed aconite used in TCM formulas with unprocessed aconite roots, or attempt to prepare this as a tincture of equal parts. **Caution:** *Aconitum* is a highly toxic herb and is specially prepared in China in a manner that changes its qualities and toxicity.

Aconitum carmichaelii	2 ounces (60 g)
Zingiber officinale	2 ounces (60 g)
Glycyrrhiza uralensis	2 ounces (60 g)
Cinnamomum cassia	2 ounces (60 g)

Combine the dry herbs, simmer using 1 teaspoon per cup of water, and then strain. Drink 2 or 3 cups each day.

Herbs for Angina

The following herbs are folklorically recommended for angina, as are beta-blocking herbs, nervine herbs, and platelet antiaggregators.

Adonis vernalis	*Lobelia inflata*
Ammi visnaga	*Phoradendron*
Arnica montana	*chrysocladon*
Crataegus oxyacantha	(also known as
Ginkgo biloba	*Viscum flavens*)
Leonurus cardiaca	*Strophanthus hispidus*

An Apple a Day

Apples (*Malus domesticus*) are among the most commonly consumed fruits in the United States, and nutritional assesments have shown children's apple consumption to correlate with increased fiber, magnesium, and potassium intake and decreased intake of total fat, saturated fat, sugar, and sodium, compared to those who eat few or no apples.[51] Epidemiological studies suggest that frequent apple consumption reduces the risk of chronic diseases such as cardiovascular disease. Apples are a rich source of polyphenols and fiber and support colonic health via a positive influence on desireable intestinal flora. Some of the beneficial effects of apples on cardiovascular health may be via supportive effects on the intestinal microbiome,[52] accomplished via local and systemic effects on lipid metabolism and vascular inflammation. Apple pectin improves glucose regulation, helps lower cholesterol, and offers important polysaccharides that may reduce atherosclerotic processes and protect against myocardial injury. Apples provide cellulose, an insoluble fiber, and pectin, a soluble fiber. Pectin is fermented by intestinal flora into short chain fatty acids, with local and systemic benefits. Pectin resists degradation by gastric acid, and it reaches the intestines and acts as a gelling agent, positively affecting transit time, intestinal enzymes, and the gut microbiota. Dietary fiber such as pectin is associated with a reduced risk of coronay heart disease, delaying occlusion by ameliorating the risk factors. Apple pectin protects myocardial cells against ischemic infects by reducing the underlying factors that lead to occlusion. Apple pectin decreases LDL, total cholesterol, and C-reactive protein levels and reduces blood pressure, helping to mitigate contributors to myocardial injury.[53] Through exerting positive effects on gut flora, apple pectin supports a healthy microbiome in the intestines, improves gut barrier finction, and reduces endotoxemia-driven inflammation.[54]

Malus domesticus, apple

Tincture for Angina with Acute Squeezing Pain

Lobelia is a natural β-nicotinic receptor ligand that may help treat coronary vasospasm and angina. *Angelica* also relaxes many muscle types in the body via effects on calcium channels, plus it is a platelet antiaggregator and general anti-inflammatory blood mover. This formula may be more effective in alleviating acute angina than the Tincture for Angina, which is more nourishing and tonifying. In the upper dosing ranges for acute angina, *Lobelia* may promote salivation and nausea and can be dosed only as tolerated. Taking the dose in mint or chamomile tea may alleviate this problem, but there would not always be time to prepare tea in acute situations.

Angelica sinensis
Lobelia inflata
Valeriana officinalis

Combine in equal parts and take 1 or 2 teaspoons every 5 minutes for acute chest pain, reducing as symptoms improve. Take 1 or 2 teaspoons 3 times a day for maintenance dosing.

Estrogen, Women, and Heart Disease

Low estrogen levels induce oxidative stress and accelerate cardiovascular disease. Men have higher levels of heart disease than women, until women reach the age of menopause. Past that age range, the markers of cardiovascular disease equalize across the genders. *Crataegus* berry medications may have cardioprotective effects and help protect the heart and blood vessels from menopausal changes in tissue status and lipids.[55]

Post-MI Recovery Tincture

The herbs in this formula are noted to improve cardiac enzyme levels, lipid levels, and antioxidant levels in cardiac muscle cells to support recovery.

Terminalia arjuna
Selenicereus grandiflorus
Crataegus oxyacantha
Piper nigrum

Combine in equal parts and take 1 or 2 dropperfuls of the combined tincture 2 or more times per day.

Rhodiola Capsules for Ischemic Heart Disease

Rhodiola is a traditional medicine for improving energy, stamina, mood, and athletic performance, and the plant has been used in formulas for angina pectoris, coronary artery disease, and heart failure. Numerous clinical studies suggest that the plant improves EKG readings and lessens the symptoms of ischemic heart disease. As *Rhodiola* has mild stimulating effects, it is best used for deficient patients, opting for other adaptogenic herbs for those prone to anxiety or nervous tension.

Rhodiola rosea capsules

Take 2 or 3 capsules, 2 or 3 times a day.

Crataegus Berry Syrup for Heart Support

Crataegus species have cardioprotective effects. Hawthorn berry has oligomeric proanthocyanadins that have vasodilatory actions and positive cardiac inotropic effect. Clinical trials suggest an ability to support normal heart function in heart failure patients. Hawthorn berries are available as a solid extract; the berries are cooked down into a flavonoid- and pectin-rich syrup suitable for use in the daily diet.

Crataegus solid extract

Consume 1 teaspoon or more each day by adding to blender drinks, by diluting with water to prepare mocktails and drinks, to drizzle over oatmeal, or to use instead of honey when making salad dressings.

Classic Chinese Duo for Protecting the Heart

Angelica sinensis and *Ligusticum striatum* (also known as *L. chuanxiong*) are a traditional duo used to support circulation and protect the vasculature. Numerous mechanisms of action have been proposed, but the pair are featured here due to their ability to inhibit the proliferation of vascular smooth muscle from oxidative and hypertensive stress and promote vasodilation via nitric oxide and intracellular calcium ions.[56]

Ligusticum striatum
Angelica sinensis

Combine in equal parts and take 1 or 2 dropperfuls of the combined tincture 3 or 4 times daily.

Formulas for Hypertension

Hypertension is a worldwide health problem, and it is a major contributor to coronary and peripheral artery disease, stroke, heart failure, and atherosclerosis. Routine blood pressure screening is part of health exams to attempt to catch hypertension and treat as early as possible in order to prevent the damage. Hypertension is especially damaging when it occurs in tandem with diabetes and hyperlipidemia, exacerbating the damage to blood vessels.

Hypertension is defined as having a systolic blood pressure greater than 140 mmHg and a diastolic blood pressure greater than 90 mmHg; and every 20 mmHg increase correlates to a 25 percent increase in relative risk for a cardiovascular event. Those with systolic above 160 and diastolic above 120 should seek immediate medical attention to prevent damage to the blood vessels and decrease the risk of stroke as quickly as possible. People with borderline hypertension are likely to become fully hypertensive over a few years' time and should be motivated to lose weight, quit smoking, control their stress, and take other lifestyle measures wherever possible.

Hypertension is classified as either primary/essential or secondary. Secondary hypertension represents around 5 to 10 percent of cases and is defined as being due to an identifiable cause, such as diabetes or renal damage, and such causes should be addressed where possible. Essential hypertension is typically multifactorial; varied causes such as age, lifestyle, diet, stress, and other factors intersect in a nebulous manner that elevates the blood pressure. This type of hypertension is more difficult to manage.

There are many classes of drugs used to treat hypertention including diuretics, angiotensin-converting enzyme (ACE) inhibitors, angiotensin receptor blockers, calcium channel blockers, α-adrenergic blockers, β-adrenergic blockers, vasodilators, and sympatholytic drugs.

Vascular smooth muscles may proliferate under the influence of increased resistance of hypertension. Agents that help the endothelium resist proliferation and inhibit fibrosis and sclerosis can help protect against the blood vessel damage. Nitric oxide protects against vascular smooth muscle proliferation, and numerous herbs are noted to promote nitric oxide release. The endothelium is highly sensitive and releases its own vasoactive substances in response to stimulation of a variety of cell surface receptors. The endothelium can help regulate vascular tone, releasing nitric oxide,

prostacyclin, thromboxane, and factors derived directly from the endothelium itself. Imbalances in these substances contribute to endothelial dysfunction and hypertension. Herbal therapies may help correct the endothelial-derived vascular modulators and help treat hypertension, as well as protect the blood vessels.

Allium is hypotensive due to its ability to increase nitric oxide, an endogenous compound with vasodilating effects. *Allium* also inhibits platelet aggregation, which can improve blood flow. *Crataegus* makes a nutritious base in formulas for all aspects of heart disease. *Crataegus* species were known as heart medicines, both to ancient

Using Nervine Herbs in Hypertension Formulas

Nervine herbs exert hypotensive effects by calming stress and anxiety, and many of them also have relaxing effects on vascular smooth muscle. *Lavandula*, for example, inhibits Ca2+ channel activity, inducing vasorelaxation. *Valeriana* and *Passiflora* are both vasorelaxant nervines, appropriate to include in formulas for hypertension, heart palpitations, and stress-related spikes in blood pressure. *Ocimum sanctum* (holy basil) is traditional for mood-enhancing effects and shown to exert vasorelaxant effects. *Melissa officinalis* (lemon balm) is an appropriate nervine for cardiovascular formulas due to anxiolytic[57] and vascular affects. Animal studies show the herb to be protective against heart damage from ventricular arrhythmias.[58] Human clinical studies show lemon balm to improve heart palpitations, especially when due to anxiety.[59] *Melissa* is a natural angiotensin-converting enzyme (ACE) inhibitor[60] and also inhibits excessive corticosterone levels.[61] *Melissa* inhibits GABA transaminase, the enzyme that metabolizes GABA, allowing higher levels to persist in the brain.[62]

Nitric Oxide Promotors

Nitric oxide is produced endogenously and plays a role in vascular tone independent of sympathetic innervation. The release of nitric oxide causes prompt vasodilation, serving to reduce blood pressure and increase circulation. Nitric oxide also helps protect the vein and artery walls—the endothelium—from low oxygen and other stressors. Herbs that promote nitric oxide can be used in hypotension formulas, as well as help protect endothelial cell from damage.

Angelica sinensis has been shown to promote vasodilation via nitric oxide promotion, as have *A. dahurica* and other species.[63]

Apium graveolens may promote nitric oxide signaling as well as affect calcium channels on vascular smooth muscles and lead to vasodilation.[64]

Cnidium monnieri is a traditional Chinese herb widely used to treat cardiovascular disorders including stroke, hypertension, and arrhythmia via numerous mechanisms of action including the promotion of nitric oxide.[65]

Trifolium pratense has been shown to activate nitric oxide synthase in endothelial cells through genomic influences that involve the beta type of estrogen receptors. *Trifolium* isoflavones appear to act synergistically with 17 β-estradiol to increase endothelial nitric oxide synthase activity and expression.[66]

European herbalists and in Traditional Chinese Medicine (TCM). The berries are high in flavonoids credited with antihypertensive and vascular protective effects. *Andrographis paniculata* contains andrographolide, a labdane diterpenoid credited with hypotensive effects via ACE-inhibiting effects.[67] *Apium graveolens* has hypotensive effects via numerous mechanisms including blockade of corticosterone stimulation, reduction of catecholamines, and reduced vascular resistance. Although

Camellia sinensis contains caffeinelike compounds, regular consumption is reported to have a hypotensive effect, largely credited to the catechins, such as epicatechin and epigallocatechin. *Hibiscus sabdariffa* is a popular tea due to its sour flavor and beautiful pink color and is credited with lowering blood sugar and having hypotensive effects. *Rauvolfia serpentina* is one of the most powerful herbs to bring down elevated blood pressure quickly. See "*Rauvolfia* for Hypertension" on page 45.

General Tea for Hypertension

This tea contains tasty, common, and multipurpose herbs, suitable to enjoy long term.

Camellia sinensis	8 ounces (240 g)
Berberis aquifolium shredded root	8 ounces (240 g)
Hibiscus sabdariffa	4 ounces (120 g)
Apium graveolens seed powder	2 ounces (60 g)

Combine the herbs, steep 1 heaping tablespoon of the mixture per cup of hot water, and then strain. Drink freely, as desired.

Tincture for Hypertension with Stress and Vascular Tension

Mental and emotional stress can both promote tension in the body and increase vascular tension, and it can worsen hypertension. Nervine herbs are appropriate for both stress and hypertension, especially in formulas where psychic stress is noted to exacerbate blood pressure. In addition to being nervines, *Tilia* and *Leonurus* are both gentle beta blockers with vasodilating action.

Passiflora incarnata	15 ml
Tilia × europaea	15 ml
Leonurus cardiaca	15 ml
Withania somnifera	15 ml

Take 1 or 2 dropperfuls of the combined tincture 3 to 6 times daily.

Garlic Spice or Capsules for Hypertension

Garlic has numerous medicinal virtues including hypotensive and hypocholesteremic properties. The organic sulfur-containing constituents, including alliin and allicin, are credited with much medicinal activity and can be used as a medicinal food or included in an herbal tincture in formulas for hypertension and high cholesterol.

Garlic capsules

Take 1 or 2 capsules at a time, 2 or 3 times daily.

Seed Oils for Heart Health

Essential fatty acids (EFAs) are "essential" because humans cannot synthesize them in the body and must receive them from outside sources. Fish, nuts, and seeds are the richest sources. Essential fatty acids are not very stable, so heating and cooking with seed oils destroys the delicate fatty acids, and they are even labile with exposure to light and room temperature storage. Therefore, storing nuts in the freezer and expensive seed oils in the refrigerator is appropriate. Unlike most animal fats, EFAs have numerous anti-inflammatory benefits for people with diabetes and metabolic syndrome, improving blood lipids and protecting against blood vessel and nerve damage.[68] Omega-3 polyunsaturated fatty acids include eicosapentaenoic, docosahexaenoic, and alpha-linolenic acids, shown to reduce C-reactive protein levels, an important marker of cardiovascular disease risk.[69]

Borago officinalis. Borage seeds are a rich source of gamma-linolenic acid (GLA).

Hippophae rhamnoides. Sea buckthorn fruits are used to produce an orange-red oil high in vitamins, flavonoids, and important minerals including selenium and zinc, and may act as a natural calcium channel blocker to lower blood pressure.

Linum ussitatissimum. Flaxseed oil contains high values of omega-3 and omega-6 fatty acids and is the richest known source of alpha-linolenic acid. Flaxseed lignans and their metabolites have beneficial hormonal effects. Flaxseed oil has antioxidant- and cardiovascular-protectant effects and is noted to reduce C-reactive protein levels and reduce cardiovascular inflammation.

Nigella sativa. Black cumin or black seed oil is shown to have hypotensive effects credited to thymoquinone.

Oenothera biennis. The tiny seeds of evening primrose are a source of GLA, available in capsules or bulk oil. GLA may be used in the body to produce the eicosanoid Dihomo-γ-linolenic acid (DGLA), which may have antithrombotic effects.

Olea europaea. Olive oil can be used in cooking and medicinal foods and is shown to reduce atherosclerotic lesions and reduce damage to the tunica intima in animal studies. Molecular investigations report benefits via effects on vascular adhesion inflammatory mediators. Oleuropein may reduce infarct damage in situations of ischemia, as well as protect against cardiotoxic drugs.

Ribes species. Black currants are especially seen in the marketplace, used for allergy and inflammation and to improve lipid ratios. Black currant (*Ribes nigrum*) seed can be used to produce an oil high in GLA.

Tea for Stress and Hypertension

Similar in concept to the above tincture, this tea features palatable nervines to use as a calmative agent to complement tinctures and pills in a broader protocol for hypertension.

Passiflora incarnata
Scutellaria lateriflora
Avena sativa
Crataegus spp. berries, finely chopped
Rosa canina hips, finely chopped
Hibiscus rosa-sinensis
Stevia rebaudiana

Combine the herbs in equal parts, such as 2 ounces (60 grams) of each, and blend. Steep 1 heaping tablespoon of the blend per cup of hot water for 10 minutes and then strain. Drink 3 or more cups daily, as desired.

Berberine Capsules for Hypertension

Coptis chinensis (goldthread) is widely used in Chinese folk medicine and is the source of berberine, credited with numerous metabolic benefits in diabetes, hyperlipidemia, and metabolic syndrome. Hypotensive properties are also reported.[70] *Berberis* is another source of berberine from North America and is also used to support liver and metabolic function.

Berberine capsules

Take 1 or 2 capsules at a time, 2 or 3 times a day.

All-Purpose Tincture for Hypertension

This formula covers all the bases—agents to relax the blood vessels, nervines to help address underlying stress, agents to lower lipids, agents that help to protect the blood vessels from the damaging effects of hypertension, and vascular trophorestoratives. Add or subtract to make your formula more specific for your patient. Compare this formula to those immediately following for examples of how to tweak a formula for a variety of patients and presentations.

Rauvolfia serpentina	10 ml
Leonurus cardiaca	10 ml
Angelica sinensis	10 ml
Valeriana officinalis	10 ml
Allium sativum	10 ml
Crataegus monogyna	10 ml

Take 1 dropperful of the combined tincture 3 times a day. For urgent situations, take 1 dropperful every 30 minutes, reducing as pressure declines.

Rauvolfia Tincture for Hypertension with Headache

While some people are unable to discern any physical symptoms when their blood pressure is elevated, others may experience headaches, dizziness, or throbbing pain at the temples or vertex. This formula contains a larger amount of *Rauvolfia* than the formula above and is aimed at bringing down hypertension with vascular pain as rapidly as possible. The formula also contains *Atropa belladonna*, so this formula should be used by

---

Folkloric Herbs for Hypertension

These herbs are all possibilities to use in formulas to address hypertension. See "Specific Indications," beginning on page 89, for tips on how these herbs may be best used. Diuretic herbs, such as *Equisetum*, and alterative and liver herbs can also be used in formulas for hypertension.

Aesculus hippocastanum	Phoradendron
Allium sativum	chrysocladon
Angelica sinensis	Tilia × europaea
Crataegus oxyacantha	Valeriana sitchensis,
Ginkgo biloba	officinalis
Leonurus cardiaca	Veratrum viride
Rauvolfia serpentina	Viburnum prunifolium

---

experienced clinicians only. The belladonna may also be removed from the formula.

Rauvolfia serpentina	20 ml
Angelica sinensis	20 ml
Valeriana officinalis	15 ml
Atropa belladonna	5 ml

Take 1 dropperful of the combined tincture every 15 to 30 minutes, reducing as symptoms abate.

Hypertension Tincture Based on R. F. Weiss

The German physician Rudolf F. Weiss used belladonna, a known vascular sympathomimetic, in his formulas for hypertension. This formula is similar to one published in Weiss's *Herbal Medicine* but altered for the North American clinician and products available. Small amounts of belladonna and homeopathic preparations are often emphasized for hyperemic situations and when there is a bounding pulse or throbbing head pain. Note that belladonna is a potentially toxic botanical and to be used by experienced clinicians only. Signs of belladonna toxicity are distorted vision and seeing halos around objects. Patients should be encouraged to contact their prescriber at once if they experience any visual disturbances. A dry mouth can also result from the use of belladonna. Use this formula for hypertensive headaches, and move on to a more nourishing tonic once the acute issue is remedied.

Valeriana officinalis	45 ml
Atropa belladonna	15 ml

Take 30 drops of the combined tincture 3 times a day.

Tincture for Hypertension with Exhaustion

Cases where hypertension is associated with long-term stress, nervous exhaustion, and what alternative medical practitioners call adrenal fatigue may respond better to herbal adaptogens than to beta blockers or simple nervines. When patients don't sleep well, can't concentrate, or complain of fatigue and exhaustion, herbal adaptogens are an appropriate option. *Glycrrhiza* tea would be complementary to this formula, especially when a relaxing ritual of daily teatime can be implemented.

Eleutherococcus senticosus	20 ml
Leonurus cardiaca	20 ml
Panax ginseng	20 ml

Take 1 teaspoon of the combined tincture 3 or 4 times a day for at least 3 months.

Rauvolfia for Hypertension

Rauvolfia (Indian snakeroot) is a traditional medicine in India, used for snakebites, as the common name implies, but more commonly used in the treatment of anxiety, headaches, dementia, agitated psychosis, insanity, and hypertension over the last 1,000 years or longer in Ayurvedic medicine. Gandhi reportedly drank *Rauvolfia* on a regular basis for its calming effects.

Rauvolfia contains powerful alkaloids, the most researched being reserpine, which was developed into a pharmaceutical hypotensive medicine as early as the 1950s. Reserpine and related indoles alkaloids, including ajmaline and ajmalicine, are credited with hypotensive effects[71] and work via a central nervous system mechanism. Reserpine favors parasympathetic tone in the central nervous system and may interfere with catecholamine stores in peripheral vesicles, both helping to reduce stress-related hypertension.[72]

Several human clinical trials have shown efficacy in treating hypertension.[73] My own clinical experience with the plant has taught me to use *Rauvolfia* for people who are anxious, tense, and even manic, but to avoid in patients with lethargy and depression, where favoring parasympathetic dominance could be too weakening or overly sedating. Choose *Rauvolfia* for vigorous people who have a full bounding pulse and who are made worse by stress, poor sleep, and vigorous workloads and schedules. *Rauvolfia* will yield the best results in those who are robust with a hot, excess constitution and may weaken, depress, or overly suppress and sedate asthenic, deficient, exhausted individuals. Due to interfering with catecholamine stores, reserpine can lead to depletion of dopamine and should not used with Parkinson's patients.[74] Depletion of catecholamines can also cause depression in susceptible individuals. Green tea used in tandem with reserpine may modulate some of these effects.[75]

Rauvolfia serpentina, Indian snakeroot

Tincture for Hypertension with High Lipids

This condition is an extremely common scenario seen in general family practice offices, and formulas such as this might be dispensed on an almost daily basis. *Opuntia* would be a good complement to this formula and can be prepared as a daily beverage (see *Opuntia* Mocktail for Metabolic Syndrome on page 30).

Allium is useful for both hyperlipidemia and hypertension, and *Commiphora mukul* (guggul) supports basal metabolism, helping to "burn" fats and thus lower lipid levels. Cinnamon would be a good substitute in this formula for any patient who does not tolerate garlic.

Allium sativum
Angelica sinensis
Salvia miltiorrhiza
Commiphora mukul

Combine in equal parts and take 1 dropperful of the combined tincture a minimum of 3 times a day. Exercise should also be encouraged in all such cases.

Tincture for Hypertension with Heart Palpitations

Leonurus is specific for heart palpitations and would make a good complement to nervines such as *Tilia* and general cardiovascular support herbs such as *Allium* and *Crataegus*. Calcium and magnesium are also relaxing to heart excitability and are complementary to this formula. Heart palpitations may be due to hyperthyroidism, stress, and menopause, and *Leonurus* can help in all such endocrine situations. Heart palpitations due to cardiomyopathies, on the other hand, are better treated with cardiac glycosides such as *Convallaria* or *Selenicereus*. See also "Formulas for Congestive Heart Failure and Cardiomyopathy," starting on page 47.

Leonurus cardiaca	20 ml
Tilia × europaea	15 ml
Crataegus monogyna	15 ml
Allium sativum	10 ml

Take 1 dropperful of the combined tincture 3 or more times daily.

Relaxing Tincture for Hypertension

This formula, like the one above, includes nervines and is aimed at relaxing the blood vessels and body and supporting quality sleep as a means of reducing hypertension. *Piper methysticum* (kava kava) may be added for acute muscle spasms.

Eschscholzia californica
Corydalis cava
Viburnum prunifolium
Tilia × europaea
Withania somnifera

Combine in equal parts and take 1 dropperful of the combined tincture 3 or more times daily.

Thromboxane Inhibitors

Thromboxane is powerful vasoconstrictor that, once activated, has effects that persist for several days. Thromboxane is increased by smoking, essential fatty acid deficiency, and a high fat diet and is deterred by these and other herbs. These herbs are commonly thought of as "blood movers," and this is one more mechanism by which they offer cardiovascular benefits.

Allium sativum[76]	*Piper methysticum*
Curcuma longa	*Cinnamomum* spp.
Foeniculum vulgare[77]	Nattokinase
Zingiber officinale	

Tincture for Hypertension with Stress

When hypertension is more associated with stress than it is with elevated lipids or atherosclerosis, nervines might be a helpful basis of a formula. *Leonurus* and *Tilia* are nervines with a mild beta-blocking effect, making them especially appropriate in hypertension formulas. *Valerian* and *Viburnum* can also relax the muscular layer of the blood vessels, lending them vasodilating properties. This formula could be complemented by a nervine tea containing *Matricaria*, *Avena*, *Passiflora*, and other calming and nourishing agents.

Leonurus cardiaca	15 ml
Tilia × europaea	15 ml
Melissa officinalis	15 ml
Valeriana officinalis	15 ml

Take 1 dropperful of the combined tincture 3 to 4 times daily for general stress and hypertension. For urgent cases, it is not harmful to increase the dosage to hourly and then reduce frequency as the blood pressure declines.

Tincture for Hypertension with Hyperthyroidism

For hypertension and heart palpitations due to hyperthyroidism, this combination of herbs is often helpful. These herbs all contain rosmarinic acid, and this organic

Calcium Channel Blockers

Calcium channel blockers inhibit the entry of calcium into cardiac muscle cells, reducing muscle contraction and lowering blood pressure via vasodilation. Calcium channel antagonist drugs are used to treat hypertension, and a number of herbs traditionally emphasized in the folklore for high blood pressure are now found to be calcium channel blockers. Osthol, a coumarin in the Apicacea family, has Ca2+-channel blocking properties,[78] and *Apium*, *Angelica*, *Cnidium*, and *Ammi visnaga* are hypotensive agents in part via their calcium-blocking coumarins. The following herbs are also calcium-channel blockers.

Allium sativum	*Cinnamomum verum*[81]
Ammi visnaga[79]	*Cnidium monnieri*[82]
Angelica spp.[80]	*Daucus carota*[83]
Apium graveolens	*Zingiber officinale*[84]

acid has been shown to help reduce excessive stimulation of the thyroid gland.

Leonurus cardiaca	20 ml
Lycopus virginicus	20 ml
Melissa officinalis	20 ml

Take 1 dropperful of the combined tincture 3 times daily or as often as hourly as needed for acute symptoms of hyperthyroidism.

Tincture for Hyperlipidemia with Poor Diet

One all-purpose approach to initial therapy for hyperlipidemia would be to support the liver's ability to emulsify fats by including herbs such as *Silybum* or *Berberis* and to support general metabolic rate, especially if the patient has a sluggish metabolism, with an agent such as *Commiphora*. *Allium* and *Zingiber* are blood movers known to improve cholesterol and triglycerides, but the use of both of them, as in this formula, may be too "hot" for some patients. Compare with the formulas that follow to see how we can alter the formula to address specific situations.

Silybum marianum	20 ml
Commiphora mukul	20 ml
Allium sativum	10 ml
Zingiber officinale	10 ml

Take 5 millileters 3 times daily, combined with dietary or other approaches, to help control elevated lipids.

Formulas for Congestive Heart Failure and Cardiomyopathy

Cardiomyopathy is a disease of the heart muscle itself where the pumping strength is compromised, making it difficult to pump blood and therefore deliver oxygen to the body and tissues. Cardiomyopathy carries a poor prognosis. Few people survive past ten years following diagnosis, and for those with congestive heart failure, just five years. Heart failure is the term used when the heart becomes so weak that it fails to pump blood at all, resulting in arrhythmias, fluid in the lungs, edema in the lower limbs, and oxygen deprivation. Heart failure is the number one cause of all hospital admissions and is increasing in prevalence in tandem with metabolic syndrome and diabetes. Herbal medicines and healthy lifestyle choices may help extend the life span, slow the progression of heart failure, and reduce the need for heroic lifesaving surgical interventions. The herbs that are appropropriate for cardiomyopathy are the same as the herbs used to treat other cardiovascular conditions such as hyperlipidemia, atherosclerosis, coronary artery disease, and hypertension. The exception is cardiac glycosides, which can help the heart muscle to squeeze more powerfully. In severe cases, heroic measures such as pacemakers, replacement of failing valves, and even heart transplants may be needed. There are several types or categories of cardiomyopathy, but in all types, the heart becomes significantly enlarged and often becomes thick and stiff. In some cases, individual muscle cells become replaced by scar tissue. The tendency to cardiomyopathy can be inherited, as genetic traits leave the heart vulnerable to oxidative and other stresses. In most cases, cardiomyopathy emerges slowly following long-standing hypertension and coronary artery disease. In other cases, cardiomyopathy may result from acute or chronic toxicity, such as drug reactions, toxin exposure, alcoholism, and other cardiotoxic contributors. Cardiomyopathy may also occur in children, but such cases are outside the scope of this book.

Diuretics may be offered to patients with edema but may cause loss of calcium, potassium, and magnesium, which can further challenge cardiac muscle contraction and arrhythmia. Beta blockers may be prescribed for patients presenting with hypertension, or when the overworked heart muscle starts to display tachyarrhythmias, which are actually a compensatory mechanism in response to poor perfusion. The use of beta blockers and other hypotensive agents often makes such patients so weak that they can barely walk across the room, but they are the standard of care in congestive heart failure (CHF) because they have been shown to reduce mortality. Herbal medicines may be complementary to or help reduce the required dosage of these agents. When there is reduced ejection fraction in cases of heart failure, cardiac glycosides and *Coleus* may be complementary to beta blockers. *Coleus forskohlii* is a traditional Ayurvedic plant and the source of forskolin, a labdane diterpenoid and a potent adenylyl cyclase activator. Forskolin increases intracellular levels of cAMP leading to activation of protein kinase, which in turn induces relaxation of vascular smooth muscle cells. *Coleus* can help lower blood pressure, especially in deficient, lethargic patients. Cardiac glycosides are capable of helping the heart contract more powerfully—the classic example being *Digitalis*. *Viscum album* (mistletoe) is a traditional Western herb for arrhythmias and the symptoms of congestive heart failure, and *Viscum articulatum* is a traditional Chinese herb for hypertension; North American species may be substituted if unavailable. *Viscum* may have diuretic effects, helping to reduce blood pressure, along with an ability to promote plasma nitric oxide, and to protect against glucocorticoid stimulation of hypertension.

Sheng Mai San for Heart Failure

This traditional Chinese formula has been used for heart disease, coronary artery disease, and diabetes to support recovery following myocardial infarction, hypotension, altitude sickness, arrhythmia, angina pectoris, myocarditis and viral infections of the heart, rheumatic heart disease, and other cardiovascular disorders. Sheng mai san is specifically indicated for shortness of breath,

weakness, and cardiac coughs. Animal research shows the blend to improve reactive oxygen and nitrogen species and to have significant antioxidant effects.[85]

Panax ginseng	3 ounces (90 g)
Ophiopogon japonicus tuber	3 ounces (90 g)
Schisandra chinensis fruit	2 ounces (60 g)

Combine the herbs. Decoct the mixture for 15 minutes, let stand for 15 minutes more, and drink 3 or more cups per day.

Cardiac Glycoside Tincture for Heart Failure

Apocynum cannabinum contains cardiac glycosides and is traditionally used to treat the symptoms of heart failure. Animal studies suggest the plant to improve blood pressure via vasorelaxation and antioxidant effects. Enhanced renal circulation may also improve the excretion of retained fluid and contribute to the antihypertensive effects. *Plectranthus forskohlii* (also known as

Options for Treating Congestive Heart Failure

These herbs and nutrients are some of the most specific options for treating heart failure. Note how some formulas in this section of the chapter combine cardiac glycoside–containing herbs with diuretics and other specifics.

HERB OR NUTRIENT	MODE OF ACTION
Angelica sinensis	Can serve as a blood-moving, anti-inflammatory, and antithrombotic base in heart formulas.
Convallaria majalis	Contains the cardiac glycoside convallarin, among many others.
Crataegus species	Can act as a nourishing base in all types of heart formulas and is said to improve nutrition to the heart muscle itself.
Digitalis purpurea	Contains the cardiac glycoside digoxin, among many others.
Drimia maritima (also known as *Urginea maritima*)	Contains the cardiac glycoside strophanthin, among others.
Strophanthus hispidus	Contains the cardiac glycoside strophanthin ouabain, among many others.
Diuretics	Helpful additions to formulas for heart failure to rid the body of dependent fluid.
CoQ10	A potent antioxidant well researched to support circulation in the heart and gums.
Magnesium	Important to inotropic currents of the heart and especially useful for arrhythmias.
Calcium	Important to inotropic currents of the heart and especially useful for arrhythmias.
Taurine	Found in high amounts in the heart muscle, where it is involved in cell integrity and resistance and helps protect against damage from elevated glucose and lipids.
Carnitine	Helps heart muscle and liver cells use free fatty acids as a fuel.

Cardiac Glycosides for Cardiomyopathy

Cardiac glycosides are steroidlike molecules used in the treatment of congestive heart failure and cardiac arrhythmia. These glycosides are found in several plant families, but also in some animals, such as Monarch butterflies, whose larvae feed exclusively on the *Asclepias* (milkweed) family. Their bodies become high in cardiac glycosides as a result, deterring predation. Cardiac glycosides strengthen heart rate and rhythm in situations of heart failure with weak and rapid heartbeat. Cardiac glycosides are the most long-standing cardiac drugs, prescribed for a positive inotropic action, increased cardiac output, and reduced pulmonary pressure, without having a negative effect on systemic blood pressure.[86] Cardiac glycosides are also known as cardenolides and have been long noted to have a positive inotropic effect on the heart. All cardiac glycosides affect ion transport across cardiac muscle cell membranes via effects on Na(+)/K(+)-ATPase enzymes, leading to an increase in intracellular sodium, that in turn increases intracellular calcium and enhances cardiac muscle contractility. Due to positive effects on ion flow, cardiac glycosides can help convert atrial flutter and fibrillation to regular sinus rhythm. Cardiac glycosides enhance cardiac contractility by increasing the concentration of calcium in heart muscle cells via effects on the transport of sodium and potassium. Effects on sodium transport in the renal tubules may also contribute the beneficial effects that cardiac glycosides have on blood pressure.[87]

Digitalis (foxglove) is the most well-known plant source of cardiac glycosides, but *Convallaria* (lily of the valley), *Apocynum* (dogbane), and other plants also contain cardiac glycosides. Plant cardiac glycosides include ouabain and digoxin. Digoxin from the foxglove plant is used clinically, whereas ouabain is used only experimentally due to its extremely high potency. Plants that contain cardiac glycosides are primarily used to treat cardiac failure, increasing cardiac output by improving the force of heart muscle contractility. Animal studies have shown *Convallaria* to increase potassium ions in the atria and increase atrial stroke volume.[88] While convallatoxin is vasoconstrictive, the sum total of all cardiac glycosides and other constituents may have a more vasotonic action and enhance circulation and coronary flow.

Digitalis purpurea,
foxglove

Coleus forskohlii) contains forskolin, a unique diterpene credited with positive inotropic effects via promotion of intracellular cyclic AMP. Many tincture companies still sell the product under the name *Coleus*.

Apocynum cannabinum
Coleus forskolii
Viscum album
Crataegus spp.

Combine equal parts and take 1 to 2 dropperfuls of the combined tincture 3 or 4 times per day.

Cactus and *Convallaria* Tincture for Cardiomyopathy

Convallaria contains cardiac glycosides and has been used traditionally for heart failure and arrhythmias. Many Eclectic herbal authors regard the plant as having a broader therapeutic window than *Digitalis* and therefore as safer and easier to work with. *Selenicereus*, a cactus having gone by numerous names over the century, is specific for an enlarged heart that results from a weak heart muscle. *Crataegus* and *Salvia miltiorrhiza* provide a nourishing base, supporting the connective tissues of the vasculature, and enhancing oxygen delivery and micro-circulation to the heart muscle, reducing inflammation.

Selenicereus grandiflorus
Convallaria majalis
Crataegus oxyacantha
Salvia miltiorrhiza

Combine in equal parts and take 1 or 2 dropperfuls of the combined tincture 3 to 4 times a day.

General Diuretic Tea for Cardiomyopathy

Equisetum, *Urtica*, and *Taraxacum* promote fluid excretion via the kidneys, offering ample minerals to create suitable osmotic gradients. These diuretic herbs are complemented by the circulatory-enhancing, cardioprotective herbs mentioned throughout this chapter.

Urtica spp.
Taraxacum officinale
Equisetum spp.
Salvia miltiorrhiza
Angelica sinensis
Ginkgo biloba
Mentha piperita

Herbs for Heart Failure from Folklore

Several herbs are used in formulas throughout this chapter as medicines capable of strengthening heart action either by acting as positive inotropic agents or by improving oxygen delivery and circulation to the heart muscle itself.

Convallaria majalis, *Digitalis purpurea*, *Ginkgo biloba*, and *Lepidium meyenii* can all be used for heart failure formulas in general, as well as for heart failure with peripheral edema or dyspnea. *Rhodiola rosea* is also useful for heart failure in general. For heart failure with peripheral edema, *Juniperus communis*, *Thuja plicata*, and *Urtica* species are also appropriate as diuretics when paired with a vascular tonic or a cardiac glycoside–containing herb.

Combine equal parts of the dry herbs, or add more mint if desired for flavor. Steep 1 heaping tablespoon per cup water, and then strain. Drink 5 or 6 cups per day.

Diuretic Tincture for Weak Heart with Edema

Cardiomyopathy can make it difficult for venous and interstitial fluids to return to general circulation and fluids can pool in the lower limbs. The use of cardiac glycosides, such as *Convallaria*, can help the heart pump fluids about the body, while gentle diuretics can assist the kidneys to diurese the dependent fluid. Using herbs that contain cardiac glycosides to treat heart failure, however, is best left to experienced herbalists and physicians. This formula contains *Petroselinum* and *Juniperus* as diuretics.

Convallaria majalis	30 ml
Ginkgo biloba	15 ml
Petroselinum crispum	10 ml
Juniperus communis	5 ml

Take ½ to 1 teaspoon of the combined tincture 3 to 6 times daily.

Selenicereus grandiflorus for Congestive Heart Failure

This plant has had its Latin binomial changed several times over the last century and has gone by the names *Cereus grandiflorus* and *Cactus grandiflorus*. One common name is night-blooming cactus. This is a folkloric plant for treating the symptoms of congestive heart failure, helping to stabilize blood pressure and strengthen heart action. It was thought to help strengthen the heart sufficiently to help restore function, via nutritional and trophorestorative effects. *Selenicereus* was specifically indicated in early American herbalism for a weak and enlarged heart, irregular heart action, aortic regurgitation, and angina, especially when associated with depression and lethargy. There has been very little scientific research on the plant, but the amino acid tyramine is found in the plant and credited with a positive inotropic action.[89]

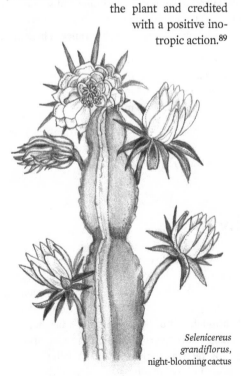

Selenicereus grandiflorus, night-blooming cactus

Convallaria majalis for Heart Failure

Lily of the valley bulbs have been used as a folkloric remedy for heart failure for at least several centuries. *Convallaria* was used as a cardiotonic to improve heart muscle function and was specifically indicated for weak circulation with hypotension, but it may also be used for hypertension when due to underlying cardiomyopathy. Modern herbalists believe the plant capable of improving ejection fraction, cardiac output, and pulmonary hypertension; they also use *Convallaria* for cardiac arrhythmias and irregular rhythms associated with valve disease such as mitral prolapse. Many of the medicinal effects of *Convallaria* are credited to the cardiac glycosides including convallarin, convallatoxin,[90] convallasaponin,[91] strophanthidin, cannogenol, and sarmentogenin.[92] Herbalists believe *Convallaria* to be better tolerated than digoxin from *Digitalis*, due to a broader therapeutic window, but like all cardiac glycosides, it can produce bradycardia in overdose.

Convallaria majalis, lily of the valley

Formulas for Heart Stress at High Altitudes

Many healthy people can feel dizzy at high altitudes and become easily winded with even mild exercise due to low oxygen levels in the ambient atmosphere. I have seen healthy people faint after eating a meal at high altitudes, as blood is shunted into the digestive organs, leaving even less oxygen for the brain. In those with cardiac disease, the symptoms can be severe and necessitate descending to a lower altitude. Several herbs may help prepare the body for high altitudes, but they will not remedy acute altitude sickness once it has already occurred.

Exercise Tolerance and Endurance Tea for High Altitudes

Lepidium is traditionally used in the Andes as a remedy against altitude sickness and is said to improve strength and stamina by supporting heart action at high altitudes. Two options for beverages containing *Lepidium* are given here.

Lepidium meyenii powder
Salvia miltiorrhiza root
Glycyrrhiza glabra shredded root
Astragalus membranaceus shredded root

Combine equal parts of the dry herbs. Gently simmer 1 teaspoon of the herb blend per cup of hot water for 10 minutes in a covered saucepan. Remove from the heat and let stand 10 minutes more, and then strain. Drink freely, 3 or more cups per day.

Variation: Prepare the *Glycyrrhiza*, *Salvia*, and *Astragalus* only into a tea. Place 1 cup of the finished filtered tea in a blender with 2 tablespoons *Lepidium* powder and 2 cups almond, soy, or other milk. Drink once or twice a day prior to departing for high altitudes and continue while residing at higher elevations. The use of a CoQ10 supplement would complement these formulas.

Tincture for Prevention of Altitude Sickness

Mountain climbers and those arriving by plane to high altitudes can use these herbs to prevent altitude sickness and improve athletic performance. *Lepidium meyenii* (maca) grows only at high elevations in the Andes, and modern research supports the traditional claim that the plant enhances exercise ability and stamina at high altitudes. *Lepidium* may exert hypotensive effects by enhancing urinary elimination of sodium potassium and chloride. *Lepidium* also has significant antioxidant and anti-inflammatory activity contributing to hypotensive effects. Altitude sickness involves pulmonary and visceral vascular congestion, and *Ginkgo* and *Silybum* are also appropriate choices because they are circulation-enhancing herbs. *Silybum* is most well known as a hepatotonic, but it is also noted to protect numerous organs from ischemia and to support glutathione and superoxide dismutase levels in tissues. The adaptogenic herb *Rhodiola* may also enhance energy and endurance at high altitudes.

Rhodiola rosea	15 ml
Lepidium meyenii	15 ml
Ginkgo biloba	15 ml
Silybum marianum	15 ml

Take 1 teaspoon of the combined tincture 3 or more times a day. The use of a CoQ10 supplement would be complementary to this tincture.

Formulas for Arrhythmias

While some mild arrhythmias, like the occasional premature ventricular contraction (PVC), are benign, those associated with chest pain and dyspea are usually associated with underlying heart diease and require immediate medical attention. Arrhythmias can emerge when electrical conduction is unable to orchestrate effective heart contraction in situations of heart failure; they are typically managed with beta blockers, to help calm tachycardia or other antiarrhythmics aimed at improving the ability of the heart muscle to contract and respond to electrical signals. Because blood may pool and clot when not effectively pumped by the heart, many patients will also be put on anticoagulant medication for life. Some patients may benefit from pacemakers for severe bradycardia or ablation procedures for tachyarrhythmias and fibrillation. Valvular disease or failure may be addressed with valve repair or replacement.

There are also several causes of arrhythmias that are not due to electrical disruption in the heart itself, for example irregular heart action that results from an anxiety disorder, such as panic attacks. Hyperthyroidism may cause hypertension and tachyarrhythmias. Other hormonal fluctuations such as perimenopause can be associated with transient heart irregularities. Anemia, dehydration, and electrolyte imbalances can cause arrhythmias and can be ruled out with simple blood tests. Blood tests are also useful to monitor the underlying causes of heart disease such as hyperlipidemia and hyperglycemia and, along with EKGs and echocardiograms, can help monitor heart disease. Hypertension can also provoke arrhythmias in some people. Amphetamine use can, of course, promote heart palpitations, as can psychotropic drugs.

Herbal therapies for heart disease are discussed throughout this chapter, with several more examples following that are more specific for irregular heart rhythm. Other herbal formulas to address heart irregularities associated with functional issues such as thyroid, menopausal, or anxiety disorders are also exemplified in this section. Alcohol, caffeine, and smoking can also worsen cardiac arrhythmias and should be avoided. Weight loss and improved cardiovascular fitness will improve heart function for many people in the early stages of heart disease.

Tincture for Atrial Fibrillation

Atrial fibrillation is a type of cardiac arrhythmia that is extremely rapid, disorganized, and irregular and significantly increases the risk of a stroke. Atrial fibrillation increases in occurrence with advancing age and is typically preceded by hypertension, hyperlipidemia, and diabetes. Due to an increased risk of stroke, herbal platelet antiaggregators can be useful, but because such patients are usually on pharmaceutical anticoagulants,[93] herbal antiaggregators should be introduced slowly while monitoring with appropriate labwork. Magnesium supplementation is also appropriate for helping maintain heart rhythm itself. See also "Platelet Antiaggregators and PAF Inhibitors" on page 77.

Ligusticum porteri
Allium sativum
Ginkgo biloba
Salvia miltiorrhiza
Crataegus spp.

Combine in equal parts and take 3 dropperfuls of the combined tincture per day, long term, to support heart rhythm and reduce the risk of thrombi or ischemic events.

TCM Formula for Paroxysmal Atrial Fibrillation

This variation on a formula from TCM is reported to enhance blood circulation and "clear heart fire" to reduce stasis in the heart and the risk of thrombi and ischemia.

Panax ginseng
Ophiopogon japonicus
Salvia miltiorrhiza
Ziziphus spinosa
Paeonia lactiflora
Coptis chinensis
Schisandra chinensis

Combine the herbs in equal parts and decoct 1 teaspoon of the mixture per cup of hot water, and consume 3 or more cups per day.

Leonurus Simple for Palpitations with Menopause

In most cases, heart palpitations emerging at the climacteric are due to metabolic regulation of the heart, more so than heart weakness or hyperexcitability. *Leonurus* is very specific for this complaint and often quickly effective. Magnesium supplementation is also helpful and complementary to a *Leonurus* simple.

Leonurus cardiaca   60 ml

Take ½ to 1 teaspoon of the tincture 3 or more times daily.

Tincture for Palpitations with Hyperthyroidism

Tachycardia is a common symptom of hyperthyroidism and a metabolic, rather than a true cardiovascular, disorder. Therefore, endocrine herbs are more indicated than vascular tonics. Many of the herbs in this formula contain rosmarinic acid, which may reduce overstimulation of the thyroid by inhibiting autoantibodies from binding to the TSH receptors and by reducing oxidative stress in the inflamed thyroid gland.

Leonurus cardiaca   30 ml
Matricaria chamomilla   30 ml
Melissa officinalis   30 ml
Lycopus virginicus   30 ml

Take ½ to 1 teaspoon of the combined tincture at least 3 times a day. For urgent cases or severe tachycardia, the formula may be taken hourly for several days, reducing as symptoms abate.

Viscum album for the Heart

Mistletoe was a revered plant of the Druids, used for magical and ritualistic purposes and for a variety of physical ailments as a panacea. Over the centuries, *Viscum* has been emphasized for the symptoms of congestive heart failure, arrhythmia, and cardiac weakness and has been developed into an anticancer medicine, including intravenous and subcutaneous preparations used for leukemias, lymphoma, and blood cancers. *Viscum* was traditionally used as a cardiotonic specifically indicated for hypotensive patients with weak hearts and a feeble pulse. *Viscum* may, however, be included in formulas for hypertension when vascular inflammation and atherosclerosis are present.[94] Overall, *Viscum* acts as a vagal nerve tonic and strengthens the heart and cardiac rhythm, normalizing both tachyarrhythmia and bradycardia and reducing nerve hyperexcitability. *Viscum* may support optimal electrical conduction through the Purkinje fibers in the heart and decrease ventricular excitability.[95] *Viscum* flavonoids can reduce tachyarrhythmias,[96] and the flavonoid kalopanaxin reduces adrenalin-induced contraction in the aorta.[97] The herb is indicated for enlarged hearts and incompetent values and for all of the symptoms of congestive heart failure, including dyspnea, heart arrhythmia, edema, cardiac coughs that are worse lying down, exercise intolerance, and angina. *Viscum* lowers blood pressure via calcium channel inhibition,[98] nitric oxide promotion,[99] and muscarinic antagonism in the jugulars.[100] *Viscum* has a positive inotropic effect via mechanisms that include optimization of β-adrenergic innervation[101] and may enhance microcirculation in the heart muscle

itself.[102] *Viscum* may also support microcirculation in general, as well as protect the vasculature from the damaging effects of chemotherapy.[103] *Viscum* phenylpropanoid glycosides, coniferin and syringin,[104]as well as lectins inhibit excessive platelet aggregation. The lectins in *Viscum* have numerous immunomodulating and anti-inflammatory actions[105] and decrease ventricular excitability, and some *Viscum* lectins have a particular affinity for endothelial cells.[106]

Viscum also improves carbohydrate metabolism.[107] Animal studies have shown *Viscum* to support basic metabolism, improve lipid and glucose utilization, prevent lipid elevations in animal models of menopause,[108] and enhance insulin secretion from pancreatic β-cells.[109] Because

Viscum album, mistletoe

metabolic syndrome and diabetes are leading contributors to heart disease, agents that improve glycemic control and vascular inflammation are important therapeutic goals. *Viscum*'s antioxidant and hypoglycemic effects are additional mechanisms of managing chronic heart disease, and they support repair of the vasculature. One human study reported *Viscum* tincture to reduce hypertension and serum triglycerides at a dose of 10 drops, 3 times per day.[110]

Tincture for Palpitations with Hypoglycemia

Low blood sugar can promote heart palpitations as the adrenal glands release adrenaline in an attempt to convert stored fuel into glucose. In addition to dietary therapies and other measures that address the underlying dysglycemia, both *Lepidium* and *Panax* are known to help regulate heart rhythm when due to metabolic disorders.

Lepidium meyenii	60 ml
Panax ginseng	60 ml

Take ½ to 1 teaspoon of the combined tincture 3 or more times a day.

Tincture for Palpitation with Cardiac Damage

When a critical amount of heart muscle or conductive fibers have been damaged due to vascular inflammation, long-term smoking, and small infarctions, life-threatening atrial fibrillation and other arrhythmias may result as the electrical fibers attempt to stimulate damaged or nonconductive tissue. Most patients with such serious pathologies will be managed by a cardiologist with antiarrhythmic drugs. The following herbs are all nontoxic and nourishing and could be considered as trophorestoratives to the heart. Such a formula may be used in tandem with pharmaceutical drugs, but it is recommended to be under a physician's care to ensure that the dose of the pharmaceutical is appropriate. Dosage may need to be reduced as the herbs gently strengthen the heart.

Crataegus spp.	15 ml
Ginkgo biloba	15 ml
Viscum album	15 ml
Selenicereus grandiflorus	15 ml

Take 1 teaspoon of the combined tincture at least 3 times a day. Continue for at least 1 year.

Simple for Palpitations with Stress

Eleutherococcus is specific for heart palpitations due to stress and may be implemented as a simple or crafted into a more complex formula given individual presentations. *Eleutherococcus* is available as a solid extract and may be used in this form or as a tea or tincture.

Eleutherococcus solid extract

Take ¼ teaspoon 2 to 4 times daily, reducing as symptoms improve.

Tincture Formula for Bradycardia

Bradycardia may result from sluggish metabolism as with hypothyroidism, as well as due to impaired electrical conduction. This formula is broad acting and is appropriate for both these situations. *Ginkgo* can improve circulation through the coronary arteries and support heart action, and it also may reduce clot risk associated with slow-moving blood. *Coleus* is a muscle tonic, supporting the metabolic function and contractile ability of individual muscle cells, and *Commiphora* supports metabolic rate and helps reduce hyperlipidemia. *Crataegus* supports a nourishing trophorestorative base.

Ginkgo biloba	15 ml
Coleus forskohlii	15 ml
Crataegus spp.	15 ml
Commiphora mukul	15 ml

Take ½ to 1 teaspoon of the combined tincture at least 3 times a day, increasing the dose and frequency until results are achieved.

Tincture for Stress-Related Palpitations

The following three herbs are specific for tics, twitches, anxiety, and restlessness related to stress and overwork. In some individuals the muscle hyperexcitability may extend to the heart muscle. The use of calcium and magnesium supplements would complement this formula. Ensure adequate sleep and implement self-nurturing, stress-relieving lifestyle practices in all such cases.

Leonurus cardiaca	20 ml
Valeriana officinalis	20 ml
Hypericum perforatum	20 ml

Take ½ to 1 teaspoon of the combined tincture at least 3 times a day. If no improvements are seen with dosing

---

Herbs for Tachycardia with Heart Disease

The following herbs are emphasized in both the folkloric literature and modern research arenas for helping to restore normal heart rates and rhythm when electrical conduction in the heart is impaired due to fibrosis and pathology in cardiac tissue.

Allium sativum	*Lepidium meyenii*
Angelica sinensis	*Panax ginseng*
Crataegus spp.	*Viscum album*
Ginkgo biloba	

Arrhythmia Herbs

Arrhythmias are best treated with an understanding of the underlying cause. Organic heart pathologies are most often to blame, but dehydration, electrolyte imbalances, and metabolic disorders can also be the cause. The main electrical conduction electrolytes—magnesium, calcium, and potassium—are appropriate for all cardiac arrhythmias. Magnesium is sometimes given intravenously in a hospital setting for cardiac emergencies, and many people are deficient in this important mineral, contributing to arrhythmias.

Allium sativum. Garlic may protect against ischemia/reperfusion insult and reduce arrhythmias, and allicin has gentle calcium blocking effects.

Crataegus monogyna, C. oxyacantha. *Crataegus* is reported to improve nutrition to the myocardium and acts as a natural calcium channel blocker. It can inhibit the inward potassium channels and prolong the effective action potential having an antiarrhythmic effect. *Crataegus* may be used as a supportive herb in any formula for arrhythmia.

Convallaria majalis. This herb contains the cardiac glycoside convallarin which the Eclectic authors report is superior to *Digitalis*, less cumulative, and has a wider dosage window.

Cytisus scoparius. This uncommon herb is mentioned in Eclectic herbals for arrhythmias.

Digitalis purpurea. *Digitalis* contains steroidal cardiac glycosides that have a positive inotropic effect on the heart and is specific for tachyarrhythmias, slowing and strengthening the heartbeat.

Gelsemium sempervirens. A potentially toxic herb for experienced clinicians only, *Gelsimium* is a profound nerve sedative. Medical herbalists may use it in minute dosages for its strong nerve suppressive effects.

Ginkgo biloba. Ginkgolide can inhibit excessive $Ca2+$ current and calm cardiac excitability. Ginkgolide may also prevent ischemia-driven arrhythmias by enhancing coronary artery flow and cardiac perfusion.

Leonurus cardiaca. *Leonurus* is especially indicated for palpitations and tachyarrhythmias occurring with hormonal disturbances such as menopause and hyperthyroidism. *Leonurus* improves microcirculation, coronary artery flow,

3 times daily after 1 week, double the dose and/or the frequency. Once results are achieved, the medicine may be gradually reduced over time.

Umbel Tea for Tachyarrhythmias

Osthol can reduce the frequency of action potential in neurons via inhibition of calcium channels. In animal studies, osthol has been shown to relax the thoracic aorta by virtue of its Ca 2+-channel blocking properties and by elevating cyclic guanosine monophosphate (cGMP) levels in vascular smooth muscle.[111] This is presently presumed to be the mechanism whereby some Apiaceae family plants such as *Cnidium, Angelica,* and *Apium* are able to reduce blood pressure. Osthol was noted to reduce hyperexcitability of the atrial electrical fibers when subjected to electrical stimulation. The protection against excessive electrical activity has been credited to calcium channel inhibition. This research suggests that Apiaceae plants might be useful for tachyarrhthmias as well as for hypertension.

Angelica archangelica
Apium graveolens seeds
Rosa canina hips
Salvia miltiorrhiza

Combine the dry herbs in equal parts. Decoct 1 heaping teaspoon per cup of hot water, strain, and drink freely.

and blood viscosity. Leonuridin and leonurine are alkaloids in *Leonurus* credited with improving circulation and reducing platelet activation. These phenylpropanoid glycosides may affect calcium channel currents and are noted to improve EKG readings in animal studies.

Phoradendron chrysocladon (also known as *Viscum flavens*). Early research on *Phoradendron* suggested possible cholinergic effects capable of lowering blood pressure and reducing cardiac output in tachyarrhythmias, yet increasing vascular resistance. Isolated viscotoxins in the plant exert a negative inotropic action, although the sum total of all the plant's constituents may lend a more tonifying effect by enhancing microcirculation and dilating the coronary arteries. Modern research has focused on anticancer effects, and cardiovascular effects have yet to be further elucidated. The *Phoradendron* genera are related to *Viscum*, and *Viscum album* is discussed in the sidebar on page 54.

***Rauvolfia serpentina*.** *Rauvolfia* indole alkaloids such as reserpine and ajmaline are powerful antihypertensive agents, able to calm arrhythmias through neurotransmitter effects. *Rauvolfia* also affects certain voltage-gated potassium channels that play an essential role in the normal electrical activity in the heart, including repolarization and termination of action potentials in excitable cardiac cells.

***Salvia miltiorrhiza*.** Red sage may be used to reduce ischemia-induced arrhythmias, and the tanshinone compounds are shown to enhance ion channel densities in individual myocytes. *Salvia miltiorrhiza* may also support the integrity of gap junction proteins in heart muscle cells protecting against inflammatory damage.

***Selenicereus grandiflorus*.** Night-blooming cactus contains cardioactive steroids used for arrhythmias with an enlarged heart. The plant was emphasized in the Eclectic literature as one of the most specific herbal choices for arrhythmia due to cardiomyopathy where there is an enlarged heart.

***Strophanthus hispidus*.** This plant contains the cardiac glycoside oubain known to upregulate Ca2+ entry into arterial smooth muscle cells, leading to vasocontriction and peripheral vascular resistance, and increase blood pressure, which may enhance perfusion in those with hypotension and circulatory weakness.

Formulas for Peripheral Vascular Insufficiency

Peripheral artery disease usually occurs in tandem with heart disease and carries the same risk for suffering myocardial infarction and stroke. The peripheral arteries can be affected by atherosclerosis and inflammation in the blood, in the same manner as the coronary and cerebral blood vessels. Hypertension, diabetes, and hyperlipidemia can lead to plaque formation, loss of elasticity, and blockage, impairing blood flow to the arms and even more so to the legs. The typical symptom of peripheral artery disease is cramping pain in the calves, referred to as intermittent claudication, and heavy and tired sensation in the legs while walking distances or climbing stairs. Pain typically abates with rest and will return with similar exertion again and again. Many people think they are suffering from muscle cramps and may not seek treatment until the condition is advanced. If left untreated, peripheral artery disease can lead to serious lack of oxygen in peripheral tissues, leading to pain at rest, followed by poor healing stasis ulcers, wounds, and even gangrene necessitating amputation of toes and limbs. Discoloration of the skin and loss of hair on the toes and lower limbs may precede more serious consequences of poor circulation. When recognized in earlier stages, smoking cessation, controlling cholesterol and blood sugar, and improving general cardiovascular health may slow the progression. Physical activity and therapies can be vitally important to promote blood flow and help enhance perfusion.

Peripheral vascular disease also includes chronic venous insufficiency where the veins are unable to optimally return blood from the lower extremities back to the heart due to valve incompetence. Weakness of the venous valves is an inheritable trait, and functioning valves may fail over a period of time due to increased venous pressure that may result from obesity, congestive heart failure, injuries to the vein such as phlebitis, or from the excessive accumulation of wastes in tissues as may occur with metabolic and catabolic diseases. As blood and fluid accumulates in the lower limbs, sensations of aching and heaviness may occur and be accompanied by progressive discoloration of the shins. Stasis ulcers may result, usually on the inner aspect of the lower limb above the ankle.

Herbal therapies may enhance perfusion, support fluid return, treat underlying abdominal pressure, and address risk factors. Formulas can be individualized for specific situations and presentations.

Tincture for Poor Circulation with "Dampness"

Long-standing poor circulation can allow metabolic waste products to remain in the lymphatic and interstitial spaces and cause congestion and a tendency to infection—sometimes referred to as "dampness." Diabetics with impaired circulation often develop thick doughy skin following many years of such vascular stasis. The following herbs may improve circulation and reduce tissue congestion.

Allium sativum	15 ml
Angelica sinensis	15 ml
Ceanothus americanus	15 ml
Iris versicolor	8 ml
Cinnamomum spp.	7 ml

Take 5 milliters of the combined tincture a minimum of 3 times daily.

Herbs for General Poor Circulation

Some of the best "blood movers" include the following herbs. Use these agents for stasis, blood clots, vascular congestion, vascular insufficiency, and hypertension.

Allium sativum	Lepidium meyenii
Angelica sinensis	Rhodiola rosea
Crataegus oxyacantha	Zingiber officinale
Ginkgo biloba	Achillea millefolium
Hibiscus rosa-sinensis	

Tincture for Poor Circulation in the Elderly

Panax is an herb with broad application for the elderly and those with weakness, deficiency, and poor circulation. Mounting research shows that Lepidium may also improve heart function. It can be used in tinctures and Lepidium powder can be added to smoothies. The general blood movers Ginkgo and Angelica may complement these heart tonics. This formula may benefit elderly patients who are generally healthy and without frank heart disease, but whose limbs fall asleep readily and who are becoming weak, dizzy, and exercise intolerant.

Panax ginseng	20 ml
Angelica sinensis	10 ml
Ginkgo biloba	10 ml
Lepidium meyenii	10 ml
Zingiber officinale	10 ml

Take 1 to 2 dropperfuls of the combined tincture 3 or more times daily.

Tincture for Poor Circulation to the Kidneys

This formula for poor circulation to the kidneys, with renal insufficiency, combines agents that support general circulation with a small amount of Thuja to act as a counterirritant. Thuja is slightly irritating to the urinary tissues and may promote glomerular filtration. It should be used only in small doses because it can irritate and inflame the kidneys.

Salvia miltiorrhiza	30 ml
Crataegus spp.	15 ml
Lepidium meyenii	10 ml
Thuja occidentalis	5 ml

Take 1 or 2 dropperfuls of the combined tincture 3 times a day or as often as hourly for conditions of renal failure.

Herbs for Poor Circulation with Diabetes

Consider the following herbs in formulas for poor circulation, high cholesterol, and high glucose related to diabetes.

Allium sativum	Syzygium jambos
Ceanothus americanus	Trigonella
Cinnamomum verum	foenum-graecum
Curcuma longa	Zingiber officinale
Opuntia ficus-indica	

Angelica for the Blood

Angelica sinensis (dong quai) is considered a "blood mover" in TCM, and modern research has identified numerous cardiovascular benefits. *Angelica* may reduce congestion in the tissues through enhanced blood and lymph circulation,[112] and vasodilating effects via both nitric oxide and calcium channel inhibition are demonstrated.[113] *Angelica sinensis* is especially vasodilating to the pelvic vasculature. *Angelica* inhibits platelet aggregation and has antihistamine, anticholine, and antiserotonin effects,[114] all helping to reduce vascular inflammation and allergic reactivity. *Angelica* also has immune-enhancing and antitumor activity. Ferulic acid is credited with many of the cardiovascular benefits, having hypolipidemic and platelet antiaggregating effects, helping to reduce blood cholesterol and triglylerides and to improve blood viscosity. Ferulic acid acts as an antioxidant in plant cells and may also protect animal tissues from oxidative damage. Coumarins and ligustilide also contribute to the platelet antiaggregating effects.[115] *Angelica* has protective and antiproliferative effects on vascular smooth muscle.[116] Coumarins from *Angelica* species have been noted to possess platelet antiaggregating effects either equal to or greatly surpassing that of aspirin. The coumarin compounds also display an ability to calm arachidonic acid and thromboxane-induced inflammatory and platelet-aggregating activity.[117] Mild hyperlipidemic effects have also been demonstrated by coumarins such as umbelliferone found in plants in the Apiaceae family.[118]

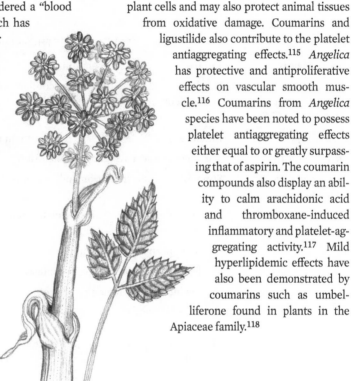

Angelica sinensis, dong quai

Tea for Poor Circulation to the Kidneys

This tea would complement the Tincture for Poor Circulation to the Kidneys. These herbs all support diuresis but are nourishing and nonirritating and can be used long term. This formula is unflavored; mint, *Stevia*, or other pleasant-tasting herbs may be added as preferred by the patient.

Ginkgo biloba	4 ounces (120 g)
Urtica spp.	4 ounces (120 g)
Petroselinum crispum	4 ounces (120 g)
Equisetum arvense or *E. hyemale*	4 ounces (120 g)

Combine the dry herbs to yield 1 pound of tea and store in a large glass jar in a dark cupboard. Steep 1 tablespoon of the blend per cup of hot water for 10 minutes and then strain. Drink 3 or more cups a day.

Tincture for Poor Circulation with Stasis Ulcer

Tibial ulcers due to venous stasis occur in the elderly, diabetics, and anyone else with poor circulation. Elevating the legs is essential to promote healing. *Echinacea* can reduce the breakdown in hyaluronic acid, helping skin lesions heal and inhibiting opportunistic infections. *Phytolacca* can improve lymphatic stasis, and *Ginkgo* improves peripheral circulation. Reduce the *Phytolacca* should any oral irritation occur.

Ginkgo biloba	20 ml
Echinacea angustifolia	20 ml
Phytolacca decandra	20 ml

Take 1 dropperful of the combined tincture every 2 or 3 hours, reducing as the ulcer heals.

Herbs for Stasis Ulcers

Stasis ulcers are common in diabetics and others with circulatory insufficiency. Blood movers, connective tissue restoratives, antimicrobial, and wound-healing agents are all helpful. Treating the underlying diabetic condition if present would also be important.

HERB	THERAPEUTIC EFFECT OR METHOD OF USE
Angelica sinensis	Improves peripheral circulation
Berberis aquifolium (also known as *Mahonia aquifolium*)	Can be used as a skin wash for secondary infections and to astringe purulent secretions
Calendula officinalis	Enhances microcirculation to the basement membrane, enhancing wound healing
Centella asiatica	Enhances connective tissue regeneration
Crataegus oxyacantha, C. monogyna	Supports general circulation and has a stabilizing effect on inflamed vasculature
Echinacea angustifolia, E. purpurea	Promotes the healing of ulcers due to supporting hyaluronic acid; inhibits secondary infections
Ginkgo biloba	Improves peripheral circulation
Hydrastis canadensis	Can be used as a skin wash for secondary infections and to astringe purulent secretions
Phytolacca decandra (also known as *Phytolacca americana*)	May be included in formulas to promote lymphatic circulation and immune response
Symphytum officinale	May be used topically and internally to promote wound healing and connective and epithelia cell repair

Topical Compress for Stasis Ulcers

Stasis ulcers are due to poor circulation, and thus topical applications alone are rarely effective. However, topical compresses and skin washes can deter opportunistic infections and be complementary to circulatory-enhancing agents taken orally. Chronic stasis ulcers can also be foul smelling so the use of essential oils is pleasant for both the doctor and the patient.

Calendula officinalis flower powder
Achillea millefolium powder
Symphytum officinale root powder
Centella asiatica powder
Equisetum spp. powder
Tea tree oil
Lavender oil

Combine 1 ounce (30 g) each of the dry powders in a resealable plastic bag or glass jar and blend well. At the time of use, remove 1 tablespoon of the powder blend and place in a small bowl. Pour 1 cup of hot water over the herbs and stir. Allow to sit for 15 minutes. Soak a soft clean cloth or sterile gauze pad in the mixture, squeeze out the excess moisture, and add 1 or 2 drops each of the essential oils to the pad. Apply to the ulcer and leave in place for at least 15 minutes. Repeat as many times per day as possible, with the legs elevated, such as 15 minutes every hour or two if possible.

Tincture for General Circulatory Weakness

General weakness and deficiency states may be constitutional or may develop secondary to stress, adrenal fatigue, illness, and other causes. General deficiency may manifest as low energy and stamina, a weak voice, and coldness. When deficiency extends specifically to the heart, exercise intolerance, general fatigue, shortness of breath, and dizziness may also accompany. The pulse may be weak or thin, the complexion pale or even grayish or bluish, and the body cold, especially the hands and feet. This formula features the so-called Chi Tonic *Panax ginseng* and combines it with circulatory tonics *Lepidium* and *Crataegus* and the warming agent *Capsicum*. See also "Formulas for Congestive Heart Failure and Cardiomyopathy," starting on page 47, for formulas for actual heart failure as the underlying cause of general circulatory weakness.

Panax ginseng	15 ml
Lepidium meyenii	15 ml
Crataegus spp.	15 ml
Capsicum annuum	15 ml

Take 5 milliliters of the combined tincture 3 or 4 times a day.

Herbal Tincture for Intermittent Claudication

Ginkgo is one of the most studied peripheral vasodilators that may enhance perfusion to the limbs. *Angelica* is another well-researched vascular tonic and a lead "blood mover" in TCM formulas for circulatory insufficiency. *Achillea* is also a peripheral vasodilator, and cinnamon is also a warming blood mover credited with improving some of the underlying factors that impair peripheral circulation, such as elevated blood glucose and vascular inflammation.

Ginkgo biloba	15 ml
Crataegus spp.	15 ml
Angelica sinensis	10 ml
Achillea millefolium	10 ml
Cinnamomun spp.	10 ml

Take 1 or 2 dropperfuls of the combined tincture 4 times a day, long term.

Tincture for Poor Circulation with Claudication

The herbs in this formula can act as peripheral vasodilators, improving blood flow to the limbs, as well as reducing abnormal clotting tendencies to improve blood viscosity. *Achillea* can act as an alterative as well as a peripheral vasodilatory, helping to improve the absorption of the formula and support general digestion. Magnesium, vitamin C, and inositol hexaniacinate are complementary nutrients to this formula.

Achillea millefolium	20 ml
Angelica sinensis	15 ml
Ginkgo biloba	15 ml
Zingiber officinale	10 ml

Take 5 milliliters of the combined tincture 3 or more times a day.

Formulas for Cerebral Vascular Insufficiency

Atherosclerosis and general inflammation in the blood and blood vessels can impair blood flow to the head as coronary arteries are gradually damaged and progressively blocked. Cerebrovascular insufficiency contributes to cognitive impairment and dementia, and improving blood flow to the brain may slow the progression and even improve cognition. Cerebral ischemia induces microcirculation disorders, but reperfusion actually induces greater damage to the brain due to the sudden increase in free radical generation. Research, therefore, investigates the chronic contributors to ischemia, as well as how to prevent reperfusion injury. Transient ischemic attacks (TIAs) can be asymptomatic or associated with symptoms. Even a few moments of ischemia can lead to permanent brain tissue damage and scarring, and immediate medical attention is important. TIAs are associated with small clots, such that the blockage of the affected blood vessel is transient and usually will not result in permanent tissue death or lasting effects. These ministrokes, however, can be harbingers of a catastrophic stroke, and if left untreated, roughly one-third of people experiencing TIAs will go on to have an actual stroke, with permanent consequences such as brain damage, loss of speech, or loss of other motor functions. Larger strokes often present with symptoms such as drooping of the face, visual disturbance, weakness or incoordination in an arm or leg, difficulty speaking, or confusion and difficulty understanding what others are saying. Strokes may be painless, although sudden acute head pain may also occur in some cases.

Ligustrazine from *Ligusticum striatum* (also known as *L. chuanxiong*) helps protect against reperfusion injury.[119] *Gastrodia elata* is used in TCM for treating stroke, epilepsy, dizziness, and dementia, and the constituent gastrodin has been shown to prevent reperfusion injury.[120] Herbs showing promise for vascular dementia

Centella asiatica against Fibrosis

Heart disease, peripheral vascular disease, and cerebrovascular disease all involve gradual loss of muscle cell contractility and connective tissue elasticity, as such cells are slowly replaced by fibroblasts. The insidious process of fibrosis leads to loss of elasticity in vein and artery walls, impaired venous return in the legs, impaired contractile strength of the heart muscle, and loss of circulation to the brain. Therefore, herbs shown to resist fibrosis and support the repair and regeneration of connective tissue can be one important prong of therapies for many cardiovascular, peripheral vascular, and cerebrovascular diseases.

Centella asiatica is a nourishing succulent leafy green native to India. It is used traditionally as a panacea for general strength and longevity and especially for wound healing including in bone, skin, joint, and other tissues. *Centella* has also been specifically recommended to support tissue regeneration for the brain and connective tissues, having supportive effects on hyaluronic acid and collagen synthesis.[121]

Centella may also have vascular protective effects, promoting genes involved with angiogenesis, and neuroprotective effects, promoting axonal regeneration,[122] and possibly protecting against ischemic and reperfusion injuries. *Centella asiatica* offers vascular protective effects to the brain, heart, and tissue by improving cutaneous microcirculation.[123]

Diabetics with impaired microcirculation show improved venoarteriolar response, blood gases, capillary permeability, and general microcirculation from *Centella*.[124]

Centella asiatica, gotu kola

and cognitive decline in the elderly include *Salvia miltiorrhiza*, *Huperzia serrata*, *Ligusticum striatum*, *Ginkgo biloba*, and *Panax ginseng*. Controlling cholesterol and inflammation in the blood is vital to controlling plaque buildup in the carotid arteries, and some severe cases will require surgical approaches to help clear the carotid arteries. Herbs that control high cholesterol and glucose, such as *Commiphora mukul* and *Allium*, may be added to the formula or taken separately as pills or other preparations to help slow the progression of atherosclerotic plaquing. Other herbs and formulas discussed in this section and throughout the chapter may help improve cerebrovascular circulation as well.

Tincture for Impaired Cognition

This formula is to enhance cerebral circulation when hypoxic symptoms such as tinnitus, frequent faintness, and dizziness are present. *Ginkgo* is a noted cerebral circulatory stimulant, and clinical trials suggest that *Ginkgo* can improve cognition in cases of vascular dementia.[125] *Rosmarinus* and *Salvia officinalis* have circulatory-enhancing effects as well as cholinergic effects, making them appropriate for memory loss and confusion related to poor circulation to the head. *Centella* and *Curcuma* are both noted to reduce fibrosis and scarring in the brain related to hypoxia, injuries, toxin exposure, and inflammation. Most of the herbs in this formula could also be prepared as a tea rather than a tincture.

Ginkgo biloba
Rosmarinus officinalis
Salvia officinalis
Centella asiatica
Curcuma longa

Combine in equal parts, and take 1 or 2 dropperfuls of the combined tincture 3 or 4 times a day, decreasing as symptoms improve.

Tincture for Dizziness or Tinnitus

Patients with carotid artery insufficiency or low blood pressure or elderly patients with poor circulation may all experience dizziness, especially with standing. The herbs in this formula all improve circulation to the head. This formula may be simplified and a smaller group of the herbs selected as preferred. Ringing in the ears,

Herbs for Poor Circulation with General Weakness

These herbs are all energizing and can help strengthen circulation and metabolism. They may be considered in circulatory formulas where hypothyroidism, adrenal fatigue, and poor energy accompany poor circulation.

Coleus forskohlii   *Smilax ornata*
Fucus vesiculosus   *Rhodiola rosea*
Lepidium meyenii   *Syzygium jambos*
Oplopanax horridus   *Commiphora mukul*
Panax ginseng

Herbs for Poor Circulation with Digestive Symptoms

Don't overlook the value of including alterative and hepatotonic herbs in formulas for circulatory disease, especially when associated with long-standing hyperlipidemia, diabetes, and metabolic syndrome. This chapter emphasizes the folkloric indications and modern research on herbs for the vasculature itself, but many of the formulas also include an alterative ingredient such as *Achillea*, *Berberis*, or *Curcuma*. Patients with diabetes, obesity, hyperlipidemia, and liver disease would benefit from additional formulas aimed at enhancing hepatic clearance and processing of fat and carbohydrate. Liver congestion also contributes to hypertension. Liver and alterative herbs might be considered in tinctures, teas, and encapsulations, especially where poor digestion, constipation, gas, bloating, and poor fat digestion accompany vascular issues. They can also be considered as one arm of a comprehensive herbal formula because of the many metabolic benefits they offer. See "Specific Indications," starting on page 89, for suggestions on when to choose one of these herbs as a supportive ingredient in circulatory formulas.

Achillea millefolium   *Rumex crispus,*
Berberis aquifolium   *R. oblongifolius*
Curcuma longa   *Taraxacum officinale*
Iris versicolor   *Zingiber officinale*

Nutraceuticals for Ischemic and Hemorrhagic Stroke

Strokes can be classified as ischemic or hemorrhagic, and 87 percent of strokes are ischemic. The remaining 13 percent are hemorrhagic stroke. For ischemic stroke, a standard therapy is anticoagulant medication, antihypertensives, and thrombolytic therapies where needed.[126] Supplementing phosphatidylcholine and related compounds may help protect neurons. Choline also supports phosphotidylcholine synthesis. A nutraceutical called citicoline (cytidine-5′-diphosphocholine) is now available in 20 countries for the treatment of ischemic stroke. Citicoline occurs endogenously where it plays a crucial role in the synthesis of phospholipids including phosphatidylcholine and phosphatidylethanolamine. Omega-3 fatty acids may also protect from neuronal damage if given promptly.

Worms, Germs, and Kinase Enzymes

Acute thrombotic stroke requires immediate infusion of thrombolytic agents that can permeate the blood brain barrier, and most patients will remain on long-term anticoagulant therapy therafter. While immediate thrombolytic drug therapy can help restore perfusion before major neuronal death occurs and may improve stroke recovery, such drugs can also cause intracranial bleeds, which can be fatal. Long-term use of such vitamin K antagonists can also cause gastrointestinal and traumatic bleeds, and they are tricky to combine with many other drugs, herbs, or even foods, because blocking the drugs may lead to new clots, and synergizing drug effects may lead to internal bleeding.

Microorganisms have been recognized as a source of thrombolytic agents, such as streptokinase from *Streptococcus hemolyticus* and staphylokinase from *Staphylococcus aureus*, and traditional foods fermented with these bacteria have been safely consumed for generations in Asia.[127] Tissue plasminogen activators streptokinase and urokinase activate plasminogen into active plasmin, which promotes degradation of fibrin in the blood clots.

However, these agents are expensive and side effects are common, especially allergic reactivity. While streptokinase doesn't have excellent penetration through the blood brain barrier,[128] it and related microbe-derived kinase enzymes are being explored as a maintenance therapy for those with an increased risk for clots, thrombi, emboli, stroke, cardiac ischemia, and infarct. Streptokinase should be administered as rapidly as possible, ideally within 3 hours, following a stroke.[129]

Similar enzymes are produced by earthworms and leeches. Earthworms of the *Lumbricus* genus have anticoagulant properties due to a serine protease enzyme they produce that has thrombolytic and fibrinolytic activity. Leech and earthworm enzymes may be able to help dissolve clots due to direct fibrinolytic effect on fibrin, and it is well absorbed in the intestines. Earthworm kinases show great potential in the management of ischemic vascular disease, preventing platelet activation, protecting neuronal tissue, and helping to dissolve thrombi. See "Nattokinase Capsules for Acute Phlebitis and Clots" on page 80 for more information.

---

tinnitus, may also occur with poor circulation to the ears, and *Ginkgo* is a traditional remedy for the complaint.

Ginkgo biloba	20 ml
Rhodiola rosea	10 ml
Rosmarinus officinalis	10 ml
Salvia officinalis	10 ml
Vinca minor	10 ml

Take 1 or 2 dropperfuls of the combined tincture 3 or more times a day, long term.

Tincture for Poor Circulation with TIAs

The small warning strokes known as transient ischemic attacks (TIAs) should be taken very seriously. TIAs are treated in the same manner as significant ischemic strokes. *Ginkgo, Rosmarinus,* and *Salvia* species may improve circulation in the cerebral circulation. These herbs and *Curcuma* may also reduce the inflammatory

processes that lead to blood reactivity and therefore impair circulation. *Centella* may help protect neurons in situations of oxidative and ischemic stress.

Ginkgo biloba	10 ml
Curcuma longa	10 ml
Centella asiatica	10 ml
Rosmarinus officinalis	10 ml
Salvia miltiorrhiza	10 ml
Salvia officinalis	10 ml

Take 1 or 2 dropperfuls of the combined tincture 3 or more times a day, long term.

Tea to Protect Against Stroke

This formula is particularly aimed at patients with heart disease or atrial fibrillation. *Ginkgo* is approved in China for the treatment of stroke. *Ginkgo* reduces capillary fragility, and terpenoids in *Ginkgo* may inhibit platelet activation, improve viscosity, and reduce the risk of stroke. Salicylates are cyclooxygenase inhibitors that inhibit platelet activation. Aspirin is widely used to reduce the risk of clots, thrombi, and stroke, but salicylate-containing herbs may also work in similar ways. *Apium graveolens* is high in butylphthalide, noted to protect again ischemic brain injury. *Salvia miltiorrhiza* is shown to reduce the impact of ischemia/reperfusion injury in the brain and support regeneration of neural stem cells following stroke.[130]

Ginkgo biloba
Salix spp.
Apium graveolens
Salvia miltiorrhiza

Combine in equal parts and take 1 dropperful of the combined tincture 3 or 4 times a day.

Traditional Chinese Herbs for Stroke Recovery

The following herbs are widely used in Asia for stroke recovery and are reported to help reduce the size of an infarct when given in the first 24 hours. They are also reported to improve cerebral blood flow, protect against reperfusion injuries, and support local angiogenesis and neurogenesis via a variety of mechanisms.[131]

Alisma plantago-aquatica (also known as *A. orientale*)	*macrocephala*
Angelica sinensis	*Ligusticum striatum* (also known as *L. chuanxiong*)
Astragalus membranaceus	*Paeonia lactiflora*
Atractylodes	*Poria cocos*
	Ziziphus spinosa

Formulas for Capillary Fragility and Telangiectasias

Formulas for telangiectasias as a dermatological condition are offered in chapter 5 of Volume I of this set of formularies, but here I cover telangiectasias as a microvascular disorder. Also known as spider veins, telangiectasias can be a normal phenomenon of aging, but may also be associated with alcoholism due to increased vascular pressure in the liver. Pregnancy may also increase vascular pressure, and connective tissue diseases such as scleroderma, lupus, and dermatomyositis are associated with increased sclerosis of blood vessels. Telangiectasias can be cosmetically improved with minor surgical procedures that cause them to collapse and be less noticeable.

Herbs high in flavonoids are noted to help stabilize the vasculature and help preserve the integrity and function of the tiny blood vessels of the dermis. High flavonoid and connective tissue–supportive herbs may help reduce the emergence of new spider veins and may help treat those who bruise easily and recover poorly from vascular traumas, such as repeatedly developing hematomas from relatively minor trauma.

Formula for Weak Blood Vessels and Dilation

Loss of integrity and elasticity in the blood vessels may underlie some cases of varicose veins, and this formula features herbs that support connective tissue integrity. Some patients who bruise easily or scar excessively

following minor injuries also may benefit from this formula. These herbs may also be used following a blood vessel infection or inflammation such as phlebitis to help the tissue fully heal without fibrosis and further

---

Microvasculature Stabilizers

Many flavonoids support capillaries, optimizing the integrity of cells, improving cell-to-cell adhesion, and preventing damage to collagen by a variety of anti-inflammatory and antioxidant actions. Those who experience easy bruising or spider veins may benefit from regular consumption of berries and medicinal herbs high in flavonoids. Because histamines increase the permeability of microcirculatory vessels, allowing white blood cells, platelets, and inflammatory compounds to squeeze through, these herbs may also improve allergic reactivity, helping to stabilize the cells and limit excessive permeability.

Calendula officinalis	*Hamamelis virginiana*
Centella asiatica	*Punica granatum*
Crataegus spp.	*Vaccinium myrtillus*
Curcuma longa	Bioflavonoids
Ginkgo biloba	Vitamin C

Herbs for Bruising and Telangiectasias

These herbs can be helpful for bruising and telangiectasias that occur as a result of microvascular fragility.

HERB	METHOD OF USE
Calendula officinalis	Use topically in compresses or include in teas and tinctures
Crataegus oxyacantha, C. monogyna	Include in teas and tinctures, or use as a solid extract
Curcuma longa	Use in teas and especially tinctures; may be used topically but can temporarily stain the skin yellow
Hamamelis virginiana	Use topically in compresses or include in teas and tinctures
Punica granatum	Use as a food, or especially a juice
Vaccinium myrtillus	Include in teas and tinctures, or use as a solid extract

loss of elasticity. *Calendula* enhances connective tissue regeneration and may enhance microcirculation in the skin. *Centella* and *Equisetum* are classic herbal medicines used to heal connective tissue. This formula may be prepared as a tea or a tincture as desired. *Calendula* and *Centella* are listed in old herbal formularies for endocarditis and phlebitis. This formula uses mint simply as a flavoring agent, and it can be omitted from the tincture.

Calendula officinalis dry petals	4 ounces (120 g)
Equisetum spp.	4 ounces (120 g)
Centella asiatica	4 ounces (120 g)
Rosa canina dry, ground hips	4 ounces (120 g)
Mentha piperita	4 ounces (120 g)

Combine the dry herbs and blend together to yield 20 ounces of dry tea mixture. Store in an airtight container in a dark, dry cupboard; the blend should keep well for 1 year in those conditions. Steep 1 tablespoon per cup of hot water and then strain. Drink 3 or more cups daily. To prepare as a tincture, combine in equal parts and take 1 dropperful of the combined tincture 3 times daily.

Therapies for Bruising and Hematomas Due to Trauma

Herbs for blood vessel trauma and connective tissue tonics would be valuable in a first aid kit or appropriate to treat patients who have suffered a fall or a motor vehicle accident. *Arnica*, *Hypericum*, *Symphytum*, and bromelain are among the agents most emphasized in folkloric and modern herbalism to speed the healing of soft tissue

trama and offer pain relief. A topical oil or ointment containing *Arnica montana* is a well-known therapy for trauma. Use as promptly as possible following injury.

Bromelain is a proteolytic enzyme from pineapples. It is available in capsule form and may help speed the healing of extensive bruising and tissue trauma. Bromelain may also reduce the risk of developing phlebitis or developing thrombi in injured blood vessels. Take 500 to 1,000 milligrams at least 3 times daily, between meals.

Use *Hypericum perforatum* orally as a tea or tincture as well as topically as a compress, oil, or salve. *Hypericum* is a folkloric classic for bruising, strains, sprains, whiplash, and trauma to innervated areas such as the neck, spine, and limbs.

Tincture for Telangiectasias

All of the herbs in this tincture contain flavonoids and are mentioned in the herbal folkore for strengthening blood vessels.

Hypericum perforatum
Curcuma longa
Aesculus hippocastanum
Crataegus oxyacantha, C. monogyna

Combine in equal parts and take 1 or 2 dropperfuls of the combined tincture 2 or 3 times a day, long term.

Tea to Support Capillary and Vascular Integrity

Similar to the Tincture for Telangiectasias, this tea also features herbs folklorically recommended for bruising, capillary enlargement, or trauma and to improve healing following vascular trauma.

Crataegus spp.
Hypericum perforatum
Hamamelis virginiana
Calendula officinalis

Combine equal parts of the dried herbs, and add an additional flavoring herb of choice if desired. Steep 1 tablespoon of the mixture per cup of hot water and then strain. Drink 2 or 3 cups each day, long term.

Herbal Juice to Strengthen Capillaries

Fruits are excellent sources of flavonoids, but even the best of them may provide too much sugar for those with underlying diabetes or heart disease. For best results, dilute the juice with herbal tea, such as the previous Tea to Support Capillary and Vascular Integrity, or with sparkling water. Flavonoids help reduce vascular disease risk factors, and dietary intake of flavonoids should be maximized for life.

Vaccinium myrtillus (blueberry juice)
Vitis vinifera (grape juice)
Prunus serotina (cherry juice)
Punica granatum (pomegranate juice)

Use a single juice at a time or a combination, as desired. These juices can be used to flavor water, make spritzers, or use as an ingredient in mocktails, salad dressings, sauces, or marinades.

Formula for Raynaud's Syndrome

Raynaud's syndrome is episodic cold-induced cutaneous vasospasms in the hands and feet symmetrically that are triggered by the body's effort to shunt blood from extremities into the trunk, giving priority to the vital organs. In this overreactivity to cold temperatures, the person's body thinks it is in danger of freezing and shunts blood to the core to spare the heart and brain, at the expense of the hands and feet. Blood vessels in the distal extremities vasoconstrict and affect hemodynamics to maintain a larger blood volume in the head and thorax. The typical symptoms are numbess in the affected digits, associated with pallor and temporary loss of sensation and motor dexterity. In its most severe form, extreme vasoconstriction leads to ischemia and can lead to cell death, and ulcerative or even gangrenous lesions. Although not common, some cases may be the first presenting symptoms of a connective tissue disease such as systemic sclerosis and precede the onset of other symptoms for many years. Females are affected to a greater degree than males, and there may be some overlap with migraine. The simplest treatment for the milder forms of this syndrome is to take care to not allow the hands or feet to get cold. In most cases, Raynaud's syndrome is not serious and merely a nuisance. However, for those who experience symptoms simply by being in a cold room or who suffer severe symptoms, taking *Angelica sinensis* and *Ginkgo* on a regular basis as a preventive may be helpful. *Angelica* is a traditional frostbite remedy in China and may attenuate cold-induced vasocontriction. Ligustilide and ferulic acid in *Angelica* are credited with balancing effect on cold regulation sensors and pathways in the tissues.[132]

Tincture for Severe Raynaud's

As there is no research on herbal medicines to treat Raynaud's syndrome, this formula is my best guess at the use of blood movers and agents that may help reduce vasospasms, including the energetically warm herbs cinnamon and ginger.

Angelica sinensis	15 ml
Ginkgo biloba	15 ml
Achillea millefolium	10 ml
Tanacetum parthenium	10 ml
Cinnamomum verum	5 ml
Zingiber officinale	5 ml

Take 1 or 2 dropperfuls of the combined tincture 3 or more times daily.

Formulas for Anemia

Many types of anemia, including the rarer types associated with blood cancers and genetic diseases, are outside of the scope of this text. Here, I offer formulas for the relatively common situations of iron deficiency anemia and vitamin B_{12} deficiency anemia.

Iron deficiency is estimated to be the most prevalent micronutrient deficiency worldwide and contributes to multiple pathologies, with children and pregnant women particularly at risk. Poor diet and/or poor digestion and absorption may limit iron reserves. Underlying contributors to anemia include blood loss, which is easily identified in women with heavy menses, but may not be readily discernible in others. In those cases, the cause should be sought out, such as by testing for occult blood in the stool. Iron supplementation is often required, yet carries its own risks because the body has no mechanism to excrete excess iron. Iron is highly toxic when present in high quantities because it facilitates the generation of highly reactive oxygen species that can damage cellular constituents. Excess iron also supports pathogenic bacteria and protozoa in the gut and susceptibility to malarial infections in some parts of the world.[133] Most bacteria have an obligate need for iron, but the species most beneficial to the gut, *Lactobacillus* and *Bifidobacterium* species require little to no iron, thus long-term use of iron can encourage intestinal dysbiosis and support more pathogenic strains of bacteria. Common side effects of taking iron supplements are constipation and intestinal irritation. Therefore, natural forms of iron that deliver the mineral in an organic foodlike matrix are preferable for long-term use. The various forms of iron present in food differ in their bioavailability, none being highly absorbed. The heme form of iron from meat, fish, and seafood is the most bioavailable. The highest sources of nonheme iron are seeds, grains, nuts, and leafy green vegetables and from 2 percent to 20 percent may be absorbed depending on the chemical type. These types include the iron salts (ferric citrate, phosphate, phytate, oxalate hydroxide) and larger molecules such as ferritin. Ferritin occurring in plants is referred to as phytoferritin and legumes are among the foods richest in iron and phytoferritins, including soybean (*Glycine max*), pea (*Pisum sativum*), and alfalfa (*Medicago sativa*). Phytates occurring in legumes and other plants, however, interfere with the absorption of minerals including iron. Phytates, also known as phytic acid, are a phosphorus-containing molecule sometimes referred to as an "antinutrient." The consumption of 5 to 10 milligrams of phytic acid can reduce iron absorption by 50 percent, and although phytic acid has numerous benefits as well, those with iron deficiency anemia might benefit from limiting the intake of these compounds. Despite their many nutrients and benefits, legumes, whole grains, and nuts are the richest sources of phytic acid. Sprouting beans, grains, and nuts greatly reduces their phytate content, as does long, slow cooking with a bit of salt and vinegar.

B-vitamin deficiency, whether from poor diet or poor absorption, can also result in another type of anemia with larger than average red blood cells, referred to as megaloblastic anemia. Pernicious anemia results from inadequate absorption of vitamin B_{12} due to lack of intrinsic factor for intestinal parietal cells. Deficiencies in both B_{12} and folate can cause megaloblastic anemia. Low B_{12} is also associated with neurologic symptoms, which are absent in folate deficiency. Low levels of vitamin B_6 are less common but can also cause anemia and may be associated with depression.

Tincture for Iron Deficiency Anemia

It is necessary to consider and treat directly the underlying cause of anemia and rule out any intestinal lesions or other source of chronic blood loss. For those with poor iron absorption, the use of iron-containing herbs combined with agents capable of enhancing its absorption will be helpful. This formula can also complement the use of iron supplements by enhancing their absorption and preventing the constipation that often accompanies as a side effect. Molasses is also an excellent source of iron. Be sure to use real molasses and not molasses-flavored corn syrup. *Rumex* is a folkloric herb for anemia, and although it is not a high-iron herb itself, it is reported to boost the absorption of iron, a mineral typically poorly absorbed.

Molasses	30 ml
Rumex spp.	30 ml
Taraxacum officinale	30 ml
Liquid iron supplement	30 ml

Pour the molasses in the bottom of a 4-ounce bottle, add the 2 tinctures and the commercial liquid iron supplement (one supplying ferrous citrate or similar iron salt), and shake vigorously. Take 1 dropperful of the combined tincture 3 or more times daily.

Ginkgo and Circulatory Enhancement

Ginkgo has been well documented to improve cerebral blood flow and provide an antioxidant action to nerves and vasculature. Ginkgolide B has been shown to act as a platelet-activitivating factor antagonist.[134] *Ginkgo* is useful in cases of arterial insufficiency, intermittent claudication, ischemic heart disease, and other cases of tissue hypoxia[135] and is being explored as a means of reducing tumor, cancer, and degenerative disease processes. *Ginkgo* will also relax the vasculature and enhance perfusion, which is one proposed mechanism by which heart muscle improves activity, memory, and senility. In cases of cerebral insufficiency, one clinical study showed cerebral blood flow to be increased 70 percent following the ingestion of *Ginkgo*.[136]

Since *Ginkgo* enhances peripheral circulation, it may benefit those with peripheral vascular disease. Several studies have shown *Ginkgo* to promote blood flow in both healthy and compromised blood vessels.[137] *Ginkgo* supplementation has been shown to reduce symptoms of tissue hypoxia in human trials.[138] One clinical trial noted comfortable walking distance was lengthened in those suffering from arterial insufficiency. Thromboxane synthesis and thrombus formation have been inhibited in animal studies. Animal studies have also shown vasospasms to be reduced with the administration of *Ginkgo*.[139] Flavone glycosides and terpenes including ginkgolides have been credited with these effects. For example ginkgolide, a flavonoid, competes with platelet-activating factor (PAF) for binding sites and minimizes the ability of this compound to promote platelet aggregation. *Ginkgo* is thought to help modulate fibrinolytic activity and improve both blood viscosity and vessel elasticity. *Ginkgo* is recommended in cases of thrombosis and phlebitis, which are other pathologies associated with atherosclerosis and loss of elasticity. *Ginkgo* is readily available in a variety of forms and formulations that are safe to use in tandem with pharmaceutical heart medicines, perhaps even anticoagulants. One very large clinical investigation found no increased risk of bleeding or other adverse effects in those using *Ginkgo* in tandem with warfarin.[140]

Ginkgo biloba, ginkgo

Tea for Iron Deficiency Anemia

These herbs all have a moderate iron content and are considered safe to ingest during pregnancy, a time when some women become anemic. *Rubus* contains many minerals and is considered to be a uterine tonic.

Mentha piperita	3 ounces (90 g)
Rubus parviflorus	1 ounce (30 g)
Taraxacum officinale leaf	1 ounce (30 g)
Verbascum thapsus	1 ounce (30 g)
Juice of 1 lemon or 1 tablespoon herbal vinegar	
Water	4 cups (960 ml)
Molasses	

Combine and blend the dry herbs. Place 3 tablespoons of the herbs in the bottom of a saucepan and add the lemon and/or herbal vinegar and allow to stand overnight. In the morning, add the water and bring to the lowest possible simmer, holding it there for 5 minutes. Allow to steep covered for 10 minutes more, and then strain. Add a ½ teaspoon of molasses to each cup and drink freely, hot or cold.

Herbs for High-Iron Teas

According to Dr. Duke's Phytochemical and Ethnobotanical Database (https://phytochem.nal.usda.gov/), these herbs are among the richest sources of iron. Because a simple infusion process does not extract minerals well, try a long overnight soak of the herbs in water prior to bringing to a brief simmer. Adding a bit of lemon juice or vinegar to the soaking water can help extract more minerals including iron. Grinding the herbs into a powder and consuming them as capsules or preparing into herbal foods can provide the most iron. The following plants are listed in order of the highest iron content first.

Taraxacum officinale (dandelion)
Artemisia vulgaris (mugwort)
Stellaria media (chickweed)
Verbascum thapsus (mullein)
Mentha pulegium (pennyroyal)*
Carthamus tinctorius (safflower)
Petasites japonicus (butterbur)*
Trifolium pratense (red clover)
Thymus vulgaris (thyme)
Camellia sinensis (green tea)
Arctium lappa (burdock)
Nepeta cataria (catnip)

* Caution: Avoid in pregnancy!

High Iron Fruit and Molasses Vinegar

Minerals including iron are better extracted with decoctions than with tinctures, and still better extracted by soaking them in vinegar. Herbal vinegar extracts are technically referred to as acetractas and are a simple method of getting more iron into the daily diet. This formula uses high mineral herbs that can be purchased or gathered from the wild.

3 cups (720 ml) apple cider vinegar
3 cups (720 ml) basalmic vinegar
1 cup (60 g) fruit (such as berries, apple or pear chunks, apricots, or plums)
1 cup (240 ml) blackstrap molasses
1 cup (10 g) fresh *Stellaria media* leaves and stems
½ cup (30 g) *Taraxacum officinale* roots and leaves
¼ cup (15 g) *Artemisia vulgaris* leaf
¼ cup (15 g) *Verbascum thapsus* leaf

Place all in a blender and blend on high until the fruit is liquefied and all is homogenized. Transfer to a large canning jar and store in a dark cupboard. Shake vigorously each day for 1 to 2 months. Filter through a wire mesh kitchen strainer, and a second time through a jelly bag (a muslin pouch). Add more vinegar or more molasses to the finished product to taste. Use the vinegar to make salad dressing and to drizzle on cooked and raw veggies. You may omit the *Stellaria* (chickweed) if unavailable or substitute fresh dandelion leaves.

Treating Hemochromatosis

Hemochromatosis is a hereditary condition of iron overload, most common in people of Celtic or Scandinavian origin, occurring in roughly 0.5 to 1 percent of such populations. Iron overload can also occur as a side effect of receiving blood transfusions. When hereditary hemochromatosis is not recognized and treated, the iron builds up in the tissues and causes muscle and joint pain, fatigue, inflammation, liver damage, and cardiac injury. It also causes other toxicities in the liver, spleen, bone marrow, pituitary, pancreas, and central nervous system. The earlier the condition is recognized and treated, the better to prevent liver and other tissue changes. The primary treatment is regular phlebotomy to help rid the body of excess iron. Deferoxamine is a pharmaceutical iron chelator use to treat iron overload, but it can cause

side effects including hepatic dysfunction, allergy, and agranulocytosis. Calcium channel blockers have become potential new therapies for reducing oxidative injury induced by iron overload in these organs. Herbs that bind or chelate iron may be helpful, as well as general anti-inflammatories to help protect cells, tissues, and organs. These herbs cause less harm than the pharmaceuticals.

Iron accumulation in the myocardium can lead to fibrotic cardiomyopathy. *Salvia miltiorrhiza* (dan shen) has been used for hundreds of years in China to treat cardiovascular diseases and may offer the heart some protection against myocardial fibrosis in cases of chronic iron overload.[141] *Salvia miltiorrhiza* is prepared into injectable medications and used to treat renal, hepatic, and cardiovascular fibrosis. Animal studies show dan shen injections can also prevent iron accumulation in the kidneys, increase antioxidant systems in the kidneys, and help normalize elevated blood urea nitrogen and creatinine.[142] Iron chelation may contribute to the anticancer and neuroprotective activity of curcumin,[143] and *Curcuma longa* may be included in formulas for hemochromatosis. *Mucuna pruriens* may help slow the progression of neurological disease, such as Parkinson's disease, in part through iron-chelating activity.[144] Baicalein and its glycoside baicalin, found in *Scutellaria baicalensis*, are natural iron chelators shown to increase antioxidant status and decrease iron content and lipid peroxidation in the liver of mice with iron-overload–induced liver oxidative injury.[145] Tetramethylpyrazine and ferulic acid, found in *Ligusticum striatum*, chelate ferrous iron and decrease the formation of hydroxyl radicals that contribute to oxidative stress, and the plant has been featured in a wide variety of cardiovascular, pulmonary, and respiratory formulas in TCM. Silymarin, the active flavonolignans in *Silybum marianum*, may help protect organs and tissues from iron overload.[146]

Protocol for Hemochromatosis

- Regular phlebotomy
- Resveratrol capsules, 2 capsules twice a day
- Quercetin capsules, 2 capsules twice a day
- Green tea capsules, 2 capsules twice a day
- *Curcuma* capsules, 2 capsules twice a day
- *Silybum marianum*, 2 capsules twice a day
- A diet rich in flavonoids—berries, grapes
- Tincture for hemochromatosis, 1 or 2 dropperfuls, 3 times a day

Tincture for Hemochromatosis

Animal studies have shown *Salvia miltiorrhiza* and *Scutellaria baicalensis* to protect tissue from iron overload, and due to their numerous antioxidant and well-established anti-inflammatory effects on the vasculature, these herbs would make useful tonics for those with hemochromatosis to consume on a regular basis. *Curcuma longa* flavonoids and flavonoids in general, including those in hawthorn berries (*Crataegus* species), may also protect the vasculature and tissues from high levels of iron.

Crataegus oxyacantha
Curcuma longa
Salvia miltiorrhiza
Ligusticum striatum
Scutellaria baicalensis

Combine in equal parts, and take 1 or 2 dropperfuls of the combined tincture 3 or more times daily.

Flavonoids and Natural Iron Chelation

Many naturally occurring phytochemicals, such as flavonoids, will bind iron and may be one of the many mechanisms whereby flavonoids protect the vascular tissues and offer other anticancer and anti-inflammatory effects. For example, the common flavonoid quercetin binds $Fe2+$ more strongly than the well-known $Fe2+$ chelator ferrozine.[147] Iron chelators reduce oxidative stress and may attenuate disease progression in various cardiovascular diseases, atherosclerosis, neurodegenerative diseases, and cancer, either through promotion of cellular growth and proliferation or through participation in redox reactions that catalyze the formation of reactive oxygen species and increase oxidative stress. Proanthocyanidins, epicatechins, flavonols, and anthocyanin contain an iron-binding group that contributes to their many antioxidant and anti-inflammatory activities. Procyanidins, such as those occurring in *Crataegus*, featured throughout this chapter, may have such versatile cardioprotective effects due to free radical scavenging activites. *Camellia sinensis* catechins have extensive research on anticancer and tissue-protective effects, and some such activity may be due to the sequestering of iron. *Vitis vinifera* contains procyanidins including resveratrol, which has a strong iron-chelating effect.[148] All such flavonols are rapidly absorbed and slowly eliminated from the plasma and are delivered to liver, heart, skin, brain, and other organs. The ability to bind iron may contribute to the broad antioxidant, cardio- and vascular-protective, hepato/renal protective, and anticancer benefits of these important dietary and herbal nutrients.

Formulas for Vascular Infections and Endocarditis

Treating sepsis and vascular infections such as endocarditis will usually require pharmaceuticals, and often intravenous antibiotics in a hospital setting, but for lesser infections or for complementary therapy, antimicrobial herbs with an ability to enter general circulation are appropriate. *Allium, Echinacea, Andrographis,* and *Curcuma* are among those herbs thought to have systemic effects. High-flavonoid herbs may help protect the blood vessels from the damaging effects of infections and inflammatory processes and may include *Salvia miltiorrhiza, Crataegus, Calendula, Hibiscus,* and *Rosa canina.* The following formulas are meant to complement the therapies offered by a cardiologist or hospitalist.

Herbs for Heart Infections and Inflammation

Cardiac infections usually require close medical supervision, antibiotic therapies, and in many cases, inpatient care. Therefore, few herbalists of the present era have much experience treating pericarditis or endocarditis with herbs alone. Patients with pre-existing mitral valve prolapse or a history of rheumatic fever are often given antibiotics when having routine dental work to reduce the risk of endocarditis, and naturopathic physicians have more experience in this regard. Antimicrobial herbs may be used in lieu of prophylactic antibiotic pharmaceuticals. The following herbs have been mentioned in folkloric literature as appropriate in cases of endocarditis and pericarditis. These can be used in tandem with antibiotic or steroidal pharmaceuticals, as well as in convalescent therapies to help protect the tissues and support recovery and repair.

Berberis aquifolium	*Echinacea* spp.
Calendula officinalis	*Hypericum perforatum*
Centella asiatica	*Salvia miltiorrhiza*
Coptis chinensis	*Symphytum officinale*
Curcuma longa	

Support Tincture for Endocarditis

This tincture might be taken in tandem with antibiotics or as a convalescent therapy following hospitalization or antibiotics.

Allium sativum
Echinacea angustifolia
Salvia miltiorrhiza
Curcuma longa
Crataegus oxyacantha

Combine in equal parts in a 2- to 4-ounce bottle. For acute infections, high doses such as 1 dropperful of the combined tincture hourly are appropriate. For convalescence and to support antibiotic therapy, a dropperful 4 or 5 times a day may be helpful. A tea of *Calendula* and rose hips would complement this tincture.

Tea to Support Recovery from Endocarditis

A cardioprotective tea such as this may help prevent fibrosis and support full recovery.

Salvia miltiorrhiza
Angelica sinensis
Rosa canina
Crataegus oxyacantha
Centella asiatica
Calendula officinalis
Glycyrrhiza glabra

Combine equal parts of the dry herbs and decoct 1 teaspoon of the herb mixture per cup of hot water. Drink 4 to 6 cups per day for several weeks to several months.

Tincture for Mitral Regurgitation

Apocynum is a cardiac glycoside–containing herb to be used by experienced clinicians only. Here it is safely dispensed by using only 15 drops of the entire formula at any one time. *Echinacea* may seem like a curious choice but was often employed by some physicians in the early 1900s for decayed and weakened connective tissue, including mitral valve prolapse.

Apocynum cannabinum	30 ml
Echinacea angustifolia	30 ml

Take 15 drops of the combined tincture every 4 hours.

Formulas for Venous Congestion and Varicosities

Varicose leg veins and hemorrhoids are common complaints associated with vascular congestion in the lower limbs. They occur in 25 to 50 percent of the US population. Lower extremity varicosity may be a cosmetic concern, or it may be associated with tingling, burning, or itching sensations or sensations of heaviness and fatigue. Leg varicosities and hemorrhoids may also ulcerate and bleed. Long-standing leg varicosities may also allow clots to form and lead to phlebitis and emboli. Varicosities in the vulvar veins may also form during pregnancy or following labor and delivery. Varicose veins in the testes may also form and result in varicocele. Vulvar varicosities will usually spontaneously resolve, while varicoceles in men are usually persistent. Varicocele may impair fertility in men and is associated with increases in reactive oxygen species and impaired sperm health,[149] in some but not all cases.

Vascular congestion in the deep leg veins, known as deep vein thrombosis (DVT), is a more serious condition and should be a concern for any patient who presents with redness, pain, and warmth in the calf without obvious cause. Such deep veins are prone to forming thrombi and may be the source of an embolus that lodges in the lungs, causing acute chest pain with dysnea. Pulmonary emboli can be fatal and immediate medical attention is required. Genetic clotting and vascular diseases may contribute to DVT, and having occurred once, the condition may become chronic in some patients related to blood-flow stasis, vessel-wall damage and increased blood viscosity. Sophisticated imaging techniques, such as compression ultrasound, assist in diagnosis, but are not always available in a timely fashion. Many clinicians begin treatment while referring or sending out for further workup due to the serious consequences of an embolus. In many cases, DVT can be asymptomatic, and many patients who have died from acute pulmonary embolism are found to have DVT upon autopsy. A dilemma in treating DVT is balancing the need for anticoagulants to help dissolve the thrombi, while avoiding initiating bleeding elsewhere in the body. Anticoagulation options include IV heparin acutely, followed by vitamin K antagonists, or the newer non-vitamin K oral anticoagulants such as rivaroxaban, dabigatran, apixaban, and edoxaban. Natural anticoagulants include nattokinase, lumbrokinase, and serrapeptase, and while not replacements for expert medical attention, these enzymes may complement standard therapy or be initiated as a daily preventive therapy, following the resolution of DVT. Because the

Fibrinolytic Botanicals

Varicosities and atherosclerosis involve a loss of elasticity in the blood vessels and may benefit from eating or supplementing with fibrinolytic botanicals. Fibrin is a tough, scar-like tissue, which is deposited within the veins and artery walls, reducing their elasticity, and interfering with normal vascular and endothelial responses. Affected blood vessels are more susceptible to thrombi formation, arteriosclerosis, and phlebitis, and fibrinolytic agents may slow the progression and the potential serious consequences. Fibrin is implicated in many heart attacks, strokes, and blood clots, because fibrin is deposited in the vasculature as a component of plaque. Fibrin can crack as the tissue becomes fragile and inelastic, allowing for new bleeding that then clots and occludes blood vessels or may break free as an embolus. Substances that break down fibrin include *Capsicum annuum*,[150] *Allium sativum*,[151] *Allium cepa*,[152] *Zingiber officinale*,[153] *Ananas comosus*,[154] and *Curcuma*.[155] These same substances also act as platelet antiaggregators because they are known to inhibit platelet-activating factor (PAF). Since ischemia can result when clotted blood in a portion of the myocardium prevents flow and tissue oxygenation, botanical agents known to prevent the blood from becoming overly viscous and clotting are of benefit for angina sufferers. Garlic, onions, ginger, and cayenne peppers have all been credited with the ability to keep blood fluid and prevent pathologic platelet aggregation. *Ginkgo*, *Leonurus*, and many botanicals inhibit platelet aggregation and reduce blood viscosity as well.

natural anticoagulants are synergistic with pharmaceuticals, expert care and regular lab evaluation is essential.

Hemorrhoids are rectal varices that can form as a result of portal hypertension or, in other cases, transient vascular pressure that results from unusual lifting and exercise. Rectal varices may be obvious external lesions or can be internal and visualized with proctoscopic or colonoscopic procedures and special ultrasound imaging techniques. Rectal varices can occasionally bleed profusely and require ligation or other interventions such as sclerotherapy, cryotherapy, and infrared photocoagulation.

Phlebitis is an inflammation of a vein, most common in varicose leg veins, or other situations of long-standing vascular stasis, but may occasionally result from trauma to the veins. Ellagitannins from *Quercus robur* and other oak species are emphasized in folklore for reducing swelling, aching, and pain in leg veins and hemorrhoids and have been shown to have numerous benefits to lymphatic circulation.[156] *Centella asiatica* contains triterpenoid saponins, collectively known as centelloids, with many medicinal effects. The plant has been used folklorically to strengthen weakened veins and promote wound healing and may improve venous insufficiency and improve varicosity-related skin lesions and ulcers.

Proanthocyanidins and their building blocks, the catechins and epicatechins, have been extensively researched for reducing vascular inflammation and are shown to inhibit the enzymes hyaluronidase, elastase, and collagenase,[157] which degrade connective tissue structures and lead to increased vascular permeability. Therefore, high doses of both dietary and supplemental flavonoids are appropriate in protocols to support venous tone and treat varicosities. Suppositories of *Commiphora mukul* have been used in Ayurvedic medicine to treat internal hemorrhoids with reported success. Those with a history of clots, phlebitis, or at high risk for strokes and heart attacks might consume a "blood-moving" tea on a regular basis as a preventive measure. Those with DVT or acute phlebitis will require more aggressive protocols. Simple varicose veins without complicating inflammation or deep vein involvement should simply be supported with flavonoids, herbs, a high-fiber diet, and an effort to avoid constipation and congestion in the pelvis, in an effort to slow the progression and the onset of complications. Aching and other discomfort may be improved with systemic herbal approaches or palliated with topical applications.

Patients on warfarin (Coumadin) or other anticoagulant drugs should attempt the following herbal formulas only under a doctor's supervision due to synergism with the drugs and risk of thinning the blood to the point of suffering bleeding. Prothrombin time tests and other lab work would be necessary in such cases. The use of herbs topically does not carry any such concern.

Varicose Vein Formula

Varicose veins are next to impossible to reverse once established, but herbs can help prevent their progression, the development of new varicosities, and any tendency to phlebitis or blood clots. It is essential to reduce obesity, liver congestion, and constipation if these concerns appear to be contributory. The use of a bromelain supplement is also helpful in those with concerns of clots or embolism. *Ginkgo* supports peripheral circulation and reduces clot risk, while *Aesculus* improves portal circulation, helping to reduce the pressure in the lower limbs. *Centella* and *Hypericum* may both support the strength and integrity in blood vessel walls.

Ginkgo biloba
Centella asiatica
Hypericum perforatum
Aesculus hippocastanum

Don't Forget Liver Herbs When Treating Varicosities

Simple liver congestion can cause biliary stasis and portal congestion and contribute to varicosities. Note symptoms of liver congestion such as aching in right upper quadrant, poor digestion with fat intolerance and constipation, or heavy exposure to hepatotoxins from alcohol to environmental toxins to pharmaceuticals. The use of alterative herbs, bitters, and lipotropic formulas can be one important approach to reducing vascular congestion for those with repeat episodes of hemorrhoids or rapidly worsening varicosities. The following herbs are also emphasized as being specific for portal congestion in the folkloric herb traditions.

Aesculus hippocastanum *Dioscorea villosa*
Angelica sinensis *Iris versicolor*
Berberis aquifolium *Quercus* spp.
Curcuma longa *Silybum marianum*
Cynara scolymus *Taraxacum officinale*

Combine equal parts of the dry herbs for a tea. Steep 1 tablespoon of the mixture per cup of hot water. You can also combine equal parts of tinctures to make a combined tincture. Take 2 dropperfuls of the combined tincture 3 or more times daily, long term.

Tincture for Varicose Veins with Aching Pain

This formula combines the all-purpose blood mover *Angelica* with herbs emphasized in the folkloric literature for vascular congestion that causes an aching pain in the labia, testes, back, or legs. Varicose veins that ache premenstrually may also improve with this formula.

Caulophyllum thalictroides	30 ml
Angelica sinensis	30 ml
Aesculus hippocastanum	30 ml
Hamamelis virginiana	30 ml

Take 1 teaspoon of the combined tincture 3 or 4 times a day for a minimum of 3 months.

Topical Paste for Varicose Leg Veins

Topical pastes of herbs may be employed for varicose veins, both to allay pain and to limit progression, and this formula features herbs emphasized in folklore[158] for varicosities and aching sensations in the legs.

Hamamelis virginiana leaves	¼ ounce (7 g)
Aesculus hippocastanum tincture	1 teaspoon (5 ml)
Hypericum perforatum oil	1 teaspoon (5 ml)
Mentha spp. essential oil	10 drops

Place the dried *Hamamelis* leaves or leaf powder in a small bowl and moisten with the *Aesculus* tincture and *Hypericum* oil, along with the mint essential oil. Blend with a fork into a paste and apply to aching leg veins. Leave in place for a half hour and rinse off with warm water. Repeat multiple times per day when varicosities are causing significant discomfort, or on a daily basis for less urgent concerns.

Topical Paste for Ulcerated Varicose Leg Veins

This paste contains more absorptive ingredients to help absorb secretions and other herbs such as *Centella* noted to heal ulcers.

Centella asiatica, dried leaves, fresh when available	1 tablespoon
Calendula officinalis succus	1 teaspoon
Hippophae rhamnoides oil	1 teaspoon
Bentonite clay	1 or 2 teaspoons

Place the *Centella* leaves in the bottom of a small bowl, saturate with *Calendula* succus and *Hippophae* oil, and blend. Add bentonite clay powder to create a paste. Apply the paste directly to ulcerated skin. Allow to air dry, and then rinse gently with warm water. Repeat 3 or more times a day for several days, until the ulcerated lesion has scabbed over.

Tincture for Testicular Varicocele

These three herbs are emphasized in the folkloric literature for treating men with varicocele.

Collinsonia canadensis
Hamamelis virginiana
Lycopus virginicus

Combine in equal parts and take 1 teaspoon of the combined tincture 3 or 4 times a day.

Tincture for Pelvic Stagnation with Vaginitis

Diabetic women or others with pelvic stagnation are sometimes prone to vaginitis. The bacterial species can shift in cases of vascular stasis, the high blood glucose of diabetics invites opportunistic infections, and intestinal flora may also shift in cases of obesity, constipation, and pelvic stagnation. All of these

Solid Extracts for Varicosities

Rosa canina, *Crataegus* species, and *Vaccinium myrtillus* are tasty high-bioflavonoid herbs available as solid extracts, a thick molasses-like form of herbal medicine. Consider using these herbs to complement a tea or tincture for varicose veins and chronic hemorrhoids. Solid extracts may also be taken by small spoonfuls or may be combined with tinctures as long as the formula is thin enough to be dispensed through a dropper. Rotate through the various options or stir them together and use all at once. These would be best used over the long term, if not for life.

conditions favor the development of chronic vaginitis or in men, jock itch and chronic fungal infections. The herbs in this formula may help address underlying pelvic stagnation and help restore healthy bacterial flora. The use of probiotics and a high-fiber diet or fiber supplement would be complementary to the herbs, and a more direct treatment for vaginitis from boric acid suppositories to vaginal douches may be necessary. *Collinsonia* is indicated for venous stasis and sluggish circulation, dilated capillaries, and muscular atony. *Collinsonia* is traditionally used for oral, rectal, and urinary mucosal irritation with atony. It may help reduce vascular congestion for varicosities, including labial varicosities, hemorrhoids, and varicose veins. A specific indication for *Collinsonia* is constipation and urinary complaints due to vascular engorgement, pelvic congestion, and prolapse.

Collinsonia canadensis
Hamamelis virginiana
Urtica urens
Allium sativum

Combine in equal parts and take 1 teaspoon of the combined tincture 3 or 4 times a day

Tincture for Pelvic Congestion with Cold Extremities

Pelvic congestion is a term used by naturopathic physicians to describe a situation of vascular and fluid stasis that may result from obesity, sedentary habits, liver congestion, biliary insufficiency, poor circulation, or a combination of these contributors. As blood pools in the pelvis, lymph circulation becomes impaired and patients may have digestive discomfort, hemorrhoids, and pitting edema in the lower limbs. Herbs considered to be "blood movers," especially those that are peripheral vasodilators, can affect hemodynamics and move blood out of the pelvis and into the limbs. Any underlying constipation and liver function are not addressed by this formula and will need to be remedied separately as appropriate. Dietary changes and exercise advice are also frequently indicated.

Achillea millefolium	30 ml
Cinnamomum verum	30 ml
Zingiber officinale	30 ml
Angelica sinensis	30 ml

Take 1 teaspoon of the combined tincture 3 or 4 times a day.

Tincture for Women with Aching Back or Pelvis

Back pain, especially premenstrual backaches, can sometimes be due to vascular congestion rather than musculoskeletal reasons. When the pain is not worse with motion and is associated with hemorrhoids, constipation, or other signs of pelvic congestion, this formula may help resolve the complaint.

Aesculus hippocastanum
Hamamelis virginiana
Lycopus virginicus

Combine in equal parts and take 1 teaspoon of the combined tincture 3 or 4 times a day

Tincture for Varicose Veins and Hemorrhoids

These four herbs are emphasized in the folkloric literature for hemorrhoids, full sensations in the rectum, and aching and sense of weight in the pelvis. Pain that is heavy and improves instantly with lying down is a sign of vascular congestion and suggests the need to use blood movers.

Aesculus hippocastanum	15 ml
Hypericum perforatum	15 ml
Collinsonia canadensis	15 ml
Hamamelis virginiana	15 ml

Take 1 teaspoon of the combined tincture 3 or 4 times daily for many months.

Tincture for Aching in the RUQ with Hemorrhoids

This formula combines herbs specific for portal congestion with general liver-supportive alterative herbs to reduce vascular congestion. Aescin from *Aesculus* seeds has anti-inflammatory and antiedema capabilities, reducing capillary permeability and promoting contractility of venous wall elastin. *Aesculus* may improve venous insufficiency[159] and reduce congestion in the lower limbs.

Aesculus hippocastanum	10 ml
Salvia miltiorrhiza	10 ml
Collinsonia canadensis	10 ml
Curcuma longa	10 ml
Taraxacum officinale	10 ml
Quercus spp.	10 ml

Take 1 teaspoon of the combined tincture 3 or more times a day for a minimum of 3 months.

Platelet Antiaggregators and PAF Inhibitors

Platelet-activating factor (PAF) is a compound released by a number of blood cells that serves several inflammatory and immunomodulating roles, especially activation of platelet-driven blood clotting, permeability of blood vessels, and white blood cell activation. Smoking, exposure to pollutants, and a variety of disease processes including diabetes may lead to an excessive release of PAF and contribute to increased risk of thrombi and vascular inflammation. A number of natural compounds have been found to act as PAF inhibitors. An umbrella term for agents shown to reduce platelet clumping and adherence, whether by PAF inhibition or other mechanisms, is "platelet antiaggregator." For example, bromelain, the proteolytic enzyme from pineapples, has been shown to be a natural PAF inhibitor, contributing to its thrombolytic effect. Many Apiaceae family members generally display platelet antiaggregating activity,[160] due to PAF-inhibiting actions of the coumarins, such as osthol, they contain. Procyanidolic oligomers (pycnogenols, leukocyanidins) are flavonoids shown to inhibit platelet aggregation. The following herbs have been shown to inhibit platelet activation.

Allium sativum. May prevent the conversion of arachidonic acid to thromboxane.

Allium cepa. May block thromboxane synthesis. Sulfur compounds are responsible for the platelet antiaggregating activity.

Angelica sinensis. Ferulic acid inhibits platelet aggregation.

Astragalus membranaceus. Astragalus inhibits platelet activation. Injectable forms of the herbal medicine are being explored in China to treat patients with diabetic nephropathy, enhancing microcirculation in the kidneys and reducing vascular inflammation.

Capsicum frutescens. Capsaicin resin inhibits platelet aggregation.

Commiphora mukul. Steroidal compounds may inhibit platelet aggregation.

Curcuma longa. Curcuma inhibits platelet aggregation by inhibiting the formation of thromboxanes and promoting the formation of prostacyclin.

Fragaria species. Lipids occurring in the leaves, fruits (achenes), and pollen of strawberry plants have been found to inhibit PAF.

Galphimia glauca. This herb inhibits PAF response.

Ginkgo biloba. Ginkgolides compete with PAF for binding sites and minimize the effects.

Glycyrrhiza glabra. Glycyrrhiza appears to inhibit platelet aggregation through thrombin inhibition and does not appear to affect collagen or PAF. Glycyrrhizin is presently thought to affect thrombin via its anion-binding exosite.

Linum usitatissimum. Flaxseed oil is high in alpha-linolenic acid known to have antiatherogenic properties. Lignins act as PAF receptor agonists.

Panax ginseng. The nonsaponin or lipophilic fraction inhibits thrombin-induced platelet aggregation. It is presently proposed that Panax acts via cGMP and thromboxane A2 to inhibit thrombin.

Picrorhiza kurroa. Picrorhiza is shown to inhibit PAF-induced bronchoconstriction in animal studies.

Syzygium jambos. Eugenol is credited with PAF-inhibiting effects.

Vaccinium myrtillis. Anthocyanosides promotes prostacyclin, which dilates blood vessels and inhibits platelet aggregation.

Vitis vinifera. Grape juice, skins, and seeds contain polyphenolic compounds known to improve microcirculation by having an antioxidant and protective effect on the vasculature. Grape products including juice and red wine appear to inhibit platelet aggregation.

Zingiber officinale. Ginger inhibits thromboxane synthetase, while promoting the formation of prostacyclin. Gingerol inhibits platelet aggregation when catalyzed by arachidonic acid and collegen but not when catalyzed by PAF or thrombin. Gingerol inhibits thromboxane B_2 and PGDz, which also inhibits platelet aggregation. Gingerol blocks secondary platelet aggregation and blocks the release of ATP for platelets when stimulated by adenosine s-diphosphate or adventure.

Coumarin-Containing Plants

Apiaceae plants produce coumarin compounds to help defend against microbes, and coumarins have numerous vascular and hormonal effects in humans. The primary coumarin coumpounds having extensive research include psoralen, umbelliferone, aesculetin, osthol, bergapten, xanthotoxin, and imperatorin. Coumarins offer many vascular benefits including hypotensive action and hypolipidemic affects. Oral absorption of coumarins is moderate due to extensive liver conversion into umbelliferone; however, enough can be absorbed from plants to be physiologically active.[161] In a concentrated and isolated form, coumarin is toxic due to internal hemorrhage and liver and kidney toxicity, and pure coumarins have been banned as a food additive for more than 50 years in the United States. Coumarin is also considered a lung carcinogen, though this is in part due to its history as a flavorant in tobacco products. However, in the organic matrix of other plant molecules, coumarin compounds appear to be safe and gentle, as with celery, carrots, fennel seeds, and other Apiaceae family plants.

Castor and Mint Compress for Hemorrhoids

Mint oil has anodyne effects and is safe to use topically. This compress is simple to make and to use even when at work or away from home.

Castor oil
Peppermint essential oil

Saturate a cotton ball or small gauze pad with castor oil. Place 5 or 6 drops of peppermint oil on one surface of the saturated compress. Place the peppermint oil side of the compress directly against the painful hemorrhoid and leave in place for at least a half hour. Repeat as often as needed for pain relief.

Sitz Bath for Hemorrhoids

Hemorrhoids can be very uncomfortable, and since it can take many weeks for systemic therapies to start to reduce vascular congestion, sitz baths can help allay the discomfort as quickly as possible. The use of a sitz bath, morning and evening, can help to astringe the tissue and speed resolution of pain and the swollen tissue. Dietary measures, improving constipation, if present, daily exercise, and the use of the Tincture for Varicose Veins and Hemorrhoids on page 76 can act more deeply over time and reduce venous pressure that contributes to hemorrhoid development.

Calendula officinalis	4 ounces (120 g)
Equisetum spp.	4 ounces (120 g)
Hamamelis virginiana	4 ounces (120 g)
Hypericum perforatum	4 ounces (120 g)

Combine the dry herbs (either finely ground or powdered) and blend to yield a pound of mixture. Place 1 cup of the mixture in a large stock pot and cover with at least a gallon of water, plus $\frac{1}{2}$ to 1 gallon more if the pan will accommodate it. Bring to a gentle simmer and then promptly remove the pot from the heat source. Allow it to sit covered for half an hour. Place a rubber tub (large enough to sit in) in a bathtub or shower stall. Pour the brew through a strainer into the tub and add more hot water to a level sufficient to immerse the buttocks. Sit in the sitz bath for at least 10 minutes and then rinse off with cool or tepid water. Pour the sitz bath brew onto a lawn or garden outdoors to avoid putting too much particulate down the household drain. Repeat as many times a day as possible.

Classic Compress for Hemorrhoids

A pain-relieving compress may also be prepared from *Hamamelis* and *Quercus*, both folkloric classics for hemorrhoids. This recipe is best used at home where the person can lie on a bed or couch.

Hamamelis virginiana, finely shredded bark	4 ounces (120 g)
Quercus spp., finely shredded bark	4 ounces (120 g)

Combine the two shredded barks. Simmer 1 heaping tablespoon of the mixture in 3 cups of water on the lowest setting for 10 minutes. Let stand in a covered pan 15 minutes more and strain. Soak a soft cloth or several layers of gauze pads in the liquid and apply to hemorrhoids. Repeat as often as possible throughout the day, until pain subsides.

Don't Confuse Coumarin with Coumadin

There is an erroneous but persistent myth in some herbal practices that coumarin has a blood-thinning effect such that it could be used instead of the brand name pharmaceutical medication called Coumadin (generic name warfarin). Some coumarins, such as those found in several *Angelica* species, have been noted to act as platelet antiaggregators but not as Coumadin-like anticoagulants. Dicoumarol, however, the agent on which pharmaceutical warfarin anticoagulants are based, has potent anticoagulant action. Dicoumarol is structurally similar to vitamin K and acts as an antagonist, competing with vitamin K and reducing the effects of this blood-clotting vitamin to promote the clotting factors VII, IX, and X. Dicoumarol may form naturally from the fermentation of coumarin in plant leaves, such as decayed *Melilotus officinalis* (sweet clover) and *Anthoxanthum odoratum* (sweet vernal grass). Therefore,

Aesculus hippocastanum, horse chestnut

anticoagulant action from dried plant material such as *Melilotus* or *Trifolium pratense* (red clover) is theoretically possible but not emphasized in historic or modern literature on coumarins except in the case of accidental poisoning of livestock. Livestock who graze on large amounts of fresh red clover or other legumes or umbels are noted to develop gas and bloating (as do humans who consume a high-legume diet) but blood-clotting difficulties are not observed. However, livestock who graze on large amounts of *spoiled* sweet clover or sweet vernal grass *are* reported to suffer hematologic and bleeding disorders, which can be fatal when a large amount of decomposed plant material is ingested, due to the conversion of coumarin to dicoumarol.[162] Unlike dicoumerol or heparin, coumarin itself is not found to be anticoagulant, but many plants high in coumarin-like compounds are noted to affect platelets, blood cells, and the vasculature, such as aesculin in *Aesculus*, umbelliferone in the Apiaceae family (the umbels), and the coumarins in *Melilotus*. These coumarins are not claimed to inhibit necessary and appropriate blood clotting, but rather to act on capillary, venule, and vein permeability, increasing fluid uptake and return from the tissues to general circulation. Coumarins may inhibit aberrant clotting by having anti-inflammatory and platelet antiaggregating effects. Such coumarin constituents appear to act in tandem with "permeability factors," such as rutin and quercetin, and improve venous return. High-rutin plants include *Ruta graveolens* (rue), *Fagopyrum esculentum* (buckwheat), and *Citrus* (citrus fruits), as well as Apiaceae plants.

Tea for Portal Congestion

Milk thistle seeds are difficult to use as a tea, but many quality and concentrated encapsulated products are available. Such capsules might be used in tandem with a liver-supportive, alterative tea to support hepatocellular function and help reduce portal congestion that can contribute to hemorrhoids and varicose veins. Various varieties of artichokes of the *Cynara* genus contain cynaropicrin, shown to improve liver function in various disease states.[163] *Cynara* is in the thistle family, and as with *Silybum marianum* (milk thistle), may help reduce portal congestion and thereby help reduce pulmonary hypertension related to liver disease. *Angelica archangelica* contains the coumarin osthol, which is noted to have a protective and restorative effect on liver function. Osthol and other coumarins may support liver detoxification pathways and protect the liver from fatty degeneration in inflammatory and high lipid situations,[164] as well as offer hepatoprotective effects against alcohol exposure,[165] hepatitis, and viral infections.[166]

Cynara scolymus leaf	4 ounces (120 g)
Quercus species shredded bark	2 ounces (60 g)
Schisandra chinensis berry	2 ounces (60 g)
Glycyrrhiza glabra, shredded	2 ounces (60 g)
Pueraria montana var. *lobata* root	2 ounces (60 g)
Angelica archangelica	2 ounces (60 g)
Salvia miltiorrhiza	2 ounces (60 g)

This formula will yield 1 pound of tea and should be stored in an airtight jar. Use 2 tablespoons in 6 cups of water, and simmer gently for 10 minutes. Let stand and strain into a thermos. Drink 1 or 2 cups of tea at a time, 2 or 3 times a day, and take 2 milk thistle capsules at each use.

Castor Oil Packs for Liver and Portal Congestion

Castor beans are pressed for their oil, yielding a thick sticky product high in immune- and circulation-enhancing lectins. Castor oil is believed to be able to penetrate the skin and affect underlying tissues, lymph and blood circulation, and immune response. When applied regularly to the liver, castor oil can help to decongest tissues, reduce pain, and be a valuable complement to other oral therapies. For best results, the oil should be covered with heat and implemented faithfully day after day.

Castor oil
Flannel or other absorbent material
Heat pack

The traditional use is to saturate layers of flannel cloth with castor oil and cover it with heat, but this is messy

Nattokinase Capsules for Acute Phlebitis and Clots

Thrombolytic drugs are used to dissolve fibrin in blood clots following myocardial infarction and thromboembolic strokes, as well as to treat deep vein thrombosis and pulmonary embolism. Such pharmaceuticals can help clear blood vessel occlusion, but there is a high incidence of reocclusion and prolonged circulatory insufficiency, necessitating long-term Coumadin, with its many concerns such as bleeding risk. Herbal and naturally sourced anticoagulants may be highly valuable to reduce the tendency to further clots or emboli and may complement or in some cases replace pharmaceutical anticoagulants.

Nattokinase from *Bacillus subtilis* var. *natto* or *Pseudomonas aeruginosa* is a potent thrombolytic agent. Natto is a fermented soy product, used as a traditional food in Asia, and the enzyme nattokinase in the food may improve blood pressure and may directly lyse thrombi in vivo, as well as enhance fibrinolytic activity in plasma and increase plasmin.[167] Nattokinase interacts with heparin and interferes with heparin's binding and activation of various clotting cascades, supports antithrombin, and limits fibroblast growth factors.[168] Nattokinase can inhibit platelet aggregation by blocking thromboxane.[169] Nattokinase may be included in protocols for reducing cardiovascular disease and improving risk factors, due to decreasing plasma levels of fibrinogen and clotting factors VII and VIII, and improving blood pressure via inhibition of angiotensin II.

Although human studies are limited, single doses of nattokinase may reduce clotting factors and increase antithrombin in healthy subjects.[170] It should be emphasized that substituting nattokinase for a pharmaceutical anticoagulant is not a simple matter[171] and should not be attempted without experience, vigilant monitoring, and expert guidance, especially in those with artificial valves or other high clot-risk situations. Nattokinase appears most appropriate to reduce blood coagulation for those with varicosities, heart disease, and cardiovascular risk factors, to improve blood circulation, and to reduce clotting risk.

and a chore to wash the fabric or wasteful to discard. A simple alternative is to simply apply the castor oil to the right upper quadrant of the torso using the palm of the hand and cover with a hot pack, or apply after a hot shower when the skin and circulation are already warm and active. Applying the oil 2 or 3 times in 5 minutes may duplicate the quantity of medicine delivered by a compress. The liver is known to repair and regenerate itself at night. Apply castor oil after an evening bath or hot shower to support liver function while you sleep.

Tincture Formula for Acute Phlebitis

Curcuma and *Salvia* in this formula can have an anti-inflammatory effect and may prevent permanent fibrosis of affected veins. *Echinacea*, *Allium*, and *Berberis* can address any underlying infectious problem, and *Allium* can also help prevent the clotting risk in inflamed vessels. This formula is best used aggressively, such as dosing hourly or even more frequently, reducing as symptoms improve. A bromelain supplement at 500 to 1,000 milligrams 5 times daily between meals would make a good complement.

Curcuma longa	15 ml
Allium sativum	15 ml
Echinacea purpurea	10 ml
Berberis aquifolium	10 ml
Salvia miltiorrhiza	10 ml

Take 1 teaspoon of the combined tincture every 30 to 60 minutes, reducing after 24 hours and as symptoms improve.

Tea for Acute Phlebitis

This formula could be prepared as a tea or tincture and would be well complemented by the use of bromelain capsules, 500 milligrams every several hours, and/or nattokinase. This formula is a tea, but in urgent cases, tinctures would allow for aggressive dosing with greater ease. Topical oils such as the Topical Oil for Acute Phlebitis that follows would also be complementary.

Angelica sinensis root	1 ounce (30 g)
Calendula officinalis	1 ounce (30 g)
Centella asiatica	1 ounce (30 g)
Equisetum arvense	1 ounce (30 g)
Hamamelis virginiana	1 ounce (30 g)

Combine the herbs, steep 1 tablespoon of the mixture per cup of hot water, allow to stand for 15 minutes in a covered saucepan, and then strain. Drink as much as possible throughout the day.

Topical Oil for Acute Phlebitis

Acutely inflamed veins will be hot and tender so this oil should be applied as gently as possible. Covering with a heat pack or an ice bag as preferred by the patient for pain management can also be helpful. Repeat hourly where possible.

Sea buckthorn oil base	1½ ounces (45 ml)
Helichrysum italicum essential oil	¼ ounce (7.5 ml)
Lavender essential oil	¼ ounce (7.5 ml)

Combine the oils in a 2-ounce bottle. Apply to the affected vein using the fingertips or a cotton ball. Reapply frequently.

Tea for History of Blood Clots

A variety of genetic disorders increase clotting tendencies. Those who have severe varicosities, are taking birth control pills, are elderly with poor circulation, are long-time smokers, and/or have suffered one blood clot may use a tea such as this on a daily basis to support healthy circulation. Those undergoing surgery or suffering from heart failure or nephrotic disease are also at increased risk for blood clots and may benefit from these herbs.

Trifolium pratense
Zingiber officinale
Cinnamomum verum
Ceanothus americanus
Quercus robur

Combine the dry herbs and blend. Gently simmer 1 teaspoon of the mixture per cup of hot water for just 1 minute. Cover the pan and let stand for 10 to 15 minutes off the heat source. Drink 3 or more cups per day.

Supportive Measures for Varicose Veins with Phlebitis

Address any underlying contributors in patients where phlebitis is recurrent, such as obesity, diabetes, platelet activation, and elevated lipids. Employ a combination of the following measures:

Angelica, *Salvia miltiorrhiza*, or *Ginkgo* as a simple tincture; 1 teaspoon, 3 or 4 times daily
Fresh pineapple or freshly made pineapple juice
Garlic and onions liberally in the diet; consider concentrated *Allium* capsules
Bromelain capsules, 500 milligrams 3 times a day, between meals

Formulas for Vasculitis

Vasculitis, also sometimes referred to as angiitis or arteritis, is an inflammation of the blood vessels due to autoimmune reactivity, rather than atherosclerotic processes. As an autoimmune condition, vasculitis accompanies lupus and other collagen vascular diseases. In such conditions vasculitis may be diffuse throughout the body or may be confined just to the skin, just the blood vessels in joints, or just the blood vessels in a particular organ, such as the kidneys or the lungs. The affected vasculature becomes inflamed and may swell, fibrose, and lose elasticity and fail to deliver oxygen and nutrients to affected tissues.

When creating herbal formulas for vasculitis, it may be valuable to address any triggers that can be identified, such as an inciting infection, drug reaction, or leaky gut. More coverage of vasculitis will be included in a future volume of this set of herbal formularies.

General Vasculitis Support Tincture

Vasculitis can be widely disseminated in various collagen vascular diseases in which the autoimmune phenomena damage connective tissue of the microvasculature. These conditions are difficult to reverse, and thus a multipronged approach is necessary. This formula would help protect affected blood vessels from becoming fibrosed. *Ginkgo* and *Angelica* are not only blood movers, but they can also help reduce inflammatory processes. *Curcuma* is noted to prevent fibrotic changes in numerous tissues while offering general anti-inflammatory and immunomodulating effects.

Ginkgo biloba
Angelica sinensis
Curcuma longa
Crataegus oxyacantha

Combine in equal parts and take 1 or 2 teaspoons of the combined tincture 3 to 6 times daily

Tea for Vasculitis

It is necessary to be aggressive when treating collagen-vascular diseases. Aim to have every beverage and food be medicinal. Select herbs from among the better-tasting options to create a tea such as this to complement the General Vasculitis Support Tincture.

Rubus spp. leaves
Centella asiatica
Ginkgo biloba
Crataegus oxyacantha leaves and flowers

Combine the dried herbs in equal parts, or adjust quantities for desired taste. Steep 1 tablespoon of the mixture per cup of hot water for 10 minutes, and then strain. Drink 3 or more cups per day

Decoction for Vasculitis

All colorful berries and high-flavonoid foods are known to prevent fibrosis of the blood vessels and the loss of functional vascular cells, including when they are dried and used for teas. *Sambucus* (elderberry), *Crataegus* (hawthorn berries), and *Rosa* (rose hips) are all readily available and reasonably affordable to include in teas throughout the year. *Hamamelis* is a folkloric herb recommended for vascular inflammation.

Sambucus canadensis berries
Crataegus spp. berries
Rosa canina hips
Glycyrrhiza glabra root
Hamamelis virginiana bark

Combine in equal parts or adjust quantities for desired flavor. Gently simmer 1 teaspoon of the mixture per cup of hot water, and then strain. Drink freely, 3 or more cups per day.

Formula for Lymphedema

Lymphedema is a collection of interstitial fluid and can occur due to lymphatic abnormalities or due to severe destruction of lymphatic channels following radiation therapy. It may also present without a clear or definable cause. Lymphedema is nonpitting, as is the edema of the lower legs related to vascular, liver, or other diseases. It

will not readily reduce with elevating the legs. Tumors may compress lymphatic channels and should be ruled out when no underlying cause can be identified.

In clinical practice, lymphedema is most often related to serious injury to lymphatic channels such as following surgery or radiation therapy, for example, when irradiation to the lungs or breasts as a cancer therapy causes lymphedema in an arm. The use of *Aloe vera* and castor oil topically prior to radiation may help to limit the damage to the fragile lymphatic tissues.

Lymphedema Tincture

The herbs in this formula are all purported to help move lymph and, although they are helpful, do not expect *too* much of them. The use of pressure bandages and vigorous exercises to the affected part can be additional supportive measures.

Hypericum perforatum	30 ml
Echinacea spp.	30 ml
Phytolacca decandra	30 ml
Centella asiatica	30 ml

Take 1 dropperful of the combined tincture every 30 to 60 minutes, reducing as symptoms improve.

Formulas for Cardiopulmonary Disease

Lung diseases such as asthma, chronic bronchitis, and chronic obstructive pulmonary disorder (COPD) are covered in chapter 3 but there is some overlap between them and vascular disease, in cases where pulmonary blood vessel damage causes lung disease. For example, pulmonary hypertension, characterized by increased arterial blood pressure and vascular resistance, leads to thickening of the small pulmonary blood vessels, referred to as pulmonary fibrosis. Pulmonary hypertension and fibrosis are progressive and fatal disorders. Underlying fibrotic diseases, such as scleroderma or other autoimmune disorders, may contribute, due to chronic inflammation and excessive collagen deposition in the airways.

COPD may also gradually lead to fibrotic changes in the pulmonary vessels, typically following decades of chronic bronchitis associated with smoking or chronic smoke exposure, such as when cooking over a fire in a poorly ventilated space. Shortness of breath (especially during physical activities), tightness in the chest with wheezing, and excessive morning mucus expectoration may precede the onset of the more serious symptoms by a decade or more. As circulation to the lungs is gradually impaired with the progression of pulmonary hypertension and fibrosis, pallor, dyspnea, cyanosis, and edema in the lower limbs may ensue. COPD is highly associated with injury to systemic inflammation in general. Therefore, vascular protectants and anti-inflammatories discussed throughout this chapter are appropriate for pulmonary hypertension and fibrosis.

Pharmaceutical treatments for pulmonary fibrosis include immunosuppressants such as cyclophosphamide and antifibrotic drugs such as colchicine, but these can't be highly recommended due to lack of significant benefits and the possibilities of severe side effects.

Herbal medicines noted to reduce fibrotic processes are *Curcuma longa, Centella asiatica, Salvia miltiorrhiza*, bromelain, and antioxidants in general. *Curcuma* improves inflammatory markers in patients with COPD.[172] *Salvia miltiorrhiza* has an antifibrotic effect on blood vessels, and research suggests it may also limit fibrotic processes in the lungs, which limits cell changes and collagen deposition in the alveoli and supportive cells.[173] *Salvia miltiorrhiza's* salvianolic acid is used in China to attenuate pulmonary fibrosis in scleroderma patients, and it can slow or even reverse fibrosis of the liver in chronic hepatitis patients. Only a portion of lifelong tobacco smokers develop COPD, and other factors contribute to the fibrotic progression. Pulmonary fibrosis may be more common in those with underlying hypersensitivity disorders such as asthma and various respiratory allergies. Elevated eosinophil levels are associated with an increased risk of pulmonary disease[174] and therefore allergic individuals, especially smokers, should make an effort to protect their lungs. Antioxidants, flavonoids, and vascular protectants may reduce harm in smokers.

Simple for Cardiomyopathy with Arrhythmia

Selenicereus is a traditional herb for an enlarged heart prone to arrhythmias and episodes of palpitations. Additional specific indications for *Selenicereus* are dyspnea and pallor, emphysema, and mitral valve regurgitation. *Selenicereus* may also benefit women who develop heart palpitations and worsening heart function after menopause, as well as long-term smokers who are developing frank heart disease, beginning with excitable and tumultuous heart action.

Selenicereus grandiflora tincture

Take 1 or 2 dropperfuls 3 to 6 times a day.

COPD Long-Term Lung Support Tincture

Herbs that protect the elasticity of connective tissue are appropriate for pulmonary diseases such as COPD. *Equisetum* and *Centella* are folkloric remedies for protecting and even regenerating injured tissue. *Curcuma* may help prevent fibrosis. *Ginkgo* and *Panax* may enhance oxygen utilization. The more, the better, when it comes to connective tissue support. In addition to this tincture, see also the Connective Tissue Tonic Tea for COPD on page 150, which would complement this formula.

Ginkgo biloba	6 ml
Equisetum arvense, E. hyemale	6 ml
Panax ginseng	6 ml
Centella asiatica	6 ml
Curcuma longa	6 ml

Take 1 teaspoon of the combined tincture 3 to 6 times daily.

Diuretic Tea for Pulmonary Hypertension

Minerals are needed to help the kidneys create the osmotic gradients necessary to excrete water. *Equisetum*, *Taraxacum* and *Urtica* are among the classic herbal diuretics and are all nourishing mineral-rich herbs.

Equisetum spp.	1 ounce (30 g)
Centella asiatica	1 ounce (30 g)
Medicago sativa	1 ounce (30 g)
Taraxacum officinale leaf	1 ounce (30 g)
Urtica spp.	1 ounce (30 g)
Mentha piperita	3 ounces (90 g)
Lemon juice or herbal vinegar	

This formula will yield ½ pound of tea, which should be stored tightly sealed in a dark cupboard. Place 4 tablespoons of the herb mixture in a saucepan. Add the juice of 1 lemon or a tablespoon of herbal vinegar and let stand for several hours or overnight. In the morning, add 4 cups of water, bring to the lowest possible simmer, and simmer for 10 to 15 minutes in a covered pan. Remove from heat, let stand 10 minutes more, and then strain. Drink freely, 3 or more cups per day.

Diuretic Tincture for Pulmonary Hypertension

This tincture uses *Convallaria* instead of *Selenicereus* as a cardiotonic. *Selenicereus* is more specific for cardiac arrhythmia and enlargement, while *Convallaria* is more specific for heart weakness with a weak, irregular pulse. Both formulas use high-mineral green leafy herbs to support renal diuresis. Because the use of cardiac glycosides such as *Convallaria* may potentiate cardiac pharmaceuticals, close medical supervision is needed when using drugs and herbal cardiac glycosides in tandem.

Urtica urens, U. dioica	30 ml
Equisetum arvense, E. hymenale	30 ml
Convallaria majalis	30 ml
Ginkgo biloba	30 ml

Begin with 1 dropperful of the combined tincture 3 times a day. If there are no changes in 2 weeks, double the dose to 2 dropperfuls 3 times a day before changing to different therapeutic options. If heart function improves after several months' time, it may be possible to reduce the dose to just once or twice a day, as the effects may be maintained at a lesser dose.

Inotropic Tincture for Cardiomyopathy with Pulmonary Hypertension

These three herbs have positive inotropic effects on cardiac muscle, improving the electrical conduction and thereby contractile strength of individual muscle cells. Inotropic agents also help shorten the recovery time following contraction so that the cell can repolarize and then contract again. People who are taking digoxin or prescription drugs for heart failure should use this formula only under professional supervision.

Ginkgo biloba	25 ml
Crataegus spp.	25 ml
Convallaria majalis	10 ml

Take 1 teaspoon of the combined tincture 3 or more times day.

Supportive Formula for Pulmonary Hypertension

Pulmonary hypertension is a serious pathology, best managed with the help of a pulmonologist or cardiologist. This formula can complement pharmaceutical approaches. The amounts listed here are for a tincture.

Ginkgo biloba	15 ml
Panax ginseng	15 ml
Crataegus oxyacantha	15 ml
Vaccinium myrtillus	15 ml

Take 1 teaspoon of the combined tincture 3 or 4 times per day.

A variation on this formula would be to take 2 *Ginkgo* capsules 2 or 3 times a day, use 1 tablespoon *Panax* powder in smoothies each day, and use a teaspoon of the blended solid extracts of *Crataegus* and *Vaccinium* wherever possible in smoothies, drinking water, salad dressings, and medicinal truffles.

Vascular Elasticity Tincture for Pulmonary Hypertension

Pulmonary hypertension has a poor prognosis because the high pressure congests the heart and can damage the blood vessels and lead to rapidly progressing fibrosis. This formula will not cure pulmonary hypertension but may help protect the blood vessels from fibrosis, slow the progression of the disease, and prolong life.

Centella asiatica	30 ml
Vaccinium myrtillus	30 ml
Crataegus spp.	30 ml
Curcuma longa	30 ml

Take 1 teaspoon of the combined tincture 3 or more times a day.

Tincture for Cardiac Cough

Fluid in the lungs due to weak heart action can trigger the cough reflex, which is referred to as a cardiac cough. Agents that support heart muscle contractility can help decongest the lungs and thereby treat the cough. Patients who must sleep propped up on pillows to prevent a suffocating sensation may also benefit from this formula. This formula may potentiate cardiac drugs and should be used under a physician's guidance. *Ammi visnaga* contains visnagin and visnadin, and both have been shown to prevent inflammation-induced vasoconstriction, especially in the airways. See also "*Ammi visnaga* for Allergic Airways" on page 146.

Ammi visnaga	15 ml
Ginkgo biloba	15 ml
Crataegus oxyacantha	15 ml
Convallaria majalis	15 ml

Take 1 teaspoon of the combined tincture 3 or more times a day.

Tincture Tonic for Cor Pulmonale

This tincture is aimed at reducing stress on the heart and oxidative inflammatory stress in the vasculature at large. It features the few herbs clinically studied and shown to help deter acute respiratory stress, fatal clots, and sepsis. The *Crataegus* and *Ginkgo* support circulation and oxygen delivery in the heart and lungs while *Curcuma* reduces fibrosis and inflammation, and *Allium* can help reduce pulmonary hypertension and support healthy blood viscosity.

Crataegus spp.	10 ml
Ginkgo biloba	10 ml
Curcuma longa	10 ml
Allium sativum	10 ml
Rheum officinale	10 ml
Melilotus suaveolens	10 ml

Take 1 teaspoon of the combined tincture 3 to 6 times daily.

Tincture for Hepatopulmonary Syndrome

When the liver and biliary tree are congested, blood pressure in the vessels both up and downstream will become elevated. This combination of pulmonary hypertension associated with portal hypertension is called hepatopulmonary syndrome. If hepatic congestion is contributing to pulmonary hypertension, the herbs in this tincture may be more specific than the Tincture for Cardiac Cough. *Curcuma* is a liver-supportive herb and is also known to prevent fibrosis. *Aesculus* is specifically indicated for portal congestion, and *Silybum* is an all-purpose liver-supportive agent, while *Ginkgo* is an all-purpose vascular anti-inflammatory and circulation-enhancing herb. Expert care is essential.

Curcuma longa	30 ml
Aesculus hippocastanum	30 ml
Ginkgo biloba	30 ml
Silybum marianum	30 ml

Take 1 teaspoon of the combined tincture 3 or more times a day.

Formulas for Hypotension

Hypotension can be a constitutional tendency, a sign of a high state of fitness, or a symptom of underlying diseases. Diastolic readings below 50 constitute low pressure, but may have few serious consequences. Sudden drops in blood pressure, such as transient dips from 70 to 50, can cause dizziness, however, and may result in serious injury if a person faints. Pregnancy can cause hypotension due to increased vascularity; this is not usually a problem as long as fainting does not occur. Clinicians must rule out heart disease and endocrine imbalances in cases of hypotension. Heart disease such as severe bradycardia and heart value problems, as well as frank heart failure, may be associated with hypotension. Adrenal insufficiency, parathyroid insufficiency, and hypoglycemia can result in hypotension. Blood loss, shock, and dehydration can cause dizziness and hypotension. Hypovolemic shock is a life-threatening consequence of severe dehydration in which the blood volume and pressure are so low that adequate oxygen can't be delivered to the tissues. Even mild dehydration can cause systemic symptoms, and use of diuretics or strenuous exercise can also deplete electrolytes and interfere with optimal renal and vascular function. Serious infections spreading through the blood and tissues can cause septic shock, and anaphylactic reactions to foods, venom, and drugs may cause acute hypotension. Anemia may also promote hypotension. Therapies for hypotension should be aimed at any underlying cause, which could include formulas directed at hormonal imbalances, electrolyte imbalance, or heart disease.

Tincture for Orthostatic Hypotension

Some "normal" people—especially tall, thin women—may experience momentary hypotension when they first stand up. This is called orthostatic hypotension, and it can happen even in the absence of adrenal insufficiency or heart disease. When the condition is problematic and causing frequent dizziness or faintness, the following herbs may improve the situation. *Lepidium* is a general metabolic and circulatory tonic, while *Rhodiola* and *Glycyrrhiza* may support adrenal response, even in the absence of frank adrenal deficiency.

Coleus forskohlii
Lepidum meyenii
Rhodiola rosea
Glycyrrhiza glabra

Combine equal parts and take ½ to 1 teaspoon of the combined tincture 3 or more times daily.

Tincture for Hypotension with Heart Failure

When the heart muscle and tension in the blood vessels are so lax and weak that hypotension results, the same herbs that are used for heart failure may be appropriate therapies. Hypotension can also be due to blood loss, to neurologic disorders where the autonomic regulation of the blood pressure is impaired, or to acid/base pH imbalances and other systemic disease such as Parkinson's or diabetes, each of which would need to be addressed directly to have an impact on hypotension. When hypotension is due to heart disease, a formula such as this may be helpful. Carnitine and CoQ10 are complementary therapies.

Coleus forskohlii	15 ml
Convallaria majalis	15 ml
Crataegus oxyacantha	15 ml
Ginkgo biloba	15 ml

Take ½ to 1 teaspoon of the combined tincture a minimum of 3 times daily.

Tincture for Hypotension with Fatigue

Patients with hypotension and/or severe adrenal deficiency conditions may suffer from faintness upon standing and general fatigue and debility. *Coleus* can help individual muscle cells, including heart muscle, in basic metabolic activity and contractility. *Crataegus* can act as a nourishing trophorestorative base, and *Glycyrrhiza* is specific for low aldosterone and adrenal insufficiency conditions.

Crataegus oxyacantha	20 ml
Glycyrrhiza glabra	20 ml
Coleus forskohlii	20 ml

Take ½ to 1 teaspoon of the combined tincture 3 or more times daily.

Broth for Hypotension with Electrolyte Imbalance

To help correct the effects of inappropriate use of diuretics, or for those with acute fluid loss from excessive sweating or nausea and vomiting, a broth such as this may help provide the needed sodium, potassium, magnesium, and fluids.

5 or 6 cups dandelion leaf tea
2 tablespoons miso paste
2 tablespoons honey
1 orange
2 calcium-magnesium capsules

Prepare the dandelion tea, strain, and return the tea to the saucepan. Add the miso paste and honey and stir until they dissolve. Soy sauce or regular sea salt may be used instead of miso for ease of speedy preparation. Squeeze the juice from the orange and add it to the tea. Open up the calcium-magnesium capsules, add their contents to the tea, and stir vigorously. Heat until well blended. Drink in small sips until finished. Then make a second batch and drink it in small sips.

Formulas for Impotence

Impotence most commonly results from peripheral vascular disease, and thus it can be treated with the formulas for circulatory support found throughout this chapter. Several herbs have been emphasized in the herbal folklore for impotence, and some of them have been found to support testosterone and endocrine balance in addition to vascular effects. Impotence will also be covered in more detail in a future volume of these herbal formularies.

Tincture for Impotence in Healthy Older Men

Impotence is primarily a problem of poor circulation. Treat any underlying diabetes or hypothyroidism and consider a general blood-moving and hormonal-support formula such as this one. The amino acid arginine may be used in tandem to support the synthesis of nitric oxide, an endogenous compound that supports vasodilation and erectile function.

Ginkgo biloba	15 ml
Lepidium meyenii	15 ml
Panax ginseng	15 ml
Rhodiola rosea	15 ml

Take ½ to 1 teaspoon of the combined tincture at least 3 times a day, increasing as needed.

Cnidium Simple for Impotence

Cnidium tincture and the isolated coumarins osthol, imperatorin, xanthotoxin, and isopimpinellin have been shown to exhibit vasodilating effects in the penis, reversing vasoconstriction of the corpus cavernosum induced by epinephrine compounds.[175] This research has implications for the treatment of hypertension in general as well as for the treatment for impotence. *Cnidium* has been used traditionally in China to improve male sexual function. Some research has suggested that osthol may promote the release of nitric oxide, promoting a vasodilating effect within the corpus cavernosum. The release of nitric oxide may be mediated via phosphodiesterase inhibition, which prolongs the activity of cGMP and cAMP intracellularly.[176]

Cnidium monnieri

Take 1 or 2 dropperfuls of the tincture 3 times per day, long term.

Tincture for Impotence with Diabetes

When impotence occurs in tandem with diabetes, herbs that enhance cellular reception of glucose may be combined with herbs that enhance circulation. It is highly advisable for all patients with diabetes to begin vascular-supportive agents as discussed thoughout this chapter as soon as diagnosed, in an effort to protect the vasculature. This formula is aimed at treating underlying diabetes and would be complemented by PAF inhibitors and an herbal tea.

Commiphora mukul	15 ml
Fucus vesiculosus	15 ml
Coleus forskohlii	15 ml
Ginkgo biloba	15 ml

Take ½ to 1 teaspoon of the combined tincture at least 3 times a day, increasing as needed.

Cnidium monnieri for Vascular Support

Cnidium monnieri contains a variety of coumarin compounds including osthol, volatile oil, monoterpene polyols, glucides, and sesquiterpenes. A broad review of the literature on *Cnidium* reports hepatoprotective[177] and hypolipidemic effects,[178] which may help reduce vascular inflammation and damage. *Cnidium* has general vasodilating, blood-moving, and hypotensive effects and may be considered in formulas for impotence due to vasodilating effects in the penis[179] via promotion of nitric oxide pathways. *Cnidium* may also improve impotence via hormonal activity because osthol also has gonadotropin-like effects. Animal studies have shown testosterone, luteinizing hormone (LH), and follicle stimulating hormone (FSH) levels to all increase in 20 days' time following daily consumption.[180]

Treating Sleep Apnea

Obstructive sleep apnea is a common sleep disorder, affecting 2 to 4 percent of the adult population, associated with an increased risk of cardiocerebrovascular disease and organ injury. Sleep apnea is more common in obese patients and can contribute to the pathogenesis of hypertension, coronary heart disease, cerebrovascular diseases, diabetes, and cor pulmonale. The consequences of sleep apnea are not just poor sleep and daytime drowsiness; those with sleep apnea are at increased risk of traffic accidents and of mortality in general, the 5-year mortality rate being as high as 11 to 13 percent.[181] The high-mortality rate is due to damage to multiple organs that, when deprived of oxygen, experience inflammatory and oxidative damage. The primary medical treatment is the use of a continuous positive airway pressure (CPAP) device, and no pharmaceutical therapies exist. Weight loss will often improve the patency of the airways in obese patients, and lifestyle interventions such as walking and exercise programs, abstaining from alcohol, and quitting smoking may be helpful. Herbs that improve cardiovascular health may be helpful, although not curative.

Protocol for Sleep Apnea

The primary therapy for sleep apnea is the use of a CPAP device. While effective in improving airway patency, many patients find the device uncomfortable and to interfere with sleep itself. CPAP devices may be covered by some insurance plans, but if denied, some patients may not be able to afford them on their own. Surgical intervention to remove excess tissue from the throat is an option but is considered extreme by many patients, and the cost may be prohibitive. Furthermore, like all surgeries, there may be complications from pain to bleeding, the procedure may not be successful or permanent in improving sleep quality, and the jaw must be wired shut for several days to ensure optimal healing.

There are other measures that can help with sleep apnea, summarized here. Along with these practices, consider using one of the general vascular-support teas and tinctures detailed throughout this chapter.

Dental devices. These devices hold the tongue in place in a manner that prevents the throat from compressing and obstructing the airways, and some patients prefer them over the face mask of CPAP machines.

Experiment with different sleeping positions. Sleep apnea occurs most often when sleeping on the back. Sleeping on the side may reduce airway obstruction. Positional pillows and foam head supports made for side sleepers can encourage maintaining the position throughout the night. Those who find themselves rolling onto their backs in the middle of the night may benefit from sewing a pocket on the back of a T-shirt in which to place a tennis ball. Rolling on the back then becomes uncomfortable and encourages staying mindful of the sleep position, even while semiconscious.

Treat underlying diabetes, heart disease, and hypertension. These conditions may contribute to vascular congestion, weight issue, or airway inflammation and obstruction. Individualized treatments for each underlying contributing factor may help achieve better sleep.

Lose weight where needed. Sleep apnea is significantly more common in obese patients, and weight loss can reduce pressure on the airways. Doctors need to be prepared to put together effective recommendations that will help patients embrace making the difficult changes required to be successful in losing weight.

Quit smoking. Smoking is a leading cause of heart disease and damage to the pulmonary vasculature. Patients should be encouraged to quit smoking for all of these reasons.

Abstain or reduce alcohol consumption. Alcohol can relax all muscles, including those of the throat. Evening drinking can increase snoring as well as sleep apnea. Abstain from alcohol to improve sleep apnea.

Exercise each day. Exercise improves heart health, blood vessel health, and oxygen saturation of the blood. Exercise also supports weight loss when continued as a regular habit. Furthermore, many people find that vigorous exercise during the day improves the quality of sleep that night, no matter the causes of sleep irregularities.

Decongest the throat with mucolytic herbs if needed. Independent of obesity or heart disease, patients with genetically small throats or airways may also be prone to sleep apnea. Congestion in the upper respiratory mucous membranes compounds the problem. Such patients should experiment with avoiding dairy products, which are notoriously mucus forming. Any food allergen may increase throat and respiratory congestion, as can hay fever and sensitivity to airborne molds, chemicals, and other antigens. Patients with atopic disorder may improve with antiallergy approaches and the use of respiratory mucolytics.

Specific Indications: Herbs for Cardiovascular, Peripheral Vascular, and Pulmonary Conditions

Herbs for vascular and pulmonary conditions include those agents capable of reducing blood pressure, such as vasodilators, and those capable of protecting the blood vessels for elevated pressure, lipids, and sugars, such as the various polyphenolic compounds in foods and herbs, such as flavonoids. Heart muscle tonics such as cardiac glycosides and cardioactive steroids are also important vascular herbs. Because many heart and peripheral vascular diseases progress insidiously over many decades' time, daily tonics to protect against fibrosis, platelet activation, and oxidative stress round out the list of important *materia medica* tools for support of vascular health. Following is a list of herbs specifically needed for the blood vessels and heart.

Achillea millefolium • Yarrow

Achillea flower buds may be used as an anti-inflammatory and antimicrobial for vascular inflammations and infections. *Achillea* is a peripheral vasodilator to include in hypertension formulas, especially when associated with constipation and liver congestion. *Achillea* is a bitter alterative herb and cholagogue with a warming quality. *Achillea* may be used as a hemostatic for urgent hemorrhage situations such as nosebleeds, uterine bleeding, and intestinal bleeding.

Aconitum napellus, A. carmichaelii • Aconite

Acontium napellus is used in Western herbalism and prepared into homeopathic aconite. It also goes by the common name wolfsbane because the roots are so toxic that they have been used historically to poison large predatory animals. In TCM *Aconitum carmichaelii*, or fu zhi, is processed by soaking in alkali and then boiling or steaming to remove a large portion of the poisonous alkaloids. Even though crude *Aconitum* roots are a deadly poison, the processed root is considered indispensable in TCM, where the roots appear in more than 10 percent of traditional formulas. It is said that *Aconitum*'s toxicity is due to its extreme concentration of potent life energy, and once processed, it can help offer yang chi at the end of life or treat a variety of diseases. Fu zhi is said to be one of the most yang herbs, commonly used to treat acute myocardial infarction-induced shock, hypotension, coronary heart disease, rheumatic heart disease, and a variety of other ailments. **Caution:** Do not attempt to alter the formulas in this chapter that contain aconite into herbal tinctures, or use aconite tincture at all without expert guidance. Do not substitute processed fu zhi for unprocessed *Aconitum napellus*.

Adonis vernalis • Pheasant's Eye

Caution: This is a potentially toxic botanical due to the presence of glycosides with cardiostimulant effects. It is indicated for low blood pressure, weak feeble heart, valve disease, and edema. Specific indications include a feeble and irregular pulse. *Adonis* is also used for cardiac inflammation such as endocarditis and myocarditis. *Adonis* can improve edema by promoting renal blood flow. The Lloyd Brothers pharmacy recommended that their *Adonis* tincture be diluted into a 2X preparation for endocarditis, or that one-half to one-third of a drop be taken every 3 hours for weak, irregular heart action. Another method of dosing the tincture is to place just 10 drops in 4 ounces of water, which may then be taken by the teaspoon hourly for acute heart failure.

Aesculus hippocastanum • Horse Chestnut

The ripe nuts from horse chestnut trees are specific for poor circulation in the abdomen, especially associated with liver congestion, portal congestion, hemorrhoids, and varicose veins. *Aesculus* is indicated for aching and weight in varicosities and pelvis, as well as for rectal prolapse. *Aesculus* is useful for back pain that is due to vascular congestion rather than musculoskeletal problems and for right upper quadrant pain. Many old texts and homeopathic *materia medicas* mention *Aesculus* for sharp sticking pains in the anus and rectum, as if splinters were present.

Alisma plantago-aquatica • Asian Water Plantain

The rhizomes contain aconitine-like alkaloids that are potentially dangerous and hepatotoxic, but have also been used in small doses for heart disease and stroke recovery. The plant is emphasized for diuretic effects and to improve renal function in formulas for poor circulation in the kidneys. *Alisma* may improve dizziness and tinnitus related to circulatory disorders, and research suggests the medicine to lower lipids and blood sugar. This species is also known as *A. orientale*.

Allium sativum • Garlic

Garlic is an all-purpose vascular herb with cholesterol-lowering, hypotensive, antithrombotic, platelet-stabilizing, nitric-oxide enhancing, glucose-lowering, antioxidant, and antimicrobial properties. The well-researched sulfur compounds, such as allicin, are credited with much of the medicinal effects. *Allium* has been the subject of several human clinical trials showing efficacy in treating hypertension. The consumption of garlic on a regular basis may offer some cardioprotection by reducing inflammatory processes and supporting antioxidant levels.

Ammi visnaga • Khella

Khella seeds are used as a vasodilator for hypertension and as an antiallergy herb to treat vasculitis and systemic inflammatory burden. The extracts have also traditionally been used for angina pectoris because of the vasodilatory action, and they are commonly included in formulas for cardiovascular and pulmonary inflammation, spasms, and congestion. Several synthetic drugs have been developed from molecules in khella, including amiodarone for hypertension. Khella is sometimes confused with its less-used relative, bishop's weed (*Ammi majus*), but the two species do have common chemical constituents and pharmacological effects in common.

Angelica sinensis • Dong Quai

Angelica root is an all-purpose blood mover for hypertension, vascular inflammation, and cardiac and peripheral vascular diseases. *Angelica* may reduce the tendency to clots, as an ingredient in formulas for acute thrombi and phlebitis. *Angelica* may improve anginal pain and pain related to blood congestion and is indicated for vascular congestion, pelvic stagnation, portal congestion, and allergic and inflammatory vascular phenomena. *Angelica sinensis* may also help with stroke recovery.

Anthoxanthum odoratum • Sweet Vernal Grass

This grass contains coumarin, noted for numerous anti-inflammatory, hormonal, and circulation-enhancing properties. The characteristic pleasant aroma of freshly cut grass is due to the coumarins. Plants named "sweet clover" or "sweet grass" are so called because of these compounds. The aerial parts have been used folklorically for vascular congestion. The grass is fed to livestock and reported to contribute to excellent quality meat, but rotten sweet grass can form dicoumerols associated with anticoagulant affects and cause possible hemorrhage.

Apium graveolens • Celery

Celery seeds and other celery root and stalk preparations are traditional remedies for high blood pressure in China. Research suggests that they may both support liver function and have nutritive diuretic effects on the kidneys. Specific indications include dizziness, headaches, and shoulder pain. Celery stalk juice may be used on a daily basis to help treat hypertension and is considered safe for pregnant and ill individuals.

Apocynum cannabinum • Dogbane

Caution: This is a potentially toxic botanical due to the cardiac glycosides occurring in parts of the plant and should be limited to use by experienced clinicians only. Native American tribes used *Apocynum* roots for a variety of complaints including heart and respiratory diseases, and weak teas were reported to be cardiotonic and to quiet coughs. Dogbane is indicated where tissues are full and there is a history of nervous debility and atony of the nervous system. *Apocynum* might help heavy smokers with cardiac irritation and will improve chest pains and general edema. Eclectic physician and author H. W. Felter reported *Apocynum* to be specific for circulatory atony and dropsy and classified it as a cardiac tonic and sedative. According to the Lloyd Brothers, the dose is 10 drops of tincture placed in 1 ounce of water, which is then taken by the teaspoon every 3 hours.

Arctium lappa • Burdock

Burdock root may be used as a supportive, alterative ingredient in vascular formulas associated with digestive, biliary, or hepatic disorders. Burdock roots are high in iron and may be included in teas for those with anemia.

Arnica montana • Leopard's Bane

The dried flowers of *Arnica* are most often used to prepare oils for musculoskeletal pain and trauma and to prepare homeopathic medicine; they are less commonly included in tinctures for internal use. *Arnica* may be used orally, in small quantities only, for anginal pain that is sore, bruised, and aching in quality. Arnica may improve feeble circulation. *Arnica* is said to be specific for weak circulation, for the advanced stages of disease where innervation is impaired, and for a sense of weight and pressure in the chest that is worse with exertion.

Astragalus membranaceus • Milk Vetch

Astragalus roots are commonly combined with *Angelica* and used for a wide variety of vascular complaints and diseases. *Astragalus* has been shown to support stroke recovery, heart disease, and inflammatory disease of the vasculature, whether induced by diabetes, hyperlipidemia, or simple old age. *Astragalus* polysaccharides are noted to protect endothelial cells, cardiomyocytes, and mitochondria from a variety of oxidative stressors and toxic assaults. *Astragalus* is a bland-tasting legume suitable for teas, and the powder can be included in smoothies and medicinal foods. *Astragalus* can be used for acute infectious or inflammatory crises as well as in long-term formulations to stall the progression of fibrosis of tissues, atherosclerotic plaques, and biofilm establishment in the heart and pulmonary tissues. *Astragalus* may help protect collagen matrices of the heart and airways, preserving elasticity and reducing permanent tissue damage.

Atractylodes species • Bai Zhu

Atractylodes lancea and *A. macrocephala* are Asteraceae family plants whose dried rhizomes are used in TCM for digestive issues and as a general tonic following stroke or other serious illness. *Atractylodes* roots are often processed by stir-frying with wheat bran with the aim of reducing its dryness and increasing its tonifying function. The plant may improve lipid metabolism and have antiobesity effects. Like other aster family plants, some of the anti-inflammatory effects are credited to sesquiterpenoid compounds. Other major constituents credited with medicinal effects are atractylodin and β-eudesmol.

Avena sativa • Oats

The ripe groats and oatmeal are useful for heart health. Eating oatmeal and oat bran provides high-quality fiber and beta glucans to help control elevated lipids and glucose. "Milky" oats are high in tryptophan and credited with a calming effect on the nerves, which may help reduce stress-induced hypertension.

Berberis aquifolium • Oregon Grape

Berberis roots contain berberine, which is useful to support metabolism and liver function, and may be a foundation alterative ingredient in formulas for hypertension, diabetes, and hyperlipidemia. *Berberis* powders, compresses, and skin washes may be used topically for stasis ulcers, specifically for thick purulent mucous accumulations. *Berberis* is specifically indicated for various cardiovascular pathologies associated with digestive or liver disorders and portal congestion. This species is also known as *Mahonia aquifolium*.

Borago officinalis • Borage

The seed oil is a rich source of gamma-linolenic acid (GLA) that may improve lipid profiles and reduce cardiovascular disease risk factors. GLA is used in the body to produce antiinflammatory prostaglandins, which may help protect the nerves, lungs, heart, and other tissues form oxidative and inflammatory stressors. Borage seed oil may be helpful in larger protocols for diabetes, elevated lipids and heart disease, respiratory distress syndrome, and neuropathy.

Calendula officinalis • Pot Marigold

Calendula flowers and/or leaf medications are useful topically for phlebitis and stasis ulcers and internally to support connective tissue regeneration and prevention of fibrin deposition in convalescent formulas for MI, endocarditis, and pericarditis. Despite its popularity, *Calendula* has not been extensively investigated, but research suggest that it may support healing and tissue regeneration through support of microcirculation. *Calendula* supports angiogenesis in injured tissue and promotes complete healing rather than fibrosis. *Calendula* may limit the formation of advanced glycation end products (AGEs), where sugars and proteins fuse and attach to cell surfaces, contributing to aging and the accelerated progression of diabetes and heart disease.

Camellia sinensis • Green Tea

Green tea is an antioxidant and gentle metabolic stimulant for inflammation. Epidemiologic studies suggest the regular consumption of green tea may reduce the risk of cancer and many inflammatory diseases. *Camellia* may benefit the heart, but present research suggests only the unfermented or partially fermented green and oolong teas benefit hypertension, and these benefits may not apply to "black" or fermented *Camellia* teas. *Camellia* contains iron and can be used in tea formulas to treat anemia.

Cassia • See *Senna*

Castanospermum australe • Blackbean Tree

This South Pacific leguminous tree contains phytosterol saponins credited with hypotensive, hypoglycemic, and cardioprotective effects. The seeds are poisonous but are processed to leach the toxic compounds and may be made edible. The alkaloid castanospermine inhibits α and β-glucosidase, interfering with the synthesis of glycoprotein receptors on some cell and endoplasmic reticular membranes including blood and endothelial cells, and thereby preventing the adhesion of various substances. Castanospermine may offer protective effects on the vasculature and extracellular matrix of the heart and blood vessels by inhibiting endothelial glycoproteins binding. *Castospermum* may also deter viral endocarditis by interfering with viral proteins.

Ceanothus americanus • Red Root

The roots of *Ceanothus* are indicated for chronic fluid stasis in the obese and for those with poor circulation, splenomegaly, and diabetics. *Ceanothus* is also helpful for the doughy skin often seen in long-term diabetics. *Ceanothus* is specific for "lymphatic" and "damp" constitutions in various folkloric traditions. Very little research on *Ceanothus* has yet been conducted.

Centella asiatica • Gotu Kola

Gotu kola is a tissue-protecting and building herb that may be used topically for phlebitis and stasis ulcers, internally to support connective tissue regeneration, and in convalescent formulas for MI, pericarditis, phlebitis, stasis ulcers, and trauma recovery to prevent fibrin deposition. The entire plant contains the triterpenoids madecassic and asiatic acids, which may protect endothelial cells from hypoxic stress and resist damage to endothelial barrier function in various inflammatory situations. Animal studies suggest that the saponin asiaticoside attenuates pulmonary hypertension and resultant cardiac hypertrophy by protecting against hypoxia-induced proliferation of arterial smooth muscle cells. A human pilot study suggested *Centella* to reduce collagen deposition in carotid artery plaques compared to placebo.[182]

Cereus grandiflorus • See *Selenicereus*

Cinnamomum species • Cinnamon

Cinnamomum verum is commonly called true or Ceylon cinnamon; the species is also known as *C. zeylanicum*. *Cinnamomum cassia* is called cassia, and it is the most common commercial form of cinnamon. *C. cassia* is also known as *C. aromaticum*. Both species are used for culinary and medicinal purposes and credited with hypotensive and antiatherogenic and antidiabetic effects. Various species of cinnamon may regulate glucose transporters, improve glucose uptake and fasting blood glucose, inhibit α-glucosidase activity, and increase insulin sensitivity in diabetes and obesity. Cinnamon bark is a warming blood mover for cold constitutions, and it reduces platelet aggregation, reducing the risk of clotting and thrombi. Cinnamon moves blood out of the pelvis and into the limbs. Cinnamon in formulas may help reduce high cholesterol and glucose. Cinnamon is an effective hemostatic for postpartum hemorrhage, menorrhagia due to fibroids and polyps, and heavy menses that are worse with motion and jarring. Both of these species contain cinnamaldehyde and eugenol credited with antimicrobial, anti-inflammatory, and other activities. Cinnamon is a general circulatory effector with warming properties in the skin and gastrointestinal tract. Cinnamon may irritate the intestines in those with irritable bowel syndrome.

Cnicus benedictus • Blessed Thistle

The leaves of this thistle are specific for sluggish liver function and are a useful alterative to improve the liver's processing of hormones, carbohydrates, and lipids. The plant has been used as a wild food, a galactagogue, and hepatonic in traditional herbalism. *Cnicus* may be used in formulas as a gentle tonic and base ingredient when treating circulatory issues related to underlying diabetes and hyperlipidemia.

Cnidium monnieri • Snow Parsley

Cnidium seeds are widely used in the traditional herbalism of China, Vietnam, and Japan, and more than 350 physiological compounds have been identified including coumarins, common to Apiaceae family plants. Among the broad antimicrobial, antiallergy, and anti-inflammatory compounds, *Cnidium*'s blood-moving effects are particularily emphasized in folkloric traditions. *Cnidium* could be used in formulas for hypertension, vasocongestion, and vasculary hypersensitivy disorders such as hives or headaches and also to support both male and female reproductive conditions, including infertility and erectile dysfunction.

Coleus forskohlii • See *Plectranthus forskohlii*

Collinsonia canadensis • Stoneroot

Collinsonia is specific for peripheral vascular congestion and varicosities associated with fluid congestion in the surrounding mucous membranes and also for hemorrhoids with a sense of weight and aching and concomitant with intestinal mucous membrane congestion. *Collinsonia* is also specific for long-standing mucosal congestion in the throat leading to a hoarse voice and full sensation or sense of constriction in the larynx or pharynx, as well as a tickling sensation in the throat. *Collinsonia* may improve vascular congestion in the throat due to overuse of the voice. *Collinsonia* is also specific for a sense of constriction in other tissues, such as hemorrhoids with the sense of splinters and sticking pains or a sense of foreign body in the anus. Felter reported *Collinsonia* to sedate vagal nerve activities when given in small frequent doses and recommended it for asthma, cough, and heart ailments especially when associated with a sense of weight and tightness in the chest.

Commiphora mukul • Guggul

Commiphora mukul is specific for elevated lipids due to diabetes, hypothyroidism, or slow metabolism. It is a warming herb that is traditional in India for obesity, diabetes, and hyperlipidemia.

Convallaria majalis • Lily of the Valley

Convallaria contains cardiac glycosides including convallarin and is specific for weak heart action causing tissue congestion and dyspnea. *Convallaria* is also recommended for valvular diseases and mechanical problems, palpitations and arrhythmias, and pitting and echymoses of the tissues due to poor venous return. The Eclectic physicians of the late 1800s thought *Convallaria* was safer than digitalis and noncumulative and used it extensively for heart diseases. A dose of 5 to 20 drops of the fluid extract was taken in a single dose; 2 to 10 milliliters was placed in 4 ounces of water and dispensed by the teaspoon, every 3 to 6 hours. **Caution:** *Convallaria majalis* has a *Digitalis*-like action and should be reserved for use by experienced herbalists and clinicians only.

Coptis chinensis • Goldthread

Coptis is "gold" due to the presence of berberine in the roots, a well-studied isoquinoline alkaloid credited with many medicinal activities. *Coptis* has antimicrobial effects and may be included in formulas for infectious endocarditis and phlebitis. Berberine supports metabolism and liver function and is also appropriate to include in formulas for cardiovascular disease related to elevated lipids, diabetes, and hypertension. Due to its many metabolic effects, *Coptis* may help deter endothelial atherosclerotic plaque deposition and may help protect the brain against neurodegenerative effects due to metabolic syndrome. *Coptis* also supports liver metabolism, improving carbohydrate and lipid processing,[183] and offers cardioprotection.[184] *Coptis chinensis* is one of the 50 fundamental herbs used in TCM, where it is called duan e huang lian.

Crataegus monogyna, C. oxyacantha • Hawthorn

Leaves, flower buds, and ripe fruits are all used medicinally, and hawthorn is a general circulatory-enhancing herb with versatile applications in all manner of cardiovascular formulas. *Crataegus* may improve perfusion, and the Eclectic era physicians claimed that it improved nutrition to the heart, making it appropriate for all types of cardiomyopathy. *Crataegus* makes a useful trophorestorative base ingredient in formulas to improve blood vessel health and integrity and as a vascular anti-inflammatory for all manner of cardiovascular and peripheral vascular conditions. *Crataegus* may improve circulatory insufficiency and help protect the blood vessels in cases of hypertension, hyperglycemia, and hyperlipidemia. *Crataegus* is not terribly specific for anything and yet is appropriate for everything. *Crataegus* is appropriate in inflammatory processes associated with hyperglycemia and hyperlipidemia and can be a foundation herb upon which to rest more specific herbs to help prevent vascular damage for long-term use in treating chronic conditions. It is appropriate in cases of both hypertension and hypotension, bradycardia and tachycardia, and sluggish circulation and tendency to clots, and it is not contraindicated during Coumadin therapy or for diseases associated with a tendency to hemorrhage.

Crinum glaucum • Swamp Lily

Like other Amaryllidaceae family bulbs, the swamp or river lily may have physiologic effects via nervous regulation. *Crinum* may be specific for hypertension with shallow breathing and heart weakness with dyspnea.

Curcuma longa • Turmeric

Turmeric roots may be used as an antioxidant and anti-inflammatory ingredient in formulas for vascular inflammation, especially when associated with liver congestion, systemic inflammation, and chronic infections. *Curcuma* is also helpful to improve lipid and carbohydrate metabolism in cases of diabetes and hyperlipidemia. *Curcuma* may also be included in formulas for hemochromatosis due to its ability to protect tissues from the damaging effects of iron overload.

Cynara scolymus • Artichoke

Artichokes and related thistles are nourishing and useful as medicinal foods or as alterative ingredients in formulas for high cholesterol and vascular complaints related to liver or digestive disorders. *Cynara* is a liver tonic and may improve cholesterol and hormonal imbalance by supporting the hepatic processing of these substances. *Cynara* improves glucose and lipid processing and is useful for hyperglycemia and hyperlipidemia. Human clinical trials have shown leaf medications to help improve liver function in patients with nonalcoholic steatohepatitis.[185]

Cytisus scoparius • Scot's Broom

Cytisus flowering tops contain the alkaloid sparteine, which is a general diuretic that may help diurese congestive heart failure patients. Sparteine may also have an oxytocic effect, explaining the plant's traditional use for stalled labor. The Lloyd Brothers pharmacy of the late 1800s and early 1900s recommended 10 to 30 drops of the prepared tincture in a given dose, repeated 2 or 3 times a day for circulatory support. The plant has not been rigorously researched but is reported to contain common flavonoids, such as rutin, quercetin, quercitrin, isorhamnetin, and kaempferol, and common legume isoflavones including genistein and sarothamnoside. Unique compounds include quinolizidine alkaloids, namely sparteine as well as arothamine and lupanine, and benzenoid compounds including tyramine. Lectins including galactosamine have also been identified.

Desmodium styracifolium • Desmodium

This legume is used to clear heat and resolve dampness in TCM, and the plant is credited with an ability to diurese and disperse toxins. The entire plant is used medicinally and may be an adjuvant in formulas for the heart and lungs; it is also used as a tonic that may lower blood pressure through promoting cholinergic activity and blocking adrenergic activity. *Desmodium styracifolium* shows fibrinolytic activity, explaining its traditional use in circulatory formulas.[186]

Digitalis purpurea • Foxglove

Caution: This classic cardiotonic is a potentially toxic botanical due to the cardiac glycosides it contains. *Digitalis* leaves act as a general circulatory stimulant, which is best for weak, feeble, and degenerated heart conditions. The publications of both the Lloyd Brothers and the German physician R. F. Weiss have reported that *Digitalis* does not always have to be continued lifelong, as is the current norm, but that it can correct and strengthen a weakened heart and then be withdrawn. Dose is 10 drops to 2 milliliters placed in 4 ounces of water and taken by the teaspoon every hour (per the Lloyd Brothers), or 1 to 3 drops of the fluid extract or 5 to 20 drops tincture (per Finley Ellingwood). Felter stated in the 1920s that *Digitalis* could be useful to treat delirium tremens and for the early states of heart failure, in which case he recommended combining it with *Selenicereus grandiflorus* (known in Felter's era as *Cactus grandiflorus*). *Digitalis* is said to first stimulate the vagal nerve and then to depress it, but when combined with *Selenicereus*, no such depression tends to occur. The goal of therapy for heart failure is to use very small doses that stimulate the heart, but not push the dose too high or for too long and risk further weakening the heart. *Digitalis* cardiac glycosides such as digoxin have been well studied and developed into inotropic pharmaceuticals used to treat cardiomyopathy and tachyarrhythmias such as atrial fibrillation.

Echinacea species • Coneflowers

Echinacea purpurea and *E. angustifolia* root medicines may be included as antimicrobial ingredients in formulas for infectious complaints of the cardio and peripheral vasculature. *Echinacea* may be a supportive herb in cardiovascular formulas, acting as an aid in moving lymphatic fluid in cases of cellulitis and lymphedema. *Echinacea* is specifically indicated for decay of the tissues, such as with stasis ulcers in circulatory insufficiency and even gangrene.

Eleutherococcus senticosus • Siberian Ginseng

Eleutherococcus is indicated for hypertension and heart palpitations associated with stress and anxiety. The roots of *Eleutherococcus* act as an adrenal tonic useful for long-term stress resulting in nervous symptoms and fatigue. *Eleutherococcus* is appropriate to include in formulas for endocrine imbalances and immune insufficiency, especially when related to stress and overwork and associated with fatigue, weakness, and emotional instability.

Epimedium brevicornu• Horny Goatweed

Epimedium is a traditional Chinese herb said to invigorate yang and often used in skin, bone, and reproductive hormonal-balancing formulas. Other *Epimedium* species are also used medicinally, with *E. grandiflorum* being most emphasized in TCM as a kidney yang tonic. *Epimedium* may also enhance peripheral nerve regeneration when damaged due to poor circulation, and animal studies support the traditional usage for improving motor function following a stroke and helping to recover normal nerve conductivity. The aerial parts contain the flavonoid icariin, shown to improve blood lipids. Icariin is shown to be a phosphodieterase inhibitor, which serves to improve blood vessel tone, yet has vasodilating effects via nitric oxide promotion in endothelial cells. *Epimedium* is a large genus, and numerous published research papers cite species names that have later been found to be synonymous. For example, *E. brevicornu* is also known as *E. brevicornum* and *E. brevicornis*. *E. koreanum*, *E. grandiflorum*, *E. rotundatum*, *E. sagittatum*, and *E. brevicornum* all go by the common name horny goatweed, are used to improve penile circulation in cases of impotence, and are shown to have general circulatory-enhancing effects.

Equisetum species • Horsetail

The aerial parts of *Equisetum* may be used as a supportive herb in cardio and peripheral vascular formulas. *Equisetum* is not specifically a vascular herb, but may be a possible synergist herb to include in vascular formulas where there is acute inflammation of blood vessels and surrounding soft tissues, such as cellulitis. *Equisetum* may support full healing of inflamed blood vessels and tissue, reduce scarring and hyperinflammatory reactions, reduce keloid formation in the skin, and prevent valve damage in the heart that may result from acute infectious and inflammatory processes. *Equisetum arvense*, *E. hyemale*, and other species are used interchangeably.

Eschscholzia californica • California Poppy

The entire plant is used as an anodyne and nervine medicine. *Eschscholzia* may act as a complementary component in formulas for hypertension-associated adrenal exhaustion due to long-term stress. *Eschscholzia* is specific for symptoms of anxiety, muscular tension, and poor sleep.

Fucus vesiculosus • Kelp

Kelp, also called bladderwrack, is indicated for hypotension associated with hypothyroidism and is specific for obesity, goiter, exopthalmia, constipation, flatulence (all of which are symptoms of hypothyroidism), and impotence in hypothyroid men. The entire seaweed is dried or prepared into medicines and medicinal foods. Because it is so high in minerals, *Fucus* is also useful for mineral imbalances and nutritional deficiencies. *Fucus* contains sulfated polysaccharides with immunomodulating effect, and the whole plant may support the integrity of connective tissue, both internally and when topically applied.

Gastrodia elata • Tian Ma

Gastrodia is an orchid that requires symbiotic association with two different fungi in order to grow. As with many orchids, *Gastrodia* is endangered and should be obtained only from cultivated sources. Because the *Armillaria* fungus that populates *Gastrodia* roots has been credited with many of *Gastrodia*'s medicinal effects, it is now cultivated as a substitute for *Gastrodia* and referred to as gastrodia mushroom. *Gastrodia* was listed in ancient herbals to guard health and protect longevity. *Gastrodia* is a traditional Chinese medicine for treating stroke, epilepsy, tinnitus, headaches, transient ischemic episodes, dizziness, and dementia, and the constituent gastrodin has been shown to prevent reperfusion injury.[187] *Armillaria* fermentation products also appear to have circulatory-enhancing effects, useful to improve cerebrovascular circulation, treat Meniere's syndrome or disease, epilepsy, and paresthesia related to diabetes.

Ginkgo biloba • Maidenhair Tree

Ginkgo leaves are one of the most studied and prescribed herbal medicines. Due to its many anti-inflammatory and antioxidant effects, *Ginkgo* may be included in herbal formulas for a wide variety of vascular complaints including hypertension, coronary artery disease, cerebrovascular insufficiency, arterio and veno-insufficiency, Raynaud's, and tissue edema related to heart failure. See "*Ginkgo* and Circulatory Enhancement" on page 69.

Glycine max • Soybean

Soybeans are well known to contain phytosterols and are specific to use in a foodlike way for hyperlipidemia, hyperglycemia, hypertension, and menopausal symptoms including osteopenia. Soy's hypotensive effects may be weak, but its overall nutritive and protective effects may be significant. The regular consumption of soy throughout a lifetime is associated with a decreased risk of cardiovascular disease parameters.

Glycyrrhiza glabra • Licorice

The roots of *Glycyrrhiza* species may be used as a supportive adaptogenic ingredient in cases of weak heart action related to endocrine insufficiency and imbalances. *Glycyrrhiza* is appropriate for viral infections, weak immunity, and general inflammation. *Glycyhrriza* is also indicated to support the adrenal glands when withdrawing from steroids, such as in convalescent formulas for acute endocarditis and pericarditis. *Glycyrrhiza uralensis* is used in many TCM formulas for heart disease, vascular inflammation, and many other conditions.

Gymnema sylvestre • Gurmar

Also called the sugar destroyer, *Gymnema* is indicated for arteriosclerosis and atherosclerosis associated with diabetes. The plant has been used in India for such complaints for thousands of years. The plant has earned the name "sugar destroyer" because chewing the leaves destroys the ability to taste sweetness in foods for an hour or two afterward. It was therefore traditionally used to curb sweet cravings and to support diabetics with weight loss and dietary changes. The plant has also been used for vascular disease and to treat both high and low blood pressure. Large doses of the plant may be contraindicated in those with rapid hypoglycemic reactivity, because it may cause blood sugar to drop quickly. A group of triterpenoid saponins called gymnemic acids have been identified and are thought to be responsible for the observed antidiabetic activity. Another group of saponins, the *Gymnema* saponins, are credited with the antisweat phenomenon.

Hamamelis virginiana • Witch Hazel

The leaf, bark, and young twigs of witch hazel have all been used traditionally to make herbal medicines, and the plant has been listed in many early editions of the US *Pharmacopeia*. Native American peoples used the plant topically and internally. *Hamamelis* is specific for varicosities and peripheral vascular inflammation. *Hamamelis* has been called the "arnica of the veins" because it is specific for vein trauma, just as *Arnica* is specific for musculoskeletal trauma. *Hamamelis* increases vasomotor tone, helping to improve varicosities and reduce passive hemorrhages. *Hamamelis* may be used both topically and internally in preparation for vein inflammation and pain and is specific for heavy, aching, distensive pains in the vasculature. For men with varicocele, combine *Hamamelis* with *Ginkgo* and *Aesculus*. For women with perineal and heavy, aching genital pain associated with premenstrual vascular congestion, combine with *Caulophyllum* and/or *Angelica*. *Hamamelis* is also specifically indicated for heavy menses due to excessive endometrial buildup or for abnormal shedding of the uterine lining, as is the case with fibroids and polyps, where *Hamamelis* is appropriately combined with *Achillea* and *Cinnamomum*. *Hamamelis* is credited with stypic and astringent properties and is also indicated for blood in the sputum (hemoptysis), bleeding from the bowels, or blood-tinged diarrhea. The Eclectic authors emphasized *Hamamelis* for atonic and flabby mucous membranes due to poor vascular tone.

Hibiscus sabdariffa • Roselle

Also called flor de Jamaica, this herb is a Caribbean, South American, and African traditional medicine for cardiovascular disease. Rich in flavonoids, *Hibiscus* sepals are best used as a sour tea, taken as often as possible to prevent the occurrence or at least slow the progression of heart disease. *Hibiscus* may be used broadly for all manner of cardiovascular diseases and appears safe for all patients to use in tandem with other herbs and pharmaceuticals. *Hibiscus* is the richest known plant source of chromium and may improve insulin response and metabolism of glucose. Therefore, *Hibiscus* might be specifically indicated for hypertension, atherosclerosis, type II diabetes, and metabolic syndrome. While the sepals of roselle are especially valuable as medicine and medicinal food, the dried petals of *Hibiscus rosa-sinensis* are also used medicinally in Asia. The plant is also called Chinese hibiscus.

Hippophae rhamnoides • Sea Buckthorn

Sea buckthorn fruits are used to produce an orange-red oil high in vitamins, flavonoids, and important minerals including selenium and zinc and may act as a natural calcium channel blocker to lower blood pressure. The seeds may be used to extract a pale thin oil high in linoleic and alpha-linolenic fatty acid, and the fruit pulp may be used to create a bright orange thick oil high in palmitic fatty acid. Both are high in vitamin E and carotenoids and can be used topically and internally for medicinal and cosmetic purposes. The oil may be consumed orally to improve lipid profiles, as well as used to help control vascular inflammation, support repair and regeneration of tissue, and reduce allergic activity.

Huperzia serrata • Chinese Club Moss

Known as qian ceng ta in China, the *Huperzia* club moss was at one time known as *Lycopodium*, long recommended for mental weakness and senility. *Huperzia* contains a group of alkaloids known as huperzines, shown to act as cholinesterase inhibitors. The entire moss has been used traditionally in China, while modern medicine isolates and concentrates the huperzines. *Huperzia*, and especially concentrated products, may be used in formulas for vascular dementia and cognitive decline in the elderly. *Huperzia* is also used for vascular injuries such as bruising, vomiting blood, hemorrhoids, and lung abcesses, and older herbals suggest it can be useful for recurrent fevers that are worse each afternoon and for congestive headache, dizziness and fainting, cough with bloody expectoration, and lung disease.

Hydrastis canadensis • Goldenseal

While not a lead herb in vascular formulas, *Hydrastis* roots may be useful as a supportive herb in some situations of vascular stasis associated with the breakdown of skin and mucosal barriers, associated with opportunistic infections. *Hydrastis* may be used as an ingredient in topical preparations for stasis ulcers, specifically for thick purulent mucous accumulations. *Hydrastis* may also improve bleeding intestinal ulcers associated with feeble circulation, loss of intestinal tone, and mucosal atrophy.

Hypericum perforatum • St. Johnswort

St. Johnswort flower buds may be used as an ingredient in formulas to strengthen and reduce inflammation in all types of vascular tissue, from the heart muscle itself, to the microvasculature, and as a circulatory and neural anti-inflammatory for retinopathy, retinitis, and vascular trauma with bruising and hematomas. *Hypericum* is a traditional medicine for bruising, strains, sprains, and trauma and may also support healing of stasis ulcers in patients with diabetes and circulatory insufficiency. Animal studies suggest that *Hypericum* has the ability to enhance fibroblast proliferation and collagen synthesis and to support revascularization and wound healing. *Hypericum* has also been used folklorically for neuropathy and trauma to highly innervated areas. Internal and topical medication may also be useful for diabetics with peripheral neuropathy, and flavonoids in *Hypericum* may offer neuroprotective action.

Iris versicolor • Wild Iris

The creeping rhizomatous roots of *Iris* may be used as a supportive ingredient in formulas for hypertension and hyperlipidemia associated with portal congestion and insufficient digestive secretions. *Iris* is an overall stimulant specific for congested lymphatic tissues and body glands including lymph nodes, spleen, liver, and thyroid. *Iris* increases the secretions of salivary and digestive glands and is useful as a complementary herb in cases of hypothyroid digestive insufficiency and the poor lipid metabolism and hyperlipidemia that may accompany. *Iris* is a warming, stimulating herb; a small dose of only a few drops combined with other herbs is all that is needed to gently stimulate the glands. Other common names are blue flag and vegetable mercury. Mercury toxicity is known to promote excessive salivation, and because *Iris* can promote saliva flow, it acts as a "vegetable" mercury.

Juniperis communis • Juniper

Juniper berries are best used as a synergist and complementary herb in formulas for edema related to poor circulation to the kidneys and renal insufficiency. *Juniperis* is occasionally appropriate as an ingredient in formulas for edema related to congestive heart failure, though only as a palliative support. *Juniperis* is used only in those with atonic renal function and output, such as elderly men with poor urinary tone and function, and in those with a postmenopausal tendency to chronic bladder infections. *Juniperis* is a urinary stimulant and counterirritant for cases of atony in the renal system. Juniper is contraindicated for acute inflammatory disorders of the urinary system. Juniper is specific for thick urine with mucous threads and secretions and should be used in small doses only. It is most appropriate for cold, underfunctioning, atonic conditions in diabetics with poor circulation, atony of the tissues, and a tendency to chronic urinary tract infections.

Lavandula species • Lavender

Lavender is specific to include in teas or use in aromatherapy for stress and anxiety-related hypertension. Research suggests that the plant may act in part via cholinergic effects and direct effects on the limbic system. Even inhaling lavender essential oils off the palms of the hands for 10 minutes may override sympathetic nerve dominance and invoke the relaxing and hypotensive effects of the parasympathetic nervous system. Lavender may also be used orally in nervine tea blends in the treatment of hypertension, and newer products include lavender oil concentrates for oral use in the treatment of anxiety and anxiety-related complaints. *L. angustifolia* (also known as *L. officinalis*) and *L. latifolia* are the main species used medicinally and to produce essential oil.

Leonurus cardiaca • Motherwort

The young aerial parts of motherwort are indicated for for vascular complaints associated with nervousness and endocrine imbalances such as menopausal transitions and hyperthyroidism. *Leonurus* is appropriate as a supportive ingredient in formulas for hypertension when associated with anxiety, hyperthyroidism, and menopause. *Leonurus* is specific for hypertension, heart palpitations, cardiac irregularity manifesting at the climacteric, and complex endocrine disorders with a restless anxiety. *Leonurus* is also appropriate for nervous debility and restlessness and for tics, twitches, and tremors.

Lepidium latifolium • Rompe Piedras

Lepidium latifolium is specifically indicated for kidney stones because it helps the urinary tissues excrete wastes and excess water. This species of *Lepidium* may also be specific for hypertension, helping the kidneys excrete sodium, and supporting renal circulation in cases of circulatory insufficiency.

Lepidium meyenii • Maca

Maca roots may be a possible formula addition for renal hypofunction, to help excrete fluid in cases of congestive heart failure, and as a supportive measure to improve renal cellular function in cases of renal hypertension. Maca may be used to best effect as a medicinal food, such as using 1 to 2 tablespoons of the powder in smoothies, oatmeal, or other food each day.

Ligusticum striatum • Ligusticum

Ligusticum is commonly used in TCM in a wide variety of cardiovascular, pulmonary, and respiratory formulas. The herb is traditional to revitalize blood circulation in the treatment of cardiovascular diseases and is also specific for headaches and vertigo. Modern research suggests efficacy for vascular dementia and cognitive decline in the elderly and to support stroke recovery. Ligustrazine from *Ligusticum striatum* helps protect against reperfusion injury. Isolated alkaloids from *Ligusticum* and purified synthetic ligustrazine have been used in China as medicinal agents for 30 years. *Ligusticum wallichii* and *L. chuanxiong* are considered synonyms of *L. striatum*.

Linum usitatissimum • Flax

Flaxseeds are specific for hyperlipidemia, but they are also a good nonspecific remedy that may benefit all cardiovascular pathologies such as thrombotic disorders, atherosclerosis, and metabolic syndrome. Flaxseeds are effective only with daily consumption of the ground seeds and oil, and this plant is not used as tea or tincture. The seeds are the source of essential fatty acids to consume in teaspoon to tablespoon-size doses, and seed meal can act as bulk laxative and bowel tonic.

Lobelia inflata • Pukeweed

There are more than 400 species of *Lobelia*, with *L. inflata* most well known to Western herbalists, who have used the aerial parts as a bronchodilator and to ease acute vasospams in cases of angina. The young flowers and seedpods contains lobeline, a powerful emetic at high doses, and a useful expectorant at lower doses. *Lobelia* has also gone by the name Indian tobacco, because some Native American peoples used the plant in smoking mixtures to treat dyspnea. *Lobelia* has a sedating effect on the circulatory system and may relax vasospasms, calm tachyarrhythmias, and help expectorate the lungs in cases of congestive heart failure. It is not, however, a cardiotonic like *Convallaria* and has no nourishing qualities as does *Crataegus*. *Lobelia* can be valuable, nonetheless, in formulas for cardiac cough and chest pain where there are bouts of spasms, restlessness, and excitability and to calm pain of a throbbing character. *Lobelia* is specific for a full bounding pulse and strong thudding or flip-flopping sensations in the chest. *Lobelia* is also specific for hyperemia and heat stroke. Avoid *Lobelia* in weak asthenic subjects, in those with a weak slow pulse, or in those with dyspnea related to advanced heart failure.

Lycopus virginicus • Bugleweed

The aerial parts of *Lycopus* are specific for vascular excitation due to sympathetic (adrenergic) stimulation and rapid pulse and for thyroid disorders with tumultuous heart action. *Lycopus* may improve testicular pain related to vascular congestion or hormonal imbalance. Although *Lycopus* is not considered a blood-sugar-regulating herb, it was also traditionally used in some diabetes formulas, where polyuria, tachycardia, and hypertension were presenting as symptoms. *Lycopus* is also specific for dysfunctional uterine bleeding that is frequent and scant in quantity. There has not yet been much scientific research into the mechanisms of action, but scant investigations suggest that rosmarinic acid found in *Lycopus* may act as a vascular antioxidant and anti-inflammatory and that neuroendocrine effects may help treat tachycardia and arrhythmias associated with hyperthyroidism or endocrine dysfunction.

Mahonia • See Berberis

Malus domestica • Apple

Apples are rich in fiber and anti-inflammatory compounds, and regular consumption improves bowel health, lowers cholesterol, and may support weight loss. While not used as teas or tinctures, apple fiber pectin is available by the pound and may be stirred into oatmeal, smoothies, or medicinal truffles for the hypolipidemic effects. High-fiber intake also supports beneficial intestinal flora with many health benefits.

Matricaria chamomilla • Chamomile

The aerial parts of chamomile may be used as a supportive ingredient in hypertension formulas where stress, nervousness, and digestive upset accompany. *Matricaria* is specific for those who are nervous, restless, irritable, apprehensive, and hypersensitive to pain. *M. recutita* is used interchangeably.

Melilotus officinalis • Sweet Clover

Also known as yellow clover, this clover is sweet due to the presence of coumarins, giving the plant the aroma of fresh cut grass and gentle blood-moving effects. The coumarins can be converted to toxic and potentially hemorrhagic dicoumerols, however, if the foliage molds. Even though *Melilotus* contains coumarins, the plant is specifically indicated for helping to stop nosebleeds, possibly via reducing vascular congestion in the head and other hemodynamic effects. Specific indications include for headaches with throbbing congested pain and for heart palpitations, especially where circulatory stagnation underlies.

Melissa officinalis • Lemon Balm

The aerial parts of *Melissa* are a supportive ingredient in formulas for hypertension related to nerves, stress, anxiety, and hyperthyroidism. *Melissa* is an antidepressant and antianxiety agent, especially for those with endocrine imbalances, including thyroid and female hormonal disorders.

Mentha pulegium • Pennyroyal

The leaves of pennyroyal are high in iron and may be included in teas for iron deficiency anemia. Avoid pennyroyal in pregnancy due to possible emmenagogic effects.

Nigella sativa • Black Cumin

Also known as black seed, black cumin possesses small seeds used for both culinary and medicinal purposes in the Middle East and North Africa for centuries where it goes by the name of kalonji. The seeds contain both fixed and volatile oils, with hypotensive effects credited, in part, to thymoquinone in the volatile oil fraction. *Nigella* seed oil may have hypoglycemic, hepatoprotective, and neuroprotective effects, reducing insulin resistance and resisting atheromatous plaques in the liver and vasculature. *Nigella* oil may reduce oxidative stress in the vasculature and may exert hypotensive effects via both promotion of nitric oxide and by acting as an angiotensin-converting enzyme (ACE) inhibitor.

Olea europaea • Olive

Olive fruits are processed into oil, used in cooking and as a medicinal food, and research suggests the greater the intake, the less the cardiovascular disease. Olive oil consumption may also improve survival and outcome in those having suffered a myocardial infarction, lessen the occurrence of strokes, and reduce all-cause mortality.[188] Olive oil is shown to reduce atherosclerotic lesions and damage to the tunica intima in animal studies and may have protective effects on endothelial function. Molecular investigations report benefits via effects on vascular adhesion inflammatory mediators including NFκB and TNFα. Oleuropein may reduce infarct damage in situations of ischemia, as well as protect against cardiotoxic drugs.

Ophiopogon japonicus • Mai Men Dong

The tuber of *Ophiopogon japonicus* is a cardinal herb for yin deficiency in TCM. The herb is noted to nourish heart and lung yin and to quiet respiratory and heart irritability. Root polysaccharides may improve insulin sensitivity, improve blood glucose and lipids, and slow the progression of diabetic nephropathy when in the early stages. *Ophiopogon* may also support stroke recovering and limit injury from ischemia and reperfusion, with some of the beneficial effects credited to steroidal saponins including ruscogenin.

Oplopanax horridus • Devil's Club

The roots and especially root bark of this little-studied ginseng-family plant are used traditionally to treat hypertension and diabetes. *Oplopanax* has sometimes been referred to as Alaskan ginseng due to its broad tonifying effects. The plant occurs primarily in the temperate rain forests of the Pacific Northwest, and Native American people of the region used the plant as a tonic and purgative and for shamanic and ritualistic purposes.

Triterpene glycosides have been identified, along with a number of polyynes including falcarinol, falcarindiol, and oplopandiol.

Opuntia species • Prickly Pear

The fruits of the prickly pear cactus support basic metabolic function and can be included in the diet or in formulas for hyperlipdemia, hyperglycemia, and metabolic syndrome. *Opuntia* is not used as a tea, but rather as a juice from the ripe fruits.

Paeonia lactiflora • Peony

Paeonia lactiflora roots are the primary source of both white peony (bai shao yao) and red peony (chi shao yao) and have been used for thousands of years to boost the chi and reduce blood stasis. Peony is very widely used in TCM where the roots are found in formulas for a variety of problems, including stroke recovery, heart disease, and circulatory disease. Early Chinese writings emphasize the plant for patients who have yang deficiency, are weak and cold, and have poor circulation with cold hands and feet.

Panax ginseng • Ginseng

Ginseng, also known as ren shen, has broad medicinal virtues, and the triterpenoid saponins and steroidal glycosides, known as ginsenosides or panaxosides, are credited with numerous medicinal effects, including hypotensive and hypolipidemic actions. *Panax* may help lower blood pressure through calcium channel blocking effects. *Panax* is best reserved for aesthenic people and presentations because the herb is said to be a hot and energizing yang-promoting herb and because it may cause side effects or even raise blood pressure if chosen inappropriately and given to those who are already hot. *Panax* is useful for general weakness, weak heart action, exercise intolerance, and heart failure. *Panax* may benefit diabetic patients with vascular disease and those with cardiovascular disease associated with fatigue and weakness. *Panax ginseng* is also indicated for vascular dementia and cognitive decline in the elderly. *Panax* may help lower blood pressure through calcium channel blocking effects.

Passiflora incarnata • Passionflower

Passiflora is a relaxing nervine that may be included in hypertension formulas due to its vasodilating effects. *Passiflora* is specifically indicated for nervous restlessness with muscular tension and twitching.

Petroselinum crispum • Parsley

Petroselinum seeds and roots may be used as a diuretic herb for fluid retention associated with heart failure and circulatory insufficiency, and for diabetics with atonic renal function. Parsley is rich in several antioxidants including volatile oils (apiol, limonene, and eugenol), and flavonoids (luteolin, apigenin glycosides, and quercetin). A scant amount of research suggests that *Petroselinum* may protect pancreatic beta cells from apoptosis when under oxidative stress, and a similar mechanism may support renal cells and enhance diuresis in cases of heart failure. *Petroselinum crispum* is also known as *P. sativum*.

Phyllanthus amarus • Chanca Piedra

Many species of *Phyllanthus* are used medicinally in South America, India, and Africa. *Phyllanthus* species are especially specific for kidney stones and renal insufficiency, as well as for treating diabetes, liver disease, and hypertension. All parts of the plant are used medicinally, and the antioxidant flavonoids and other constituents are noted to protect the muscles and blood vessels from oxidative stress and prevent the depletion of important tissue protectants during exercise or other stressors.

Phytolacca decandra • Pokeroot

Phytolacca may be used as a supportive ingredient in formulas for lymphedema and for acute thrombophlebitis involving cellular inflammation, cellulitis, and significant tissue swelling and edema. *Phytolacca* contains lectins credited with immunomodulating actions, particularly on white blood cells and lymphatic tissues, and medicines prepared from *Phytolacca* roots are traditional for helping treat lymphatic stasis and tissue congestion. Because *Phytolacca* proteins have a pronounced stimulatory effect on lymphocytes, the term "pokeweed" mitogen is used; pokeweed is used in research to stimulate white blood cell activity as an investigational aid. *Phytolacca americana* is synonymous.

Pinus pinaster • Maritime Pine

The bark of this pine species is the source of Pycnogenol, which is concentrated for use as a powerful antioxidant supplement, but previously the plant was considered specific for venous insufficiency. Pycnogenol in the dosage range of 200 millgrams/day may lower blood pressure via an ACE-inhibiting effect.

Piper nigrum • Black Pepper

Black pepper can bring warmth to a formula when treating cold conditions and constitutions, and *Piper nigrum* is shown to increase the absorption of nutrients and many other plant constituents. Small amounts of ground pepper can be added to teas when there is lung congestion with a chill, and *Piper* tincture may be included in formulas to boost the absorption of curcumin and other compounds.

Plantago indica • Psyllium

Psyllium seeds may exert modest improvements in blood pressure when consumed on a daily basis. The best results are seen with the consumption of 15 grams or more daily (2 to 3 teaspoons), which can be difficult to do. This species is also known as *P. psyllium* and *P. arenaria*.

Plectranthus forskohlii • Coleus

Plectranthus forskohlii (also known as *Coleus forskohlii*) enhances signal transduction intracellularly and is indicated for atony and sluggishness of general metabolic functions, with excessive inflammatory response. *Coleus* may be appropriate in formulas for heart failure, hypotension, and weakness. Folklore and research suggest that *Coleus* may exert antispasmodic effects on the vasculature and GI muscles. Much of the research has focused on the diterpenes coleonol and forskolin.

Polygonum cuspidatum • See Reynoutria japonica

Polygonum multiflorum • See Reynoutria multiflora

Poria cocos • Hoelen

Also known as fu ling in China, this mushroom is credited in TCM with sedating properties and an ability to resolve phlegm. *Poria* is a polypore fungus, widely used in formulas for dampness with inflammation. Modern research suggests that the plant may support stroke recovery.

Pueraria montana var. lobata • Kudzu

This species is native to Asia and used traditionally in China for muscle relaxation and hypertension. Growing research on this and related species show isoflavones to alleviate menopausal symptoms and support bone density. Traditional and modern information suggest that *Pueraria* might be specific for menopausal hypertension and angina. *Pueraria* reduces serum lipids and improves hepatic lipid metabolism.

Punica granatum • Pomegranate

Pomegranate fruits may be used as a preventive against heart disease because, like so many other high flavonoid foods, they are antioxidant cardioprotectants. Pomegranate juice contains ellagitannin compounds, including punicalagin, shown to have antiatherosclerotic effects in human studies, significantly decreasing LDL, HDL, macrophages, and atherosclerotic lesion oxidative status. Studies also suggest that daily consumption of pomegranate juice has beneficial effects in patients with carotid artery stenosis and may even attenuate atherosclerotic plaque development in the carotid arteries. Similarly, daily consumption of pomegranate juice improves stress-induced myocardial ischemia in patients with coronary artery disease. Pomegranates may also weakly inhibit angiotensin-converting enzymes.

Quercus robur • Common Oak

The bark of various species of oak trees is used as vascular astringents with an affinity for the mucous membranes of the abdominal, liver, and portal blood vessels. *Quercus robur* (red oak), *Quercus alba* (white oak), and other oak species are emphasized in folklore for reducing swelling, aching, and pain in leg veins and hemorrhoids and have been shown to have numerous benefits to lymphatic circulation.[189]

Raphanus sativus var. niger • Spanish Black Radish

Radish tubers are high in vanadium, which may benefit glucose regulation in diabetes, and are fairly powerful bile stimulants. Radish tubers and sprouted seeds are useful for those with high lipids and glucose, especially if associated with liver congestion or poor bile flow.

Reynoutria japonica • Hu Zhang

This is one of most commonly used species of *Reynoutria* (also known as *Polygonum cuspidatum* and *Fallopia japonica*). It is similar to *Reynoutria multiflora* (fo ti or he shou wu), in that both species are used to invigorate circulation. Tinnitus, dizziness, Meniere's disease, and leukopenia are all said to respond to root decoctions. The roots are said to have a cold property and are able to clear up heat, detoxify, and disperse edema and swelling. *R. japonica* is high in resveratrol, and it is noted to help protect vein and artery walls and to help reduce cholesterol.

Reynoutria multiflora • Fo Ti

The dried root of this plant, which is also known as *Polygonum multiflorum*, has been used to treat infection and inflammation. It is famous for its alleged ability to prevent hair from turning gray. The roots were commonly boiled with soybeans to prepare a nourishing yin and blood tonic. Fo ti is commonly used to slow the aging process, such as with a rejuvenating decoction combined with *Astragalus* and *Salvia miltiorrhiza*. Additional specific indications are numbness in the extremities, dizziness, and tinnitus that may occur with poor circulation. Fo ti is used both raw and prepared, and both forms may improve serum lipid profiles. Other common names are he shou wu or Chinese knotweed.

Rheum palmatum • Chinese Rhubarb

Rheum roots contain rhein, aloe emodin, emodin, chrysophanol, and physcion, all credited with improving triglyceride lipid levels, although high doses of rhein or emodin can cause diarrhea and intestinal cramping. Rhein improves hepatic lipid metabolism and has an antiobesity and antidiabetic effect. Rhein and other constituents are anti-inflammatory, antioxidant, and antifibrotic, explaining the traditional use of *Rheum palmatum* as a neuroprotective, renoprotective, hepatotoprotective, and vasculoprotective herb. *Rheum* is used in TCM formulas for heart disease, hypoxia, circulatory insufficiency, renal failure, atherosclerosis, and ischemia-reperfusion injury. The plant has been widely researched, and numerous cardiovascular protective effects have been identified, including endothelial protective effects, mast cell–stabilizing effects, and numerous anti-inflammatory effects.

Rhodiola rosea • Arctic Rose

The astringent roots of *Rhodiola* are considered to have adaptogenic activity and are useful to support recovery from prolonged infection and for various states of debility. *Rhodiola* may be used for heart failure, fatigue, muscle weakness, inability to exert oneself, and mental fatigue. *Rhodiola* is shown to improve oxygen utilization in the heart muscle, preventing oxidative stress and improving athletic performance and exercise tolerance at high altitudes. *Rhodiola*'s ability to improve the cardiac and bronchial muscles' utilization of oxygen may also benefit patients with COPD, preventing against acute exacerbations and improving ventilation. *Rhodiola rosea* contains p-tyrosol, salidroside, rosavin, pyridine, rhodiosin, and rhodionin, which are credited with anti-inflammatory and antioxidative actions and the ability to enhance endurance. Adaptogenic effects are also reported. There are at least 200 species of *Rhodiola*, and more than 20

species have been used in a similar manner, including *R. crenulata*, *R. quadrifida*, and *R. kirilowii*.

Ribes species • Currants

The seeds of currants are rich in quality oils that can be used to improve lipids and reduce allergic and vascular inflammation. The seed oil contains vitamin E and the unsaturated fatty acids alpha-linolenic acid and gamma-linolenic acid. *Ribes* fruits are rich sources of anti-inflammatory polyphenols, including the anthocyanin delphinidin, which may offer cardioprotective effects.

Rosa canina • Dog Rose

The hips of roses are rich in vitamin C and bioflavoids, that, along with the pectin they contain, can support heart health, improve lipid profiles, and reduce inflammation in the vasculature. Use *Rosa canina* in teas and powered rose hips in smoothies and medicinal foods. Although *Rosa canina* is one of the primary species of medicinal rose hips available commercially, all rose hips contain these valuable nutrients, so all *Rosa* species may be collected for one's self and used in medicine or medicinal foods.

Ruscus aculeatus • Butcher's Broom

Ruscus is a lily-family shrub and contains ruscogenins that have vasoconstrictive and anti-inflammatory effects on veins. The plant is traditionally used both internally and topically to treat varicosities and atonic, enlarged, flabby veins. *Ruscus* has not yet been greatly researched, but ruscogenin, a steroidal sapogenin in the plant, supports cell-to-cell adhesion in a manner that may protect blood vessels from fibrotic changes and protect against diabetic nephropathy. Ruscogenin is an antielastase agent capable of decreasing capillary permeability, explaining the traditional use in the treatment of chronic venous insufficiency and vasculitis.

Salvia miltiorrhiza • Red Sage

Also known as dan shen, *Salvia miltiorrhiza* has a long history of use in China as a blood mover and vascular protectant and is used to enhance perfusion to the organs, such as to the kidneys in cases of diabetic nephropathy. Red sage is often a foundation herb in traditional formulas to treat hypertension, heart failure, coronary artery disease, endocarditis, vascular dementia of the elderly, pulmonary hypertension, and other vascular disorders. The scientific research is now quite vast, demonstrating numerous anti-inflammatory effects on the endothelium, platelets, and vascular smooth muscle, supporting the traditional uses. Include red sage in formulas for coronary artery disease and hyperlipidemia and in formulas to prevent endothelial fibrosis in hypoxic and chronic inflammatory conditions. Many of the medicinal effects are credited to tanshinone and related diterpene components.

Schisandra chinensis • Five Flavor Fruit

Also known as Chinese magnolia vine, *Schisandra* is a woody vining plant native to northern China where it goes by the name wu wei zi. The fruits have a strong taste and are sometimes referred to as the five flavor berry. Five flavor fruit is a traditional herb for liver disease in China and is one of the 50 fundamental herbs of TCM used to astringe dampness in the body, reduce sweating, and yet generate fluid. *Schisandra chinensis* is particularly credited with hepatoprotective actions and may help protect the liver from fibrotic changes. *Schizandra* may reduce lipid accumulation in the liver in cases of hyperlipidemia and metabolic stress[190] and may be hypotensive via support of nitric oxide regulation. The lignin schisandrin is cardioprotective against myocardial ischemia and may support recovery from acute myocardial infarction.

Scutellaria baicalensis • Scute

Scute is widely used in TCM as an anti-inflammatory and antiallergy agent. Known as huang qi in China, the roots are credited with lipid-lowering effects and may promote cholesterol-processing enzymes in the liver. The flavonoid baicalein and its glycoside baicalin are credited with numerous medicinal virtues, including hypotensive effects. Baicalin may affect endothelium nitric oxide pathways. Scute may help protect the cerebral, peripheral, and cardiovasculature from ischemic and reperfusion injury. Current research supports *Scutellaria*'s traditional use for stroke and infarction recovery and for neuro- and cardiovascular protective effects in a wide variety of chronic inflammatory diseases. Other common names are huang qin and Chinese scute.

Selenicereus grandiflorus • Night-Blooming Cactus

This cactus is found in both hot dry and hot tropical areas and may be used for heart failure symptoms by people local to the various areas. Night-blooming cactus is indicated for weak feeble heart action and is not as toxic as most of the other strong cardiotonics. Cactus can be used long term, though some authors have reported it to be curative and not always needed indefinitely. The Eclectic physicians reported *Selenicereus* to be nonirritating and noncumulative and to improve nutrition to the heart. Cactus is of benefit to those with valvular diseases, chronic heart failure, dyspnea with edema, and a sense of weight in the chest. Though it is sometimes used for hypertension, *Selenicereus* improves a weak heart action and will raise a low arterial pressure. *Selenicereus* is considered a cardiac normalizer and cardiotonic and may be dosed from 5 drops up to ½ a dram (2 milliliters), several times a day. The botanical name of this plant was once *Cactus grandiflora*. It then changed to *Cereus grandiflorus* and finally to *Selenicereus grandiflorus*. The common name night-blooming cactus has persisted throughout these changes.

Senna species • Senna

There are more than 600 species of *Senna* (many also known by the genus name *Cassia*), mostly found in the tropics of India and China. *Senna insularis* is a weedy shrub of the Caesalpinioideae subfamily of the legume family and is a traditional remedy in India for tachycardia and hypertension, and as with other species, it may have antidiabetic effects. Research suggests that extracts may antagonize acetylcholine and have muscle-relaxing effect through curare-like activity.[191] *Senna alata*, *Senna occidentalis* (also known as *Cassia occidentalis*), and *Senna alexandrina* (also known as *Cassia senna*) are also used medicinally. The ripe seed pods of various *Senna* species are most commonly used as medicines, especially for their laxative effects, but other plant parts such as the leaves or stem bark may occasionally be used as well.

Silybum marianum • Milk Thistle

The ripe seeds of this thistle are useful as a supportive ingredient in formulas where hypertension and hyperlipidemia occur concomitantly with liver congestion and digestive organ inflammation or difficulty. *Silybum* may also improve portal hypertension, help the liver better metabolize carbohydrates and lipids, and reduce pressure in the legs, helping to treat varicosities, and in the portal veins themselves, helping to reduce blood pressure. Milk thistle helps the liver process fats,[192] hormones, and exogenous toxins,[193] may reduce fatty and fibrotic degeneration, and may protect against cirrhosis[194] and nonalcoholic fatty liver disease.[195]

Silymarin, the active flavonolignan in *Silybum marianum*, may help protect organs and tissues from iron overload.[196]

Stellaria media • Chickweed

The plant is a wild salad green and is high in iron, useful for treating iron deficiency anemia. Fresh *Stellaria* juice and dried leaf teas have also been used to treat bronchitis and pulmonary diseases and are Ayurvedic remedies for obesity. *Stellaria media* may prevent high-fat-diet–induced fat storage in adipose tissue by inhibiting the intestinal absorption of dietary fat and carbohydrates through inhibition of digestive enzymes. *Stellaria* contains saponin, flavonoids, steroids, triterpenoids, glycosides, and other components.

Stevia rebaudiana • Sweet Leaf

Stevia is commonly used to sweeten teas and improve palatability, but also has medicinal properties of its own, enhancing insulin reception and thereby treating the diabetes and metabolic syndrome that may underlie heart disease. While most of the research has concentrated on antidiabetic actions, *Stevia* may also be cardioprotective, protecting cardiac muscle mitochondria from ischemic and reperfusion injury. Diterpenes such as isosteviol are shown to have hypotensive and hypoglycemic actions. Include *Stevia* in teas and medicinal foods to protect the vasculature in cases of diabetes, coronary artery disease, stroke recovery, and cardiomyopathy.

Strophanthus hispidus • Strophanthus

Caution: This is a potentially toxic herb that should be reserved for experienced clinicians. *Strophanthus* has been used as an arrow poison in Africa and contains the cardiac glycosides ouabain, strophanthin, and strophanthidin, which are strong cardiac glycosides whose effects can persist for weeks. *Strophanthus* is indicated for a weak and feeble heart, low pressure, and tachycardia. *Strophanthus* seeds are tinctured and the tincture is dosed 1 to 10 drops at a time. *Strophanthus* also contains lignans and cytotoxic alkaloids contributing to powerful and potentially poisonous effects. Small doses of *Strophanthus* can offer protective effects against myocardial infarction and hypertension. The plant also goes by the common name kombé.

Syzygium species • Roseapple

Syzygium is a genus of flowering plants that belongs to the myrtle family, Myrtaceae. The genus comprises more than a thousand species, and many go by the common name plum rose, roseapple, or water apple. Many *Syzygium* species are valuable as ornamental trees, and the fruits are used in jams and jellies. They are folkloric herbs for diabetes, hyperglycemia, glycosuria, weakness, thirst, and emaciation. *Syzygium aromaticum* (clove) is an important spice prepared from the unopened flower buds. Cloves and many *Syzygium* species were at one time classified in the *Eugenia* genus.

Tanacetum parthenium • Feverfew

Feverfew may reduce platelet activation and histamine release and reduce inflammatory process in the blood stream. *Tanacetum* tincture and pills are readily available and are used orally to reduce allergic vascular phenomena including migraines and hives. *Tanacetum* has also been shown to inhibit the proliferation of vascular smooth muscle, helping to slow the progression of atherosclerosis. The sesquiterpene lactone parthenolide has a protective and stabilizing effect on platelets, mast cells, endothelial cells, blood cells, and the vasculature in general.

Taraxacum officinale • Dandelion

Dandelion is used as a supportive ingredient in formulas where hypertension and hyperlipidemia occur concomitantly with liver congestion and digestive organ inflammation or difficulty. Dandelion roots contain inulin, which supports beneficial intestinal flora and provides a valuable alterative action, while the leaves are nourishing and high in minerals that can help the kidneys excrete wastes and reduce edema in cases of congestive heart failure. The leaves are also high in chromium, which may improve insulin sensitivity, and can be considered in teas for long-term use by diabetics.

Terminalia arjuna • Arjuna

Terminalia is a traditional Ayurvedic medicinal plant that has been shown to benefit atherosclerosis, anemia, angina, congestive heart failure, and cardiomyopathies.[197] The stem bark, root bark, fruits, leaves, and seeds are all used medicinally, with the stem bark being most common and most important. *Terminalia* may be used as a cardiotonic to hep reduce lipids and protect the vasculature from oxidative stress. Saponin glycosides in *T. arjuna* may offer inotropic and cardioprotective effects, while the flavonoids/phenolics are antioxidant and inhibit endothelial activation and platelet aggregation. All have been the subject of a great amount of research, supporting the traditional use of the plant as a general vascular tonic and longevity herb.

Theobroma cacao • Cacao

Cacao trees bear pods whose seeds are a rich source of flavonoids, including resveratrol. Cacao powder is the most natural form of the beans, being raw and unprocessed. Coco or cocoa powder is the roasted beans, and Dutch processed cocoa uses an alkali to render the beans less bitter and acidic and does not seem to impair the nutritional or antioxidant profile of the product. Cacao, cocoa, and even prepared chocolate, especially products that are minimally processed and low in added milk, sugar, or fats, may offer significant health benefits. Unprocessed cacao and prepared chocolate are more preventive than medicinal but may be encouraged in cases of hypertension and hyperlipidemia. Cacao's hypotensive effects relate to enhancement of endothelial nitric oxide production and inhibition of angiotensin-converting enzyme. Cacoa flavonoids may also reduce oxidative stress in the blood, improve cholesterol ratios, and inhibit platelet aggregation. When regularly consumed, cocoa may reduce the incidence of heart failure due to the antioxidant and hypotensive effects.

Thuja occidentalis • Northern White Cedar

The young leaves of the cedar tree are used as a hot, stimulating herb, which is potentially irritating to the renal passages, but also specifically indicated for urinary atony. *Thuja* may be used in small doses to stimulate poor functioning kidneys in diabetics with renal insufficiency. *Thuja* may be used as a synergist in small amounts in formulas for congestive heart failure to support diuresis.

Tilia × europea • Linden

The young leaves and flowers of linden have both nervine and moderate hypotensive actions. Linden flowers are traditionally used to treat stress, including anxiety, insomnia, and stress-related hypertension, and heart palpitations. The dried flower buds and tincture may be included in formulas for anxiety, tension, and hypertension.

Trachyspermum ammi • Ajwain

This caraway relative is an Apiaceae family plant whose seeds may help lower blood pressure and lipids, as seen with other aromatic seeds in this family. This species is also known as *Carum copticum*. Like some *Carum* species, this plant is high in hypolipidemic volatile oils. Broncho- and vasodilating properties occur via calcium channel blocking effects on muscles of the airways and vasculature. Like fennel and anise seed oils, both of the same family, ajwain oil may be used topically over the chest for tightness.

Trifolium pratense • Red Clover

Clover blossoms may be used as a "blood mover" to reduce the tendency to clots and to improve venous return of dependent fluids. Red clover is high in iron. *Trifolium* is most studied for its estrogenic effects and the isoflavones it contains, but the plant may also protect platelets from oxidative stress and reduce activation, thereby improving blood viscosity. Coumarins in clover support blood flow without interfering with appropriate clotting, but red clover is contraindicated for patients on blood thinners due to possible synergistic effects.

Uncaria rhynchophylla, U. tomentosa • Cat's Claw

The bark of the thin vines of *Uncaria* are used in South America for infectious illnesses and cancer but may also have hypotensive action. Modern research reports that the indole alkaloid hirsutine may inhibit calcium channels and lead to a reduction in intracellular calcium.

Urginea maritima • Squill

Urginea is reported to have been used by ancient druid priests/physicians as a diuretic stimulant for cardiac edema and a cardiac cough. Other common names are squill, sea onion, and maritime squill. *Urginea maritima* is also known as *Drimia maritima*. **Caution:** *Urginea* contains cardiac glycosides and is potentially toxic. It is dosed just several drops at a time.

Urtica dioica • Nettles

Nettle leaves may be used as a general mineral tonic and diuretic in cases of fluid retention and premenstrual edema. *Urtica* may reduce dependant edema due to heart failure and poor circulation to the kidneys. Nettles are a nutritive plant with many benefits, including hypotensive actions. *Urtica* may induce nitric oxide, as well as inhibit calcium ion flow, both leading to vasodilation. *Urtica* may also improve blood sugar regulation and reduce inflammation in the blood, factors that lend regular consumption of the teas or other medicines vascular protective effects.

Vaccinium myrtillus • Bilberry

Bilberry and blueberry anthocyanidins are a vascular tonic with antioxidant and anti-inflammatory effects on blood vessels. *Vaccinium* leaves, berries, jams, tinctures, and tea are specifically indicated for diabetic retinopathy to improve circulation to the eyes in cases of macular degeneration, cataracts, and failing eyesight of the elderly.

Herbs for Cardiovascular, Peripheral Vascular, and Pulmonary Conditions

Valeriana officinalis • Valerian

The roots of *Valeriana* may be used as a vasodilator in cases of hypertension, especially when associated with nervousness, hypersensitivity, obsessive thinking, symptoms from worry, stress headaches, and long menstrual cycles with scanty flow.

Vinca minor • Lesser periwinkle

Native to Eurasia, *Vinca minor* contains more than 50 alkaloids, including vincamine and reserpine, in its leaves. One nutraceutical, vinpocetine, has been developed and is a semisynthetic derivative of vincamine; it is being explored as a therapy for cerebrovascular insufficiency. Limited research suggests that the vincamines have a pronounced cerebrovasodilatory and neuroprotective activity. Clinical evidence suggests vincamine to improve cerebral circulation in the elderly, and it may be used to treat memory disturbances, vertigo, transient ischemic deficits, and headaches. Vincamine increases cerebral blood flow, oxygen consumption, and glucose utilization. *Vinca minor* also goes by the name creeping myrtle, due to the plant's vining and trailing habit.

Viscum album • Mistletoe

As a vagal nerve tonic, mistletoe may help strengthen a weak pulse, as well as slow a tachyarrhythmia or enliven a sluggish bradycardia. *Viscum* is specifically indicated for cardiac enlargement and valvular incompetence, as well as angina, shortness of breath, and general dyspnea, edema, palpitation with exertion, inability to lie down, and other symptoms of congestive heart failure.

Mistletoe was also used for seizures and epilepsy,[198] the so-called St. Vitus's dance, motor twitching, spasm, and nerve hyperactivity.

Withania somnifera • Ashwagandha

Withania is a well-known adaptogen often used to restore the nervous system and adrenal glands following long periods of stress and overwork. *Withania* may be included in cardiopulmonary formulas where exhaustion, anxiety, or nervous debility accompany symptoms.

Zingiber officinale • Ginger

Ginger roots are a warming blood mover, having hypolipidemic and hypoglycemic actions. *Zingiber* is specific for those with cold constitutions, cold hands and feet, poor sluggish digestion, inflammation, and arthritic complaints.

Ziziphus spinosa • Suan Zao Ren

Ziziphus is used in TCM as a nourishing yin tonic, especially for the heart and blood. The plant is also said to calm the spirit and treat insomnia and is commonly included in formulas for heart palpitation, heart disease, and stroke recovery. Many traditional writings say that the plant has sedative properties, and traditional formulas for nourishing the heart and blood often combine *Ziziphus* with *Cnidium*. *Ziziphus jujuba* (red date or Chinese date) is also used medicinally. The small crab apple-like fruits are dried and eaten or used in decoctions for a variety of purposes, including as hypotensive cardiotonics.

— CHAPTER THREE —

Creating Herbal Formulas for Respiratory Conditions

Creating Topical Applications for
Lung Complaints   109

Formulas for Coughs
and Altered Breath Sounds   114

Formulas for Dyspnea   124

Formulas for Hemoptysis   125

Formulas for Allergic Rhinosinusitis   127

Formulas for Acute and Chronic Bronchitis   130

Formulas for Cystic Fibrosis   134

Formulas for Emphysema   136

Formulas for Pleurisy   138

Treating Hyperventilation Syndrome   140

Formulas for Pneumonia   140

Formulas for Asthma   142

Formulas for COPD   147

Treatment for Acute Respiratory
Distress Syndrome   153

Formulas for Cor Pulmonale   154

Formulas for Tuberculosis   155

Specific Indications:
Herbs for Respiratory Conditions   156

This chapter follows a slightly different format for reasons of logic and best use as a teaching tutorial. Diagnostic categories for respiratory conditions are not rigid, they are often overlapping or in a state of flux. Thus, I have organized formulas in this chapter by *symptom*, such as cough, altered breaths sounds, and dyspnea, as well as by classical condition names such as bronchitis or asthma. This is a practical approach because, for example, a respiratory virus can quickly turn into a bacterial infection, or a person with asthma can develop a particular quality of bronchitis, or a smoker with a chronic bronchitis can lose lung function and slowly develop chronic obstructive pulmonary disease (COPD). All of these diagnoses are covered in this chapter, but because they can all overlap, the chapter begins with symptoms that present most commonly. Because herbs are so well understood and categorized by their *action*, more so than by *diagnostic categories*, it is important for practitioners to build their skill at crafting herbal formulas with specific applications, for example, "bronchitis with a wet cough" versus "bronchitis with a dry cough," "bronchitis with wheezing," and "bronchitis in a long-term smoker." Pulmonary conditions are vascular issues as much as they are lung conditions, and you might find it helpful to review the discussion of pulmonary conditions in chapter 2.

In treating infectious and allergic lung conditions, some of the best ways to ensure that herbal medicines actually reach the lung tissues are to administer them via steams, inhalants, and topical applications. Thus I begin the chapter with a section on creating topical applications for lung conditions, and among the formulas in this chapter, I offer suggestions for steam inhalation and the use of volatile oils in the management of acute infectious respiratory disorders.

It is likely that some volatile compounds, such as sulfide compounds from *Allium* spp. and camphor or menthol-like volatile oils of *Mentha* spp. and *Eucalyptus globulus*, can effectively reach the lower airways even when ingested orally. Other than those few exceptions of agents with direct effects on the airways, however, therapy for lung conditions is most often systemic, having general antimicrobial or anti-inflammatory effects. For example, one can use herbs that elicit a systemic immune response via the blood and thereby treat pneumonia or infectious conditions. Another approach is to use herbs that act on all mast cells in the body and thereby reduce allergic reactivity. For acute and urgent lung symptoms, the use of inhalants, steams, hydrotherapy, and heat packs, together with systemic treatment, may yield the best results.

General approaches to treating respiratory issues include the use of anti-inflammatory herbs having an affinity for the lungs and upper respiratory passages. Numerous flavonoids are credited with anti-inflammatory and antioxidant activity, including numerous anti-allergic effects on the airways.[1] The flavonoids khellin, hispidulin, luteolin, and isoquercitrin are especially noted to have bronchodilatory properties. Polydatin (synonymous with piceid), a resveratrol-derived glucoside that is rich in grapes and red wine, helps protect the airways from air pollution[2] and is also found in high amount in *Reynoutria multiflora* (also known as *Polygonum cuspidatum*), for which the compound is named.[3] Paeoniflorin is a monoterpene glucoside from *Paeonia lactiflora* (peony), a plant widely used in China for blood-building and vascular effects. Studies suggest paeoniflorin to reduce oxidative stress and help protect the lungs from cigarette smoke.[4] The all-purpose anti-inflammatory herbs, such as *Curcuma longa* (turmeric), *Camellia sinensis* (green tea), and *Zingiber officinale* (ginger), and seed oils including *Linum usitatissimum* (flax) and *Oenothera biennis* (evening primrose) may be included in the diet and used in herbal formulas to reduce hyperreactivity or to slow the progression of chronic respiratory disease. These general, gentle, and nourishing anti-inflammatories are used throughout the chapter as supportive base herbs in formulas and protocols, combined with bronchodilators for wheezing, antitussives for cough, antimicrobials for acute infections, and expectorants, inhalants, immunomodulators, and demulcents as needed to craft specific formulas. These all-purpose anti-inflammatories may also be taken separately in the form of resveratrol, green tea, ginger, or turmeric capsules to complement tea and tincture formulas. Powdered herbs or flax oil can be added to smoothies or salad dressing and other medicinal foods intended for long-term daily consumption to reduce allergic reactivity or to protect against fibrosis and loss of lung function in chronic inflammatory conditions.

Herbal Bronchodilators

Modern investigation shows that many herbs folklorically emphasized as remedies for acute wheezing are, in fact, bronchodilators. The following herbs and specific molecular constituents found in them are noted to have significant bronchodilating effects.

HERB	MOLECULAR CONSTITUENTS
Albizia lebbeck	Saponins
Alstonia scholaris	Ditamine, echitamine, echitenines
Artemisia caerulescens	Quercetin, isorhamnetin
Cissampelos sympodialis	Warifteine
Clerodendron serratum	Phenolic glycoside
Elaeocarpus serratus	Glycoside, steroids, alkaloids, flavonoids
Ephedra sinica	Ephedrine
Galphimia glauca	Tetragalloylquinic acid, quercetin
Gardenia latifolia	Saponins
Ginkgo biloba	Ginkgolides
Justicia vasica	Alkaloids
Lepidium meyenii	Alkaloids, flavonoids
Lobelia inflata	Lobeline
Mikania glomerata	Coumarin
Ocimum tenuiflorum	Myrcenol, nerol, eugenol
Passiflora incarnata	Alkaloids
Picrorhiza kurroa	Androsin
Plectranthus forskohlii, also known as *Coleus forskohlii*	Forskolin
Rosmarinus officinalis	Caffeic acid (CA), rosmarinic acid
Tephrosia purpurea	Flavonoids, tephrosin
Tylophora indica	Tylophorine
Vitex negundo	Casticin, isoorientin, chrysophenol D, luteolin

Creating Topical Applications for Lung Complaints

Mustard poultices are a time-honored home remedy for alleviating lung pain and congestion, where mustard is mixed with flour to form a paste that is employed topically. Allyl isothiocyanate in ground mustard seeds is noted to diffuse passively across the skin, having mucolytic and expectorating effects in the lungs.

Nicotine is able to permeate the skin and reach systemic circulation. Nicotine patches are used not only for smoking cessation, but also in an experimental capacity for efficacy in Parkinson's disease, pulmonary sarcoidosis, schizophrenia, breast cancer, pain, alcoholism, PTSD, and ADHD. Even though smoking is associated with heart and lung disease, nicotine itself may have therapeutic value for coronary artery disease. Lobeline, an alkaloid in *Lobelia inflata*, is structurally similar to nicotine and may have activity via similar pathways, affecting β-adrenergic signaling in the coronary and respiratory vasculature. *Lobelia* plasters and the topical application of *Lobelia* oil can help alleviate bronchospasm, wheezing, coughing, and lung pain.

Transdermal Drug Delivery

Traditional medicine has used herbal baths, compresses, salves, and steams as respiratory therapies and a variety of local and systemic effects. A modern development includes the addition of strong compounds, both natural and synthetic, to creams and skin patches referred to as *transdermal medication*. Transdermal delivery of medicines has many advantages including reducing the need for painful injections, offering continuous release of substances that have a short half-life, and improving systemic delivery of compounds that are poorly assimilated or rapidly metabolized. Most transdermal drugs have been developed into skin patches, with nicotine being the most commonly employed transdermal agent.

Scopolamine was one of the first biologic medicines approved for release as a skin patch in the 1970s, followed by nitroglycerin, clonidine, and estradiol in the 1980s, and fentanyl in 1990. Physical therapists may use ultrasound and iontophoresis to enhance the transdermal delivery of lidocaine and other medicines to treat musculoskeletal disorders. Some transdermal agents may even reach the central nervous system, and the natural compound rivastigmine may help treat dementia when topically applied.

Scopolamine is a tropane alkaloid, as is atropine in *Atropa belladonna*, and stramonine in *Datura stramonium*. All three have excellent transdermal absorption. Whole plant *Atropa belladonna* leaf baths and poultices are traditional therapies to treat pain and spasm due to traumatic injuries. *Datura* leaf smoke has been inhaled to relax bronchospasm in cases of acute wheezing.

More than 300 substances have been shown to permeate the skin to reach underlying tissues, and/or systemic circulation, with more being developed each year, especially various liposomal and nanoparticle-based medications. Some such substances may not be medicinal themselves, but may enhance the absorption of other medicines, as is the case with sulfoxides such as dimethyl sulfoxide (DMSO). Alcohols, fatty acids, phospholipids (such as lecithin), amines, nicotine, and some terpene-based essential oils are all natural substances that may act as permeation enhancers. Among the absorption-enhancing essential oils are limonene (found in *Citrus* species) and carvone, which is found in *Carum carvi* (caraway) and *Pimpinella anisum* (anise).[5] Among the amines are diethanolamine (DEA), which is reported to have mood- and cognition-enhancing effects.

Excellent hydration of the skin greatly enhances skin diffusion, so bathing, steaming, and applying moist heat packs may be helpful prior to topical therapies.

Though there is little modern research on this topic, it is highly plausible that other medicines could be delivered to the lungs via the skin. Such transdermal drug delivery has evolved slowly in the pharmaceutical arena, and there is a small collection of pharmaceuticals and natural biologics shown to be effective with topical application. Clinical trials are ongoing to test new hydrophilic compounds, macromolecules, herbal compounds, and synthetic compounds for transdermal delivery.

Plaster Base for Lung Plasters

This formula can be used as a waxy base that other medicinal substances can be stirred into. This formula is not too runny and not too stiff; neither is it sticky like pure lanolin nor greasy like pure oil. Once prepared and left to cool, it will harden, but it can be reheated when needed. The beeswax is first melted in a double boiler or microwave, and the lanolin, paraffin, and oil are then stirred in. Making the plaster base is a bit of work, and you may need to search out and special order the ingredients, but if made in quantity, the base can last indefinitely.

The plaster base is melted down at the time of use and powdered herbs are added. The plaster is applied to a patient's oiled chest or back, by first painting or spooning the plaster onto a thin oiled piece of muslin, cut to the right size. The muslin is placed facedown on the patient's oiled torso and covered with plastic sheeting, a hot pack, and a towel. This traditional formula uses paraffin wax, which is a petroleum-based product. But if you wish to avoid paraffin, there are alternatives created from shea, jojoba, soy, and other oils that can also supply the needed stiff, spreadable consistency. Ideas for the powders to be stirred into the plaster are in the Topical Lung Plaster recipe below.

Beeswax	4 ounces (120 g)
Castor oil	2 ounces (60 ml)
Anhydrous lanolin	2 ounces (60 g)
Soft paraffin	8 ounces (240 g)

To melt beeswax, place it in a storage vessel set in a water bath or double boiler. Or place the beeswax in a stainless steel or ceramic container and place the container on the bottom of a large slow cooker. Add water to the slow cooker, keeping the level below the lip of the container. Beeswax can also be melted in a microwave. Once the beeswax is melted, keep it on the heat source and add the three remaining ingredients, allowing them to blend together, stirring occasionally with a disposable

tool, such as a wooden stick. Once thoroughly melted and blended together, pour into glass jars for storage. The jars can be heated in a hot water bath to melt the plaster base when needed.

Topical Lung Plaster

This formula is an example of using herbal powders in a lung plaster. *Lobelia* and *Capsicum* (cayenne) are a common duo featured in early twentieth-century herb books, and it was stated that the cayenne potentiated *Lobelia*. Don't add too much *Capsicum*, though, lest it irritate or even burn the skin. Start with just ¼ to ½ ounce.

Plaster base	16 ounces (480 g)
Lobelia powder	4 ounces (120 g)
Capsicum powder	¼ to ½ ounce (8 to 15 g)

Heat the plaster base (see the description in the Plaster Base for Lung Plasters formula above) until soft and then stir in the powdered herbs. When thoroughly blended,

Emodin and Rhein to Halt Lung Fibrosis

Emodin and rhein are naturally occurring anthraquinone glycosides responsible for the laxative effects of some medicinal plants. Research also suggests that these compounds may attenuate fibrotic processes. In Volume I, chapter 2 describes details on emodin-attenuating fibrotic processes in the pancreas, even with topical application over the abdomen. This raises the question of whether any such herbs might be used in lung plasters or driven into tissues with castor oil. Emodin found in several medicinal plants can also attenuate fibrosis of the lungs due to exposure to airborne silica.[6] Emodin occurs in *Rhamnus purshiana*, *Rheum emodi*, *R. palmatum*, *R. officinale*, *Reynoutria japonica* (also known as *Fallopia japonica* and *Polygonum cuspidatum* or *japonica*), *Rumex* spp., and *Senna* spp. (also known as *Cassia* spp.), and emodin is shown to reduce collagen deposition in the airways following silica exposure.

spread on a piece of muslin or other suitable fabric about 1 foot square in size. Apply this directly to the oiled chest or back, medicine side down, and cover with a hot towel or heating pad.

Mustard Poultice

Mustard poultices and plasters are a long-standing traditional therapy for the treatment of lung congestion, coughs, bronchitis, and pneumonia. Making a poultice is not as complicated as it sounds, especially after you've done it once. And once you've tried it, you'll understand why the mustard poultice is such a time-honored medicine. It offers immediate improvement to discomfort in the chest. In the days prior to antibiotics, it was a common treatment for lung infections.

Brassica nigra (also known as *Sinapis nigra*) seed	½ cup (50 g)
Flour	1–2 cups (100–200 g)
Castor oil	

Grind the *Brassica* (black mustard) seed in a coffee grinder or other grinder. Mix the mustard powder with the flour using 1 or 2 parts ground mustard seeds to 10 parts flour. (Use double the amount of flour when preparing the plaster for a child or someone with sensitive skin.) You may need to use a full cup of mustard powder if it is not very strong. Place the flour and mustard powder mixture in a bowl and add hot water and whisk with a fork to create a paste.

Help the person recline or lie comfortably with their feet in a bucket of water. They may find it convenient to wear a bathrobe and lounge in an easy chair or lie flat on blankets on a soft carpet with their knees bent and their feet in the hot water. Use the hottest water the person can comfortably stand and maintain the heat in the bucket throughout the treatment. Apply castor oil to their bare chest and/or back. Soak a torso-size sheet of muslin in hot water and lay this on the oiled area. Stir very hot water into the mustard paste until it is sticky and spreadable. Spread the mixture on the muslin, and cover with a sheet of plastic wrap. Top that with a second piece of dry muslin. The poultice may be covered with a heat pack, although a mustard pack and foot bath are often sufficient heating without additional hot applications. If the person is reclining, a second poultice can be applied under the back or they can lie on a heating pad, if desired. Close the bathrobe and wrap up in additional blankets if desired. Leave for 15 to 30 minutes, stopping promptly if there is any discomfort. The procedure will most likely promote much perspiration and reddening of the chest. The individual treated should be given plenty of liquids during the procedure and take a warm or cool shower, as desired, afterward, and then rest or gently stretch for another half hour.

Liniment for the Lungs

A liniment is a topical application prepared from oil and alcohol that's meant to be applied with friction. Liniments are more commonly formulated to address musculoskeletal pain, but here is a variation intended for massage into the chest and back. Applied by a health care professional or loved one, the massage becomes part of the medicine. When there is chest pain due to coughing, achiness of the flu, or underlying lung infections, such liniment massages can be very comforting. This formula requires cayenne oil, which is available commercially, or it can be prepared for one's self ahead of time. The clove and cayenne provide pain relief, as the hot sensation overrides other painful sensations. *Lobelia* and *Eucalyptus* are antispasmodic for pain due to vigorous coughing. It would be harmful to ingest it. Another interesting point about the use of *Lobelia* here is that lobeline binds to β-nicotinic receptors, and the leading pharmaceutical beta blockers propranolol and timolol are shown to control epistaxis when prepared as a nasal gel.[7] **Caution:** Because this formula uses isopropyl alcohol (which is much less expensive than vodka or other options), the formula *must* be labeled for topical use only.

Syzygium aromaticum (also known as *Eugenia caryophyllata*) powder	½ ounce (15 g)
Lobelia inflata powder	1 ounce (30 g)
Eucalyptus globulus leaves	1 ounce (30 g)
Isopropyl alcohol	2 cups (480 ml)
Capsicum spp. oil	1 cup (240 ml)
Mentha piperita essential oil	100 drops
Pimpinella anisum essential oil	100 drops

Soak the dried herbs in the 2 cups of isopropyl alcohol (rubbing alcohol) for 5 or 6 weeks. Cayenne oil may be prepared by macerating dry *Capsicum* powder in castor oil. Press out the dried herbs and combine the resulting herb-infused alcohol and castor oil with peppermint and anise essential oils. The oils and alcohol will not go into a true solution together and thus this liniment should be shaken before each use. Massage it into the back and chest with friction for wheezing or tight spastic coughs. The effects are heightened by covering the torso with heat both before and after the application of the liniment.

Eucalyptus for the Lungs

Eucalyptus species are native to Australia and have been introduced worldwide as a source of fast-growing lumber, wood pulp, and essential oils. *Eucalyptus* has antibacterial, antiviral, and antifungal properties and has been used to decongest the lungs and treat respiratory infections since antiquity. All *Eucalyptus* species contain the essential oils α-pinene and 1,8-cineol, with activity against a variety of microbes.[8] These essential oils are monoterpenes with significant secretolytic and secretomotoric effects on the airways, as well as antioxidant, anti-inflammatory and antibacterial effects. These effects all help to treat acute and chronic respiratory infections, including acute and chronic rhinosinusitis, acute and chronic bronchitis, and chronic obstructive pulmonary disease (COPD).[9] Many inflammatory airway diseases such as rhinosinusitis, COPD, and bronchial asthma involve overproduction of mucus from the epithelial surfaces of the airways. Also known as eucalyptol, 1,8-cineol may help correct mucosal hyperreactivity and reduce hypersecretion of mucus.[10] Some researchers have developed *Eucalyptus* nanoemulsions, in which tiny droplets are dispersed in a lipid base that increased absorption and medicinal effects when applied topically to wounds.[11] Eucalyptus essential oil may be prepared into inhalants for COPD and general respiratory infections, as well as used in teas, tinctures, and topical medications.

Eucalyptus globulus, eucalyptus

Camphorated Castor Oil for Topical Use

Camphor is a classic respiratory topical agent, used for generations. Most commercial products dissolve the camphor in petroleum jelly, but many herbalists prefer to avoid petrochemical products. This recipe provides an alternative. It is easy to prepare and often has instant bronchodilating and cough-relieving effects once applied.

Camphor crystals	1 cup (240 ml)
Castor oil	2 cups (480 ml)
Beeswax	2 ounces (60 g), optional

Dissolve the camphor crystals in warm castor oil. The combination may be used as is, or prepared into a semisolid salve by adding melted beeswax to it. Beeswax can be melted in the microwave, put in a small metal container and placed in a double boiler, or melted in a heatproof vessel on the stove or hot plate. Stir the melted beeswax into the warm camphor oil and quickly pour into waiting salve containers immediately, as the salve will harden rapidly once removed from the heat.

Rub the oil or salve into the entire chest, ribs, and over the back and shoulder blades. Cover with a hot moist towel, hot water bottle, or other heat pack and leave in place for a half hour. Repeat throughout the day. When treating a child, apply the oil or salve to the child's chest and back at bedtime. This will often help allay coughing to support falling asleep.

Mint Oil Topical for Cough and Burning Sensations

Mentha has been used as a bronchodilator and anodyne since ancient times and marketed in topical chest rubs since at least the mid-1800s. Menthol is found in various species of mints, and it has a reputation as providing a cooling affect. Research shows the cooling sensation to occur due to the activity of cold-sensing TRP nociceptors on sensory afferent nerves. Menthol also reduces lung irritation via inhibitory effects on sodium channels and may exert antitussive effects due to such effects on nasal mucosal innervation.[12] Menthol's antitussive effect may be fairly ephemeral and require frequent dosage for any significant relief in acute or chronic bronchitis.

| Mint essential oil | 30 ml |
| Castor oil | 30 ml |

Combine the oils in a 2-ounce (60 ml) bottle, and place ½ to 1 teaspoon in the palm of the hands and rub into the back and chest, as needed for painful coughing, and a sensation of burning or tightness in the chest. May apply as often as hourly, reducing as symptoms improve. Covering with a heat pack may increase the pain-relieving effects.

Stillingia Liniment

Harvey Wickes Felter, MD, and his contemporary John Uri Lloyd, a pharmacist, wrote *King's American Dispensatory* in 1898, and it remains a treasured herbal resource. This liniment recipe appears in the book and is recommended to treat chronic asthma, chronic cough, croup, and chest pain and may be used orally as well as topically. Various herbalists have attempted to improve upon the formula by using their own homemade *Stillingia* oil and substituting tincture for the alcohol, or glycerin for part of the alcohol. The use of *Stillingia* oil and glycerin can improve the viscosity of the liniment, helping the formula to homogenize and better cling to the skin. Possible variations to the original formula are noted below.

Stillingia sylvatica	½ ounce (15 ml)
Melaleuca spp. (tea tree) essential oil	½ ounce (15 ml)
Alcohol	3 ounces (90 ml)
or substitute 45 ml each of alcohol and glycerin	
Lobelia inflata oil	1 scant teaspoon (4 ml)

The formula will fill a 4-ounce bottle. Place 1 to 5 drops at a time on a slippery elm lozenge, or in a syrup vehicle. As more typical of liniments, this formula is also useful to rub into the moist chest or back (use ½ to 1 teaspoon). The back may be oiled ahead of time using castor oil or other body oil to facilitate the best absorption and penetration into the lungs. Repeat as often as hourly. Add a "Shake Well before Each Use" to the label and replace if the formula begins to solidify.

Nitroglycerin Patches for Chest Pain and Cough

Nitroglycerin and transdermal nitroglycerin products have long been established for treating coronary vascular spasm, such as angina, and may help to allay chronic cough by reducing hypersensitivity of the cough reflex.[13] Nitroglycerin and other agents that serve as nitric oxide donors reduce vascular pain and inflammation via promoting vasodilation and other mechanisms, including activity at particular TRP channels. (See chapter 2 for more information on nitroglycerin.)

2 percent nitroglycerin patch or ointment

The therapy is used to prevent chronic chest pain related to coronary artery disease and will not alleviate acute dyspneic episodes.

Apigenin against Lung Cancer

Apigenin is a powerful antioxidant flavonoid found in many Apiaceae family plants (parsley, fennel, anise, *Cnidium*), as well as in chamomile, rosemary, and thyme. Researchers have prepared apigenin into lipid-based medications for oral use. Because apigenin is not highly soluble in water, preparation into oils, vitamin E, lipid-based products, and liposomes may greatly enhance delivery into tissues. Researchers have reported that apigenin prepared into a liposomal medication had possible tumoricidal effects against lung cancer. In addition to its anti-inflammatory effects, apigenin has many anti-cancer mechanisms of action including beneficial effects on phosphoinositide 3-kinase signaling pathway, and upregulation of tumor necrosis factor (TNF).[14] Although there is no research on whether apigenin in crude plant material can be delivered though the skin, many such plants have been historically used in inhalants for cough and lung pain and in topical oils, plasters, and liniments for application to the back and chest to treat a variety of lung diseases. Consider including fennel or anise powders or essential oils of the other plants named above in formulas for lung oils.

Topical Applications for Lung Pain

Some relief from lung pain due to cancer and lung metastasis may be accomplished with the use of topical lidocaine plasters or creams.[15] There are 5 percent lidocaine topical medications that are approved for postherpetic neuralgia but they may be helpful here as well. High-potency 8 percent capsaicin topical patches may also be helpful. Capsaicin affects the synthesis, storage, transport, and release of substance P in nociceptive fibers. It comes in varying strengths, the stronger of which are licensed for intermittent application. A compounding pharmacist may be able to prepare a combination medicine, or each agent may be applied separately. If making such a product for one's self or with the help of a compounding pharmacist, it may be useful to include an apigenin-containing plant extract. Oiling a patient's back with castor oil prior to application may help wick the medicines deeper into the lungs for optimal pain-relieving effects.

5 percent lidocaine ointment, a 3 to 4 ounce (75 to 100 g) tube

8 percent capsaicin, a 3 to 4 ounce (75 to 100 g) tube

The simplest manner to employ this therapy would be to obtain the two separate products from a pharmacy and apply ¼ teaspoon of each to the oiled chest, repeating the application every hour or two to help allay chronic pain. Alternately, *Matricaria* oil (available commercially) and cayenne powder may be blended into a proprietary lidocaine ointment by squeezing the tube into a 4-ounce salve container and working in 2 teaspoons each of cayenne powder and *Matricaria* oil. Omit the cayenne for those with sensitive skin or if any burning sensation occurs.

Formulas for Coughs and Altered Breath Sounds

Coughs and altered breathing of various types are the primary respiratory symptoms, and they may be representative of a variety of diagnoses. In this section, I begin by addressing how an herbal clinician might create formulas that address a variety of types of cough. I follow this with a more specific discussion of the diagnostic headings that typically involve coughing and altered breath sounds. It is often more effective to create a formula that addresses the quality of such symptoms—such as hacking, dry, wet, spastic, tight, or burning—rather than the underlying diagnosis of bronchitis, pleurisy, asthma, or other condition, because many herbs are appropriate for a variety of respiratory diagnoses. It is the *quality* of the *symptoms* that helps the herbalist choose which herbs to include in a formula, as much as or more so than the diagnosis of a condition.

Coughs

Coughs may be triggered by muscle spasms in the throat and bronchi due to the presence of mucus, a chemical or particulate irritant, or an allergic irritation. Coughing may also be a nervous habit or even an unconscious tic. Each underlying cause may be treated accordingly. Coughing may be a healthy and appropriate function to rid the airways of excessive mucus. However, when coughing is vigorous and persistent, it can be exhausting and injure the ribs over time, and the cause must be addressed. In many cases, the underlying cause is a simple rhinovirus or other viral infection, often referred to as an upper respiratory infection (URI), or the "common cold." URIs are self-limiting, meaning that even with no treatment, the infection will often resolve on its own. However, herbal therapies can provide palliative comfort, shorten the duration and severity of an infection, reduce chronicity in those who suffer numerous colds each year, and help to avoid complications for those whose URIs tend to progress to ear infection, sinus infection, or bronchitis.

Folkloric herbals often mention the symptom of "hemming," referring to those who must clear their throats throughout the day due to mucous accumulation. URIs, hemming, postnasal drip, and similar presentations can be improved with immune support, avoidance of dairy products or other mucus-forming foods, and the use of mucolytic, drying antimicrobial herbs such as *Hydrastis* spp., *Rosmarinus officinalis*, *Allium* spp., *Berberis aquifolium* (also known as *Mahonia aquifolium*), *Achillea millefolium*, *Thymus vulgaris*, *Euphrasia officinalis*, or *Commiphora myrrha*. Folkloric texts also refer to "hacking" coughs, where muscular spasms in the bronchi, diaphragm, and/or intercostal muscles are involved. Hacking coughs are best treated with respiratory antispasmodics such as *Thymus*, *Lobelia inflata*, *Ammi visnaga*, and *Prunus*, all noted for their ability to affect bronchial smooth muscles.

The Cough Reflex

Sensory afferent nerves in the vagal plexus can trigger cough reflexes. Various subsets of these afferent fibers can be activated and cause immediate and explosive coughing or a tickling or irritating sensation that is relieved by a voluntary cough. Infections in the airways can cause the urge to cough, but in other cases noxious agents, fumes, and allergic reactivity can trigger the cough reflex. Vagal afferent nerves are involved in triggering the cough reflex and may contribute to dyspnea. Agents that affect vagal nerve transmission may therefore be useful in managing cough, dyspnea, and some types of lung pain. In research studies, it is well established that adenosine triphosphate (ATP) and adenosine can induce coughing in humans when inhaled, and respiratory nerve tone can also be affected by stimulating nociceptive receptors such as various subtypes of transient receptor potential (TRP) channels that respond to noxious compounds, temperature, and mechanical pressure. Type C-fibers in vagal nerves may also mediate nociceptive pain in the airways, and several different receptor types including 5-HT, nicotinic, acid-sensing, purinergic, and P2X receptors can activate C-fibers.[16]

Adrenaline-bearing nerves in the airways, called the adrenergic neurons, can activate nicotinic receptors shown to be most abundant on receptors near airway epithelium and less dense in airway smooth muscle. (There is a dizzying array of nicotinic receptor subtypes.) When inhaled, nicotine leads to a sense of irritation and constriction and initiates the cough reflex. Endogenous activation of nicotinic receptors promotes a sensation of irritation, tightness, and coughing. Therefore, blockage of adrenergic signals and nicotinic receptors can quiet coughs.

Noxious agents such as capsaicin may bind to the vanilloid subtype of TRP channels on vagal afferent C-fibers and induce the release of neuropeptides into the airways, leading to constriction of bronchial smooth muscle and triggering the cough reflex. However, prolonged activation of the vanilloid TRPs can reduce sensitivity and thereby reduce the coughing and block activation by other agonists. This may explain the folkloric use of cayenne (*Capsicum annuum*, also known as *C. frutescens*) in many respiratory formulas and the finding of a *Lobelia-Capsicum* duo in many traditional cough and dyspnea formulas. Ankyrin is another subtype of TRP channels expressed on the vast majority of vagal C-fibers. When ankyrin is activated by irritants ranging from ozone to toluene and formaldehyde, coughing may result. Natural cinnamaldehyde-based molecules and mustard oil will also activate the ankyrin type of TRP channels and may cause coughing. Menthol, found in many mints, will bind another type of TRP channel less commonly found on vagal nerves but known to be activated by cold temperature, explaining the cooling sensation produced by mint.

Lidocaine can completely anesthetize vagal afferent fibers by disabling key sodium gated ion channels and can block the cough reflex, but it can have undesirable cardiac and central nervous system (CNS) effects, a problem that could be overcome if receptor-specific medications could be developed. When used judiciously as a transdermal patch, lidocaine may allay harmful coughing or allay dyspnea for those with lung cancer or other serious lung diseases. Inflammation upregulates these channels and increases the reactivity of nerves to various irritants. Therefore, the combination of agents that block sodium channels and agents that desensitize TRP channels may help reduce vagal hyperexcitability to reduce coughing in a variety of disease states. An inhalant form of such medicines may help target the lungs and limit systemic and cardiovascular effects.

People presenting with coughs describe the quality of their cough in different ways, such as tight and dry feeling, moist with much expectoration, spastic and painful in the chest, or burning and painful in the throat. In general, a cough associated with abundant expectoration occurs in children or allergic individuals and most often at the onset of a URI or an allergic episode, such as rhinosinusitis. Coughs that are nonproductive and described as having a tight, dry sensation occur most often in chronic conditions, such as in a long-term smoker. A logical approach to creating herbal formulas for such *qualities* of coughing are to include moist demulcent herbs for hot, dry, tight sensations and drying herbs for damp, abundantly productive situations. Herbs with drying actions on the airways include *Hydrastis*, *Thymus*, *Salvia officinalis*, and *Achillea*. Mucilaginous herbs with demulcent qualities include *Glycyrrhiza glabra*, *Petasites hybridus*, *Tussilago farfara*, *Symphytum officinale*, *Ulmus fulva*, and *Althaea officinalis*. Codeine, although a classic cough syrup ingredient of natural origin and somewhat effective, is not preferred by herbalists, due to its sedating and potentially addictive effects. Children should not be treated with codeine due to the risk of respiratory suppression. Many effective nonopioid herbal options are available, and thus finely tuned cough syrups and formulas are preferable.

Wheezing and Stridor

Other altered breath sounds include wheezing, typical of asthma; stridor, typical of upper laryngeal spasms; and rhonchi, typical of bronchial tone issues. These are described in "The Cough Reflex" sidebar on page 115. When a cough or altered breath sounds do not appear to correlate with smoking, an infection, an allergy, or recent exposure to a chemical or particulate irritant, clinicians must rule out other underlying causes, which could range from a lung lesion or tumor to pulmonary emboli, especially if lung pain and systemic symptoms are present. Chest X-rays are a useful screening test to rule out lung lesions. Pulse oximetry, blood gases, and spirometry can also provide useful information.

Gastroesophageal reflux disease, commonly referred to as reflux disease or GERD, can present as a cough due to the irritating effects of stomach acid on the throat and esophagus. Some medications will induce coughs as a side effect, particularly ACE inhibitors used to treat hypertension, and all medication should be reviewed when no other obvious cause of coughs is apparent. Throat and ear lesions, and even something as simple as impacted cerumen, can trigger coughing. I once saw a patient with a troublesome cough due to a small hair stuck to the tympanic membrane; the cough resolved once I irrigated out the hair. Some cases of asthma present without wheezing, but rather with a chronic spastic cough, referred to as cough-variant asthma. Some people may attempt to mask Tourette Syndrome's vocalizations with a cough.

Wheezing is most often associated with asthma and respiratory allergies and is typically episodic, occurring in allergic individuals in obvious association with allergen exposure. However, wheezing may also occur with other conditions. A lung lesion, such as a tumor, may produce a wheezing sound in the location of an obstruction, and when alterations to the voice accompany wheezing, a vocal cord or throat lesion may be the underlying cause. Wheezing in association with circulation insufficiency indicates a cardiac pathology. When wheezing is accompanied by fever and expectoration, it's possible that an underlying infection is triggering bronchospasm.

Stridor is a high-pitched, squeaky sound that occurs with inspiration and is most often due to blockage or constriction in the throat or upper airways. Edema in the upper airways may also produce stridor; the edema may occur with anaphylaxis or other allergic phenomena, as well as upper throat, larynx, or vocal cord lesions, tumors, and abscesses. When stridor presents with a severe sore throat, a retropharyngeal abscess is possible, or with unusual salivation and drooling, epiglottitis is likely. When there has been progressive or long-standing hoarseness or vocal change, a laryngeal or vocal cord lesion is likely. An aspirated foreign body may be the underlying cause of stridor in a toddler, and some atopic individuals can develop laryngeal spasm as part of an allergic phenomenon. Diagnostic imaging may be required to further assess underlying causes of stridor.

Stridor due to allergic hyperreactivity may be improved by antiallergy herbs such as *Angelica sinensis*, *Ginkgo biloba*, *Curcuma*, *Scutellaria baicalensis*, *Glycyrrhiza*, or *Foeniculum vulgare*. When underlying infections present with enlarged tonsils, or when there is thick or choking mucus present, *Hydrastis*, *Commiphora*, or *Phytolacca* spp. may be helpful. Simple soothing agents will allay vocal cord irritation due to dust, overuse of the voice, and exposure to chemical fumes, especially the antiallergy demulcents *Tussilago* and *Petasites*. Be cautious about using undiluted tinctures for those with epiglottitis or allergic reactivity in the throat, as such tinctures could worsen laryngeal spasm and lead to acute dyspnea and airway obstruction in some sensitive individuals. Use teas or lozenges, or dilute the tinctures in a small glass of water and take in small sips over time in cases of acute epiglottitis.

Following are formulas that address various presentations, regardless of the underlying diagnosis. Such symptoms are further detailed under appropriate diagnostic headings such as asthma, bronchitis, emphysema, and laryngitis. Some formulas to address these symptoms may be all-purpose—such as formulas for general cough syrups that can be helpful in treating symptoms due to colds, bronchitis, or pneumonia. Other formulas provide a basic tutorial for choosing herbs specifically for tightness, dryness, excessive expectoration, and so on.

All-Purpose Cough Syrup

This all-purpose cough syrup features herbs most emphasized in the folklore for simple coughs. In fact, *Tussilago* means "cough dispeller" in Latin. Cherry bark (*Prunus serotina*) contains spasmolytic hydrocyanide compounds, relied upon for generations as a base ingredient in cough formulas. Licorice teas and syrups are useful for treating cough and have been included in a wide variety of traditional cough formulas.[17]

Ligusticum porteri root	4 ounces (120 g)
Prunus spp. bark	2 ounces (60 g)
Tussilago farfara root	2 ounces (60 g)
Glycyrrhiza glabra root	2 ounces (60 g)
Zingiber officinale root	2 ounces (60 g)
Lobelia inflata leaves	2 ounces (60 g)
Thymus vulgaris leaves	2 ounces (60 g)
Honey	2 quarts (2 liters)
Citrus aurantium (bitter orange) essential oil	¼ ounce (8 ml)
Foeniculum vulgare (fennel) essential oil	¼ ounce (8 ml)

On low heat, gently simmer the dry herbs in 2 gallons of water until reduced down to 1 gallon. Strain and add the honey to the liquid over low heat. Dissolve the honey fully, using a spoon to skim any scum that forms on the top. Remove from the heat, and allow to cool to room temperature before adding the bitter orange and fennel essential oils. Pour quickly into bottles before the oils have a chance to evaporate. Take by the teaspoon for coughs, as often as hourly, reducing as symptoms improve.

Ginger Cough Syrup

This is another method of making a powerful antitussive using both decocted dry herbs and fresh ginger root, stabilized with honey and herbal tinctures. Fresh ginger has a much sweeter, gentler, and aromatic quality than dried ginger and is preferred for this recipe. *Armoracia* (horseradish) is a powerful expectorant, mellowed here by the other ingredients in the formula.

Zingiber officinale fresh root	1 cup, coarsely chopped (200 g)
Armoracia rusticana fresh root	1 cup, coarsely chopped (200 g)
Tussilago farfara	2 ounces (60 g)
Sambucus spp. berries	2 ounces (60 g)
Foeniculum vulgare seeds	2 ounces (60 g)
Water	6 pints (3 L)
Honey	1 quart (945 ml)
Asclepias tuberosa tincture	4 ounces (120 ml)
Lomatium dissectum tincture	4 ounces (120 ml)

Coarsely chop or grate the fresh ginger and horseradish roots, and combine with the dry *Tussilago*, *Sambucus* (elder) berries, and *Foeniculum* (fennel) seeds, and place in a large pan and cover with the water. Bring to a gentle boil, and then reduce heat and hold at the lowest possible simmer for 20 minutes, with the lid on the pan. Remove from heat, strain, and return to the stovetop; add the honey, and bring to a brief simmer a second time, skimming any scum that surfaces with a spoon. Remove from heat again, and allow to cool to room temperature before adding the *Asclepias* and *Lomatium* tinctures. Bottle and store in the refrigerator. Dose by the tablespoon 4 times daily up to every several hours when needed for respiratory congestion.

Onion Cough Syrup Home Remedy

Onions are antimicrobial and expectorating—as anyone who has chopped a fresh onion knows—and onion cough syrup is a traditional home remedy due to the ready presence of onions in most households. The proportions are equally "homey."

2 large white or yellow onions
Honey
1 lemon
Lung tincture of choice

Antitussives for Problematic Coughs

The list of antitussives could be quite long. This short list includes my own favorites that I commonly use in tinctures, teas, and, in some cases, topical preparations.

Drosera rotundifolia	*Symphytum officinale*
Eucalyptus globulus	*Symplocarpus foetidus*
Lobelia inflata	*Thymus vulgaris*
Prunus serotina	*Tussilago farfara*
Sticta pulmonaria	*Verbascum thapsus*
Stillingia sylvatica	

Coughs and Altered Breath Sounds

Chop the onions and place in a double boiler. Pour in enough honey to barely cover. Squeeze the lemon and add the lemon juice, and cook slowly on low heat for several hours. The honey will thin down as the water from the onions is released. Strain out the onions and measure to determine the volume of the resulting honey-lemon liquid. Add an equal volume of an antimicrobial and antitussive lung tincture such as *Lomatium*, *Thymus*, or *Eucalyptus* (or a combination).

The tincture will help preserve the honey and boost the effect. Wine or other plain alcohol would be a less expensive option, if preferred. You can cut down the recipe by using one small onion instead of two large ones. If the intention is to use up the small batch within a week's time, no alcohol would be needed, and the flavor would be the most palatable for a small child. Refrigerate the onion cough syrup and take 1 teaspoon as often as hourly.

Herbs for Altered Breath Sounds

A variety of altered breath sounds may be detected via auscultation of the lungs with a stethoscope. Each type may hint at an underlying pathology or diagnosis. Each may also suggest specific herbal remedies that most specifically remedy a variety of altered breath sounds. Each of these diagnoses is discussed individually in this chapter, but this summary of herbal therapies addressing the specific symptoms will also be useful to practitioners learning to craft sophisticated and specific herbal formulas for their patients.

Crackling sounds. These can occur with vascular congestion or lung infections and disease. When due to cardiac causes, cardiac glycoside herbs such as *Digitalis purpurea* and *Convallaria majalis* are most indicated. Mucous rattles and obvious moist crackles might be best addressed with drying antimicrobials such as *Thymus*, *Eucalyptus*, *Lomatium dissectum*, *Asclepias tuberosa*, or *Salvia officinalis*.

Wheezing. The tight sound is sometimes audible without a stethoscope and is due to constriction of the bronchial muscles, as occurs with asthma, but is also a symptom that occurs with chronic obstructive pulmonary disorder (COPD), congestive heart failure, exposure to toxic gases and fumes, aspiration of a foreign body, or a vocal cord lesion. Wheezing due to bronchoconstriction responds to herbal bronchodilators including *Lobelia*, *Prunus*, *Thymus*, *Ammi*, and *Eucalyptus*.

Rhonchi. The low rubbing sounds referred to as rhonchi, typical of bronchiectasis and COPD, are best treated on a lifelong basis with lung and connective tissue tonics such as *Equisetum* spp. and *Centella asiatica*, combined with anti-inflammatory immunomodulators such as *Panax ginseng*, *Zingiber*, or *Curcuma*.

Friction rubs. The high-pitched inspiratory friction rubs and squeaks of pleurisy and pleuritis are best treated with antimicrobials and anti-inflammatories such as *Asclepias*, *Curcuma*, *Tussilago*, and *Bryonia dioica*.

Stridor. The throaty squeak of stridor suggests an airway obstruction fairly high up, such as an aspirated foreign body, epiglottitis, or vocal cord lesion. Agents that decongest the throat such as *Commiphora myrrha*, *Phytolacca decandra*, *Armoracia rusticana*, and *Allium* spp. might be appropriate.

Diminished breath sounds. These occur when there is little air moving through the airways, such as when the airways are blocked by mucus, as is the case with pneumonia, emphysema, and COPD.

Accentuated breath sounds. These sounds indicate that air is moving through the airways, but is impeded due to bronchoconstriction and edema, yet not severe enough to result in wheezing. General respiratory anti-inflammatories are appropriate here, from the inhalation of *Eucalyptus* essential oil to gentle bronchodilators such as *Thymus vulgaris* or *Prunus* tea.

Tincture for Simple URI with Cough

The most ubiquitous respiratory complaints are simple upper respiratory infections (URIs)—the common cold. A formula such as this is all-purpose, but can be altered given specific presentations, as discussed throughout this chapter.

Euphrasia officinalis	15 ml
Thymus vulgaris	15 ml
Allium sativum	15 ml
Petasites hybridus	15 ml

Take 1 teaspoon of the combined tincture every 15 minutes to an hour, reducing as symptoms improve.

Tincture for Cough from Postnasal Drip

When postnasal drip from the sinuses and throat causes hemming and coughing, these herbs may help astringe the upper airways. If postnasal drip is not due to a new onset URI, consider a dairy allergy or an airborne allergen as a possible cause of chronic postnasal drip. *Achillea*, *Thymus*, and *Commiphora myrrha* have drying effects when taken fairly frequently, as well as antimicrobial and immunomodulating effects to help treat an underlying infection.

Achillea millefolium	15 ml
Thymus vulgaris	15 ml
Echinacea spp.	15 ml
Commiphora myrrha	15 ml

Take 1 teaspoon of the combined tincture every 15 minutes to an hour, reducing as symptoms improve.

Tincture for Cough with Bronchial Spasms

This formula features herbs with effects on bronchial smooth muscles, and similar formulas are featured in the bronchitis and asthma sections of this chapter.

Thymus vulgaris	15 ml
Prunus spp.	15 ml
Ammi visnaga	10 ml
Lobelia inflata	10 ml
Foeniculum vulgare	10 ml

Take 1 teaspoon of the combined tincture every 15 minutes to an hour, reducing as symptoms improve.

Tincture for Moist Cough

This formula and the Tincture for Tight Dry Cough contrast drying versus moistening formulas. This formula employs drying herbs to check excessive secretions, and the following formula employs stimulating herbs to remedy scant expectoration. The herbs in this formula all have drying effects, with *Thymus* being a particularly effective bronchial antispasmodic. Use a formula such as this when there is abundant, watery, free-flowing expectoration.

Achillea millefolium	15 ml
Thymus vulgaris	15 ml
Commiphora myrrha	15 ml
Origanum vulgare	15 ml

Take 1 teaspoon of the combined tincture every 15 minutes to an hour, reducing as symptoms improve.

Tincture for Tight Dry Cough

The warm, stimulating resinous herbs *Capsicum*, *Grindelia*, and *Stillingia* act as counterirritants. *Lobelia* is a more powerful bronchial antispasmodic with additional expectorating actions, in a soothing base of *Petasites* to prevent excessive irritation. Use a formula such as this for a dry hacking cough with scant or no expectoration, or when the mucus is thick and rubbery.

Petasites hybridus	20 ml
Grindelia spp.	10 ml
Stillingia sylvatica	10 ml
Lobelia inflata	10 ml
Capsicum spp.	10 ml

Take 1 teaspoon of the combined tincture every 15 minutes to an hour, reducing as symptoms improve. Should any throat irritation occur, or for sensitive individuals, each dose may be put in a small sip of water.

Tincture for Mucous Rattling in the Lungs

In herbal folklore, these herbs are specifically indicated for audible mucous rattles.

Asclepias tuberosa	15 ml
Thymus vulgaris	15 ml
Lomatium dissectum	15 ml
Sambucus spp.	5 ml

Take 1 dropperful of the combined tincture as often as hourly, reducing to 3 or 4 times per day as symptoms improve.

Tincture for a Severe Spastic Cough

This formula features some of the best options to calm spastic coughing. *Tussilago* may relax bronchospasm via effects on calcium channels, while *Lobelia* has complex

Coughs and Altered Breath Sounds

actions on pressure sensors in respiratory vasculature. *Drosera* (sundew) may require some searching, but is an excellent option for tightness in the throat, croup, a barking cough, and laryngeal spasm. Because *Drosera* is not found abundantly, I reserve its use for such cases. Thyme is an excellent anti-spasmodic with the added benefit of being drying to excessive mucus and having broad anti-microbial effects. *Glycyrrhiza* (licorice) is demulcent and antiviral and helps to sweeten up the formula for children.

Tussilago farfara	15 ml
Lobelia inflata	15 ml
Glycyrrhiza glabra	15 ml
Drosera rotundifolia	8 ml
Thymus vulgaris	7 ml
Foeniculum vulgare essential oil	10 to 20 drops

Combine in a 2-ounce bottle and dose a child with 10 drops at a time, as often as every half hour. An adult may take up to 60 drops at a time. Back off on the frequency of dosing if any nausea or increased salivation occurs; these are potential side effects of repetitive doses of *Lobelia*.

Tincture for Respiratory Rhonchi

Rhonchi is sometimes poetically described as the low squeak of several stout boat ropes rubbing together. It occurs due to inflammation in the large airways. In such cases the connective tissue of the larger bronchi may benefit from anti-inflammatory and restorative support, hence the use of connective-tissue herbs *Centella* and *Equisetum* in this formula. The other herbs are chosen due to their demonstrated abilities to protect against fibrotic changes in a variety of inflammatory conditions.

Ginkgo biloba	15 ml
Curcuma longa	15 ml
Panax ginseng	14 ml
Centella asiatica	8 ml
Equisetum spp.	8 ml

Take 1 dropperful of the combined tincture as often as hourly, reducing to 3 or 4 times per day as symptoms improve.

Tincture for Inspiratory Friction Rubs

The high-pitched squeak of friction rubs occurs when there is inflammation between the pleural and visceral

More about Mucus

Mucus provides a protective barrier against pathogens, toxins, and irritants in the airways, and its liquid nature helps to move such offending agents out of the body, with the help of cilia, specialized cell appendages capable of propelling mucus along. Many respiratory diseases involve an overproduction of mucus, which triggers coughing and breathing difficulty, as well as is a breeding ground for microbes, inviting secondary bacterial infections. Mucus is secreted by goblet cells and consists of a mixture of water, ions, glycoproteins, mucins, and lipids. Naturally occurring disulfide compounds give mucus a fluid nature, breaking bonds that crosslink the glycoproteins. Oxidative stress is noted to increase mucin disulfide crosslinking in a manner that toughens mucus to an excessive degree, making it rubbery and tenacious.[18] Mucins are large glycoproteins with elastic properties; they are overproduced in cases of asthma, cystic fibrosis, and chronic bronchitis. Mucolytic agents, which are natural or synthetic compounds capable of breaking apart mucus, would always be indicated in cystic fibrosis, as well as any respiratory issue accompanied by profuse mucous production. Garlic, onions, horseradish, and radishes are all natural mucolytic agents. Due to the spicy sulfur compounds they contain, they are capable of thinning mucus and allowing it to be more easily expectorated. All can be prepared into medicinal foods. Garlic, radish, and horseradish are also available in tincture form to include in mucolytic formulas for lung disease. Garlic and Spanish black radish are also available in capsules. Bromelain, N-acetylcysteine (NAC), and other sulfur compounds capable of thinning and moving mucus are available in commercial encapsulations.

membranes of the lungs, as with pleurisy. *Asclepias* is emphasized in the folkloric literature for pleurisy, and *Bryonia* is specifically indicated for pain that is worse with motion.

Asclepias tuberosa	20 ml
Lomatium dissectum	15 ml
Thymus vulgaris	15 ml
Bryonia dioica	10 ml

Take 1 dropperful of the combined tincture as often as hourly, reducing to 3 or 4 times per day as symptoms improve.

Tincture for Stridor with Laryngeal Swelling

Stillingia, *Commiphora* (myrrh), and *Phytolacca* all have an affinity to the throat, according to folkloric herbal traditions. Myrrh and *Stillingia* are resinous plants that help to move thick mucus out of the throat. *Armoracia* (horseradish) acts as a counterirritant to provoke fresh, thin mucous production, also supporting expectoration. *Tussilago* lends a soothing element to the formula; the other herbs are all stimulating and potentially irritating.

Tussilago farfara	15 ml
Commiphora myrrha	15 ml
Phytolacca decandra	10 ml
Stillingia sylvatica	10 ml
Armoracia rusticana	10 ml

Take 1 dropperful of the combined tincture as often as hourly, reducing to 3 or 4 times per day as symptoms improve. Dilute the tincture in water and consume in small sips to avoid any further throat reactivity.

Demulcent Tincture for Acute Stridor

Stridor occurring with *severe* swelling in the pharynx or larynx is a serious concern due to possible airway obstruction; it may be due to acute allergic reactivity. Great care must be taken when using herbs, lest the hyperreactive state extend to any medication attempted. Initiate the use of herbs cautiously, just a few drops at time, diluted in water, increasing as tolerated. Dropping herbal tinctures onto a slippery elm lozenge is another option. Unlike the formula above, this formula is crafted to be gentle and soothing and contains no high volatile oil, resin, or counterirritant herbs.

Tussilago farfara	10 ml
Ulmus fulva	10 ml
Glycyrrhiza glabra	10 ml

Western Herbal Respiratory Antimicrobials

These are among my favorite herbs to include in formulas for upper and lower respiratory infections. They are all folkloric classics for bronchitis, cough, cold, pneumonia, and flu.

Allium sativum	*Lomatium dissectum*
Commiphora molmol	*Olea europaea*
Echinacea angustifolia	*Sambucus nigra*
Eucalyptus globulus	*Thymus vulgaris*
Ligusticum porteri	*Usnea barbata* or *U. hirta*

Place 2 dropperfuls of the combined tincture in a shot glass of water, and consume by taking a tiny sip every 60 seconds, reducing as normal breathing is restored. Or place 10 drops on a slippery elm lozenge, suck until the lozenge is dissolved, and repeat.

Tincture for Acute Stridor Due to Allergen Exposure

When stridor occurs in allergic individuals rather than in association with pharyngitis or another infectious trigger, allergy herbs would be more appropriate than the herbs featured in the formula above. *Scutellaria baicalensis*, *Ginkgo*, and *Curcuma* are among the all-purpose antiallergy options. Emergency services may be required for acute laryngeal edema or other serious cause of acute stridor, and the tincture that follows may be appropriate to be taken en route to the hospital. Take care to dilute potentially irritating tinctures in water for those suspected to be displaying allergic hypersensitivity reactivity.

Curcuma longa	15 ml
Ginkgo biloba	15 ml
Scutellaria baicalensis	10 ml
Foeniculum vulgare	10 ml
Glycyrrhiza glabra	10 ml

Place 1 tablespoon of the combined tincture in 2 tablespoons of warm water and take a tiny sip every 30 to 60 seconds, reducing as normal breathing is restored.

Tincture for Costochondritis

When bronchitis and pneumonia cause frequent coughing, repetitive expansive of the rib cage can lead to costal chondritis. The herbs in this formula do not treat underlying infections, but rather help quiet the cough reflex with *Eucalyptus*, while treating musculoskeletal inflammation

Coughs and Altered Breath Sounds

Lobelia inflata—A Nefarious Lung Herb

Lobelia inflata belongs to the Campanulaceae family and has gone by the common name pukeweed because of the powerful emetic effects demonstrated by ingesting the fresh buds. *L. inflata* is also called Indian tobacco, and the plant was well known and discussed in early medical writings once the European settlers learned that Native Americans used it in smoking mixtures, or through other methods of inhaling the vapors, to treat wheezing and problematic coughs. *Lobelia* was observed to relax tightness in the airways and was also included in teas for coughs, asthma, and constricted airways, as well as for skeletal muscle spasms and anxiety with body tension. *Lobelia* may help relieve severe dyspnea such as the membranous throat of diphtheria or the severe spasms of asthmatic wheezing and croup. *Lobelia* was also widely employed as a plaster directly over the chest to relieve asthmatic airway constriction, wheezing, and pain in the lungs.

Lobelia contains piperidine alkaloids, the most well known being lobeline. None are being robustly studied, and the understanding of their physiologic effects is still being elucidated.[19] Lobeline is both an agonist and antagonist to β-nicotinic receptors,[20] first stimulating respiratory tension and then relaxing it. Lobeline and related *Lobelia* alkaloids also cross the blood-brain barrier, and similar to nicotine, promote the release of the neurotransmitters dopamine and norepinephrine.[21] Another compound in *Lobelia*, beta-amyrin palmitate, reduces the response to sympathetic nerve signaling and exerts a mild sedative effect,[22] in part by inhibiting the flow of calcium ions into adrenal chromaffin cells.[23]

Lobelia inflata, pukeweed

Lobeline initially excites pulmonary afferent nerves[24] and can promote salvia and respiratory mucus, and at higher dosages, it can promote emesis and/or hyperventilation and hyperpnea.[25] The nefarious aspect of *Lobelia*'s history is its curious ability to immediately promote a sensation of constriction in the throat,[26] causing users to cough or clear their throat, often within seconds of swallowing a dropperful of tincture. When the dosage is repeated, gagging and choking sensations occur and may trigger emesis when continued. *Lobelia* has been shown to increase neural firing, including in the phrenic nerve.[27] This short-lasting effect may be due to direct effects on baroreceptors in the carotids[28] and chemoreceptors in the adrenergic system.[29] Yet *Lobelia* is a classic remedy to treat coughing, because subsequent to the simulating effects that trigger mucus expectoration, *Lobelia* relaxes the airways.

The smooth muscle of the upper airways also possesses J fibers (short for juxtapulmonary capillary receptors[30]), which are associated with the cough and gag reflex as they receive excitatory input. The fibers are activated by a variety of irritants, including capsaicin, lobeline, bradykinin, sulfur gases, and mechanical manipulation such as sticking a finger down the throat.[31] For the asthma patient or someone suffering from excessive secretions in the bronchial passages, *Lobelia* may act as a counterirritant. A counterirritant is a term unique to herbal medicine, where small doses of an irritating or caustic substance are used at the optimal physiologic dose to stimulate sluggish or deficient function. In the case of a respiratory counterirritant, lobeline stimulates respiration, enhances secretions, and supports productive coughing. For best results, *Lobelia* should be formulated with herbs that are soothing to the throat and lungs and have an antispasmodic effect to complement its stimulating effects, preventing it from being too stimulating, irritating, and potentially emetic. When used topically, *Lobelia* may simply help to calm excessive coughing and help allay the pain and spasm of respiratory and other muscles.

with *Hypericum* and *Curcuma*. *Piper* and *Bryonia* can help palliate acute rib pain. For those with severe painful coughing where the pleural membranes are the source of the pain, or when there has been such vigorous and prolonged coughing such that the ribs are becoming inflamed, wrapping several elastic bandages firmly around the rib cage can help lessen respiratory motion and reduce pain a bit. Soaking in a hot Epsom salt bath several times each day can be a comfort to those with chest pain due to lung infections and musculoskeletal conditions.

Hypericum perforatum	15 ml
Eucalyptus globulus	15 ml
Curcuma longa	10 ml
Piper methysticum	10 ml
Bryonia dioica	10 ml

Take 1 dropperful of the combined tincture every hour, reducing as symptoms improve.

Stillingia "Formula Number 176"

This formula from *King's American Dispensatory*, published in 1898, was recommended for syphilis, rheumatism, and scrophula, a tuberculosis-like collection of symptoms including swollen lymph nodes, chronic cough, and upper respiratory symptoms.

Stillingia sylvatica	10 ounces (300 g)
Corydalis cava	10 ounces (300 g)
Iris versicolor	5 ounces (150 g)
Sambucus nigra	5 ounces (150 g)
Chimaphila umbellata	1 ounce (30 g)
Coriandrum sativum	2 ounces (60 g)
Xanthoxylum americanum berries	2 ounces (60 g)
Alcohol	6 ounces (180 ml)
Water	3 ounces (90 ml)
Glycerin	3 ounces (90 ml)

Grind the dried herbs together into a moderately coarse powder, blend with the alcohol, water, and glycerin, and macerate for a month. Press out the liquid and dose roughly ½ teaspoon of the liquid at a time 1 or more times daily.

Topical *Arnica* Oil for Costal Chondritis

Topical *Arnica*, a classic herb for musculoskeletal trauma, would complement the above oral formula.

Arnica oil

Rub the oil into painful ribs and cover with heat such as a heating pad or microwavable moist heat pack. Reapply at least 3 times daily.

Tincture for Dry Cough

Both *Bryonia* and *Sanguinaria* were used by early American physicians to treat chronic hacking cough by diluting the herbs in water.

Bryonia dioica	20 drops
Sanguinaria canadensis	20 drops
Water	480 ml

Combine both tinctures in a glass with the water and take 1 sip every 15 minutes to ease coughing.

Tincture for Influenza with Cough

Cough that accompanies influenza requires immune support and palliative care. This formula from Cloyce Wilson, MD, (circa the 1920s) features *Lobelia* as a cough suppressant, along with *Echinacea* as an immunomodulator. *Actaea* may seem like an odd choice, given its present emphasis as a menopausal and "women's herb," but it does indeed do a great job of alleviating myalgia due to viral infection. Several *Actaea* species contain cimicifugin, a resinoid isolated from roots and rhizomes, which may deter respiratory syncytial virus via inhibition of viral attachment and promotion of interferon responses.[32]

Echinacea spp.	10 ml
Lobelia inflata	10 ml
Actaea racemosa	10 ml
Aconitum napellus	5 drops

Place 10 drops of the combined tincture in a sip of water. Repeat this dose once an hour, reducing frequency as the cough subsides.

Coughs and Altered Breath Sounds

Formulas for Dyspnea

Severe dyspnea is a medical emergency, and even mild dyspnea feels alarming to the person experiencing it. When a fresh episode of dyspnea occurs, those who have asthma may understand exactly what is occurring and

reach for their inhaler, but fairly sudden onset of dyspnea requires prompt evaluation because the breathing episode may be due to an acute myocardial infarction, pneumothorax, a pulmonary embolism, or aspiration of a foreign body. All cases of abrupt onset of dyspnea in individuals with no prior history are worrisome and require immediate medical attention. Atopic individuals and asthmatics may also experience fairly sudden wheezing and dyspnea following exposure to dust, toxic fumes, environmental chemicals, chlorine, smoke, or other trigger. Removing oneself from the offending environment is crucial, and treatment with bronchodilators and antiallergy agents such as antihistamines is needed. Gradual worsening of dyspnea occurs with congestive heart failure, COPD, metabolic acidosis, and emphysema as part of the natural disease progression, and herbal therapies may slow this progression and delay mortality when used long term. Advanced lung diseases and heart failure are common causes of chronic and slowly progressive dyspnea, and patients will typically require oxygen therapy at some point to allay acute dyspnea. Formulas to address dyspnea should address the underlying cause as much as possible; see "Quality of Dyspnea and Associated Diagnoses" for examples. If the underlying cause is not obvious, chest X-rays, EKGs, blood work, pulse oximetry, and other diagnostic workups may be run to assist with the diagnosis.

Quality of Dyspnea and Associated Diagnoses

Asthma. Dyspnea is associated with a sense of tightness or pressure in the chest, as if something heavy is sitting on the chest. There is a sensation of being unable to move air deeply into the lungs, prompting people to use all chest and accessory muscles.

Metabolic acidosis. Dyspnea occurs because the blood cannot "hold" the oxygen due to a metabolic effect on blood gases and pH balance. People suffering from metabolic acidosis present as if out of breath, breathing shallowly and rapidly; this is also referred to as Kussmaul breathing.

Heart failure. Dyspnea occurs with the least exertion and subsides with rest; it's associated with heartbeat irregularity or blood pressure irregularity.

Heart failure progressed to pulmonary disease. Dyspnea occurs when lying flat on the back, and those affected must sleep propped up on pillows to avoid fluid accumulation in the lungs.

Pneumonia. Dyspnea occurs along with coughing, fever, and malaise.

Hyperventilation. Dyspnea with tingling and spasm of the fingers is typical of hyperventilation.

Panic attacks. Dyspnea occurs with acute heart palpitations and extreme sense of fear and anxiety.

Anemia. Vague dyspnea occurs with dizziness, weakness, and fatigue, especially when exercising.

Tincture for Dyspnea Associated with Underlying Anemia

Treat the anemia by controlling any blood loss, and supplementing iron or B vitamins as specifically needed. This tincture supports blood building with *Medicago* and *Paeonia* and circulation to the lungs with *Ginkgo* and *Lepidium*.

Ginkgo biloba	15 ml
Lepidium meyenii	15 ml
Paeonia lactiflora	15 ml
Medicago sativa	15 ml

Take 1 teaspoon of the combined tincture 3 to 6 times daily for a month or more, reducing as symptoms improve.

Bronchodilating Tincture for Acute Asthmatic Dyspnea

This formula is aimed at opening the airways as quickly as possible and may help reduce reliance on steroidal inhalers. A separate formula is advised, which is aimed at treating hyperreactivity more deeply, using this formula as needed for acute dyspnea.

Thymus vulgaris	15 ml
Ammi visnaga	15 ml
Lobelia inflata	10 ml
Foeniculum vulgare	10 ml
Eucalyptus globulus	10 ml

Take 1 dropperful of the combined tincture every 5 to 10 minutes, reducing as normal breathing is restored.

Nervine Tincture for Hyperventilation and Panic Attacks

All of the herbs in this formula have anxiolytic effects and help attenuate cortisol and adrenaline surges that can underlie a panic disorder. A nerve tonic tea and daily exercise and/or meditation would be good complements to this tincture. A homeopathic remedy such as *Aconitum* may help acutely if taken at the onset of a panic attack.

Actaea racemosa	15 ml
Glycyrrhiza glabra	15 ml
Panax ginseng	15 ml
Melissa officinalis	15 ml

Take 1 teaspoon of the combined tincture 3 to 6 times per day, reducing as symptoms improve.

Tincture for Dyspnea with Anxiety at End of Life

The end stages of heart disease, COPD, and cor pulmonale are typified by progressive circulation failure accompanied by dyspnea. Opioid drugs may be prescribed for acutely distressing dyspnea and end of life care, where they will speed death, but make it a more comfortable, less panicked, process. *Aconitum* is a highly toxic plant due to aconitine, a powerful alkaloid. However, aconite

---

Breathing Technique for Dyspnea

To relieve dyspnea due to hyperventilation, breathe in and out of a paper bag acutely until normal breathing is restored. Follow with a nervine formula. (See the "Treating Hyperventilation Syndrome" section, starting on page 140.)

---

has been a traditional remedy to bring a spark of heat and light to the last days or hours of life, due to the belief that it is one of the most yang plants, and that heart failure is a deficiency of yang energy and chi—the vital force. *Aconitum* has been used in combination with *Glycyrrhiza* in chronic heart failure patients in China since ancient times.[33] It's important to note that opiates are contraindicated long term in respiratory diseases because they can weaken respiration and contribute to respiratory acidosis. It is unlikely that the opioid herbs *Eschscholzia* or *Corydalis* are problematic, as no adverse events have been reported. Pharmaceutical opioids may be offered as end of life care.

Eschscholzia californica	20 ml
Corydalis cava	20 ml
Glycyrrhiza glabra	18 ml
Aconitum napellus	2 ml

Take 1 dropperful of the combined tincture every 10 minutes, as part of a hospice protocol, under the care of a physician.

Formulas for Hemoptysis

Hemoptysis—expectorating blood from the respiratory system—is a grave finding that is most often due to the presence of a lung lesion, such as a tumor or granulomatous tissue in the lungs, or blood vessel damage. Less worrisome causes such as respiratory infections may involve blood-tinged sputum, but rarely frank blood. Tuberculosis and severe lung fungus such as *Aspergillus*, as well as necrotizing sorts of infections and lung lesions, are leading causes of hemoptysis. Vascular and cardiac pathologies may also cause hemoptysis in advanced stages. Clotting disorders due to anticoagulant therapy, vitamin K deficiencies, and other coagulation issues may also cause hemoptysis. The bleeding most often occurs from the bronchial, subclavian, or internal mammary arteries and is sometimes so severe as to require surgical measures, such as bronchial artery embolization.[34]

In cases of scanty quantities of blood, where no obvious cause or lesion can be detected after a thorough workup and diagnostic imaging, the term *cryptogenic hemoptysis* is used. As many as one-third of cases of hemoptysis may not identify any lesion, and it is presumed that the blood loss is from tiny diffuse lesions, occurring at a low grade, that are not yet possible to visualize. In these cases, lung anti-inflammatories, connective tissue tonics, and folkloric herbs for hemoptysis such as *Achillea, Centella, Usnea barbata,* and *Equisetum,* might be employed, in combination with herbs that are more specific to the particular case. For example, lung antimicrobials may be indicated in some cases, immunomodulators and anti-inflammatories in others, and vascular tonics in others. There are few published studies or clinical trials on herbs for hemoptysis.

Tincture for Hemoptysis Due to Pulmonary Congestion

Digitalis is primarily known as a heart tonic and ionotropic agent, but may also be useful in some cases of cough with hemoptysis when due to pulmonary congestion in weak, asthenic subjects. Cinnamon tannins and other constituents may also be synergistic, astringing capillary beds and supporting the tonic herbs in the formula.

Usnea barbata	20 ml
Stillingea sylvatica	10 ml
Digitalis purpurea	10 ml
Thymus vulgaris	10 ml
Cinnamomum verum (also known as *C. zeylanicum*)	10 ml

As a tonic for those experiencing occasional hemoptysis, take a dropperful of the combined tincture 3 times per day. For those with chronic hemoptysis, take 1 or 2 dropperfuls at a time, every 2 or 3 hours, reducing as bleeding subsides. Omit the *Digitalis* if the resting pulse drops below 55 beats per minute.

Tea for Hemoptysis Due to Tuberculosis

This tea is based on a traditional Ayurvedic formula. Human clinical trials have found this formula to complement traditional therapy and help reduce hemoptysis, cough, and dyspnea, as well as improve body weight.[35] Most, but not all, of these herbs are available from North American sources of supply, and I include this formula due to its reported success and for inspiration to create unique and custom formulas for hemoptysis due to TB. *Tinospora* species are tropical vines and *T. cordifolia* goes by the common name guduchi in Sanskrit and Ayurvedic medicine, where it is used to treat a variety of "heat" syndromes. *Phyllanthus emblica* goes by the name amalaki in Sanskrit and the tree is considered sacred to the Hindus. *Phyllanthus emblica* is used to balance all the bodily energies, the doshas in Ayurvedic medicine, and is often a leading ingredient in rejuvenating formulas.

Phyllanthus emblica
Tinospora cordifolia
Withania somnifera
Glycyrrhiza glabra
Piper longum
Hemidesmus indicus
Saussurea costus (also known as *S. lappa*)
Curcuma longa

Combine equal parts of the dry herbs. If you cannot find a source of supply for any of the herbs listed, it is fine to omit those. You can use more or less licorice to match the flavor preferences of the patient. Gently simmer 1 teaspoon per cup of hot water for 10 minutes, and let stand 10 minutes more. Strain and drink 3 or more cups per day.

Verbascum Simple for Hemoptysis Due to Tuberculosis

Verbascum thapsus (mullein) has a long folkloric history for lung complaints, cough, and tuberculosis. A trial conducted in an Irish hospital in the 1800s, prior to the advent of pharmaceutical remedies, reported *Verbascum* to help six out of seven TB patients.[36]

Verbascum thapsus

Steep 1 tablespoon dried leaves per cup of water and then strain. Drink 4 to 6 cups each day.

Styptic Tincture to Reduce Bleeding from the Lungs

A styptic refers to an agent capable of reducing bleeding due to astringent and other effects on blood vessels and capillary-rich tissues. These herbs are all emphasized folklorically for blood in the sputum or to astringe blood vessels and provide a styptic effect.

Equisetum spp.	15 ml
Centella asiatica	15 ml
Achillea millefolium	15 ml
Tussilago farfara	8 ml
Cinnamomum spp.	7 ml

Take 1 teaspoon of the combined tincture 3 to 6 times daily.

Tincture for Hemoptysis with Influenza

Lycopus is often overlooked in the treatment of influenza, but its astringent effects make it especially indicated for lung congestion with hemoptysis. *Eupatorium* is specific for respiratory virus and influenza, and both *Thymus* and *Tussilago* have antitussive effects, the former being a drying antimicrobial, and the latter being a soothing demulcent antispasmodic.

Lycopus virginicus	15 ml
Eupatorium perfoliatum	15 ml
Thymus vulgaris	15 ml
Tussilago farfara	15 ml

Take ½ to 1 teaspoon of the combined tincture 3 to 6 times per day, reducing as symptoms improve.

Topical Treatment for Hereditary Hemorrhagic Telangiectasia

Hereditary hemorrhagic telangiectasia (HHT), also called Osler-Weber-Rendu disease, is characterized by the presence of abnormally dilated blood vessels, ranging from telangiectasias in mucocutaneous tissues to large arteriovenous malformations in various organs including the lung, brain, and liver. The telangiectasias are prone to rupture, resulting in recurrent nasal or gastrointestinal bleeding that is difficult to control. When in the respiratory tissues or upper airways, such telangiectasis may be a rare cause of hemoptysis and epistaxis.

The pharmaceutical drug propranolol is a nonselective beta blocker routinely used to treat hemangiomas in infants due to its antiangiogenic effects. Propranolol may promote apoptosis in proliferated endothelial cells. In one small clinical study, all six patients with HHT had less epistaxis following the use of a propranolol nasal gel.[37] Apply a 1.5 percent propranolol nasal gel prepared by a compounding pharmacist twice daily to the nostrils for at least 12 weeks, continuing pending progress.

Although theoretical, propranolol and other beta blockers might be employed topically over the chest for hemoptysis due to vascular lesions in the lungs. *Lobelia* is often effective for respiratory spasm, dyspnea, and coughing, and thus it would do no harm to try *Lobelia* oil topically to treat hemoptysis. Lobeline, which is found in *Lobelia*, has similar absorptive and other properties as nicotine, and transdermal absorption of nicotine is excellent.

Formulas for Allergic Rhinosinusitis

Allergic rhinitis is a chronic illness that affects 10 to 40 percent of the worldwide population. Acute rhinosinusitis occurs most often in atopic individuals, and some familial tendency may be evident. Allergic rhinitis is characterized by intermittent or persistent symptoms of nasal congestion, rhinorrhea, nasal/ocular pruritus, sneezing, and postnasal drainage. These symptoms are a result of immunoglobulin E (IgE) mediated allergic inflammation in the nasal mucosa. These symptoms are all due to IgE release in response to the presence of triggering allergens including pollen, mold, dust mites, pet dander, pests (cockroach, mice), and possibly environmental chemicals. Immunoglobulin E drives inflammation, and immunomodulating agents may reduce immune-driven increases in eosinophils, mucus secretion, and cytokine responses, helping to reduce sneezing, nasal obstruction, and rhinorrhea. Intranasal corticosteroids may be prescribed, but they offer only symptomatic relief and will not correct or cure the underlying chronicity. Furthermore, steroids and other long-term medications may result in a wide variety of adverse effects ranging from mild dizziness, pain, and fatigue to more serious headaches, palpitation, nose bleeding, and hypertension and other cardiovascular problems.

Nutritional protocols can often improve underlying atopy. Such protocols include avoiding dairy products and allergens, using antioxidants such as high-flavonoid herbs and supplements, and ingesting high-quality essential fatty acids such as flaxseed oil. Pycnogenol is a trademarked name for antioxidant flavonols derived from the bark of maritime pine (*Pinus pinaster*). Pycnogenol is shown to reduce allergic inflammation.[38] Rosmarinic acid, found in *Rosmarinus*, *Nepeta cataria*, *Perilla frutescens*, and other medicinal herbs, also has antiallergy effects. One human clinical trial reported *Petasites* spp. to be helpful in treating hay fever symptoms as effectively as a prescription antihistamine, but without the drowsiness side effect.[39] *Tanacetum parthenium* (feverfew) is in the same family and has similar chemistry and is also a traditional remedy for headache, allergies, and asthma, with many medicinal effects credited to the sesquiterpene lactones such as parthenolide.[40] *Curcuma* has been used to reduce allergic reactivity with reported success.[41]

A full discussion of halting allergic hypersensitivity is beyond the scope of this volume, but for any allergic disorder, avoid environmental toxins and allergens as much as possible, support gut health, supplement with omega-3 fatty acids (with fish oils), and take various antioxidant supplements. A high intake of butter, nuts, and potatoes correlated to increased incidence of allergic rhinitis in school-age children in one study.[42] Gut and tissue microbiota have an enormous influence on immune modulation, and intestinal flora is increasingly being recognized to have a great impact on autoimmune and allergic phenomena. Gut health, digestion, and probiotics, therefore, should be considered in all cases of respiratory allergy. The complex communities of microorganisms that inhabit the nasal mucosa can also have a profound impact on allergic tendency.[43] Supplementation with *bifidobacteria* reduced allergic rhinitis in children in one clinical study.[44]

Tincture for
Hay Fever with Sinus Congestion

This is my go-to formula that often works quite quickly to improve sinus pain and reduce symptoms of allergic rhinitis. *Armoracia* promotes upper respiratory mucous flow, while *Thymus* dries excessive mucus and provides antimicrobial effects. *Petasites* and *Euphrasia* both offer antiallergy and anti-inflammatory effects and are specifically indicated for hay fever and respiratory allergies.

Petasites hybridus
Armoracia rusticana
Euphrasia officinalis
Thymus vulgaris

Combine the herbal extracts in equal parts and take 1 dropperful of the combined tincture as often as every ½ hour for acute symptoms, decreasing as symptoms improve.

Ephedra Simple Tea

Ephedra is no longer available as a tincture or concentrated capsules, due to abuse of the products. However, *Ephedra* twigs are an age-old remedy for allergic rhinitis and are very safe and gentle when consumed as a daily tea, in the folkloric tradition.

Ephedra spp. twigs

Gently simmer 1 heaping tablespoon per cup of hot water for several minutes, remove from the heat, and let stand in

Ephedra sinica for Respiratory Allergies

Ephedra, ma huang in Traditional Chinese Medicine (TCM), has been used in the treatment of respiratory diseases for more than 5,000 years. Its methylxanthine component, ephedrine, is thus the earliest antiasthmatic agent known. Ephedrine stimulates the release of endogenous catecholamines, resulting in bronchodilation. Ephedrine was commercially available until the 1980s, when *Ephedra* products were removed from the market due to their abuse—as ephedrine capsules were taken to excess as a stimulant or processed into amphetamine-like compounds. High dosages can cause hypertension and heart palpitations, but when used by skilled clinicians, the herb is quite safe. Nonetheless, ephedrine products are difficult to find, so some herbalists gather and make their own ephedrine medicines.

An herbal inhalant based on *Ephedra*, *Zingiber*, and *Syzygium aromaticum* has been used in Korea to treat lung disease, especially chronic obstructive pulmonary disease (COPD). Animal research shows the therapy to reduce inflammatory cytokines and protect against pathological tissue changes.[45]

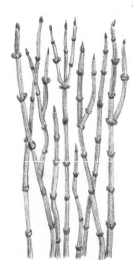

Ephedra sinica, ma huang

Petasites and Allergic Airway Diseases

Petasites (butterbur) is a folkloric plant for asthma, spastic cough, and airway reactivity, with *Petasites hybridus* being traditional in folkloric herbalism of the United States, and *Petasites japonicus* being used for respiratory disease in Asia. Modern investigations show the plant to reduce elevated nitric oxide, to help normalize eosinophil and neutrophil counts, and to inhibit leukotriene and histamine-induced bronchoconstriction, typical of allergic airway disease. *Petasites* also inhibits cyclooxygenase, an enzyme involved with the synthesis of inflammatory prostaglandins. One clinical trial showed *Petasites* to alleviate allergic rhinitis symptoms as effectively as antihistamines,[46] and in a similar trial, as effectively as the prescription medicine fexofenadine (trade name Allegra).[47] Another clinical trial also showed *Petasites* to reduce allergic rhinitis symptoms compared to placebo in as little as 7 days.[48] One small human clinical trial on asthmatics reported *Petasites hybridus* root extracts to reduce the severity, frequency, and duration of asthma attacks.[49] In those using asthma medications and inhalers, 40 percent of study participants required less medication at the end of the second month of study compared to the baseline, and improved spirometry readings could be demonstrated. A review of clinical trials on *Petasites* suggest efficacy for allergy and asthma that warrants further investigation.[50] Petasin and isopetasin in roots and leaves of *Petasites* are pyrrolizidine alkaloids with potential liver toxicity. While generally well tolerated, large and frequent doses of the plant can cause digestive upset, and long-term use may be hepatotoxic. For this reason, *Petasites* is contraindicated in those with prexisting liver disease such as hepatitis C or biliary insufficiency, and its use should be limited to a duration of just several months. *Petasites* can help reduce the severity and frequency of asthmatic episodes relatively quickly, while diet optimization and fatty acid supplementation work more slowly, but can offer long-lasting improvements. Therefore, I often use *Petasites* in the initial month of therapy. As hyperreactivity is improved, *Petasites* may be removed from a formula, relying on other herbs for long-term maintenance. *Petasites* should also be avoided during pregnancy and lactation.

Combining *Petasites* with hepatoprotective herbs such as *Curcuma* or *Silybum* may be wise. Avoid alcohol, acetaminophen, and other potentially hepatotoxic agents when consuming *Petasites*.

Petasites hybridus, butterbur

a covered pan for 10 minutes. Strain and drink 3 or more cups per day. Consuming the tea only in the early part of the day will help avoid sleep disturbance that could result from any potential amphetamine-like effects.

Tea for Allergic Rhinitis

This tea is based on a traditional Korean formula called So-Cheong-Ryong-Tang, also known as Sho-seiryo-to or Xiao-Qing-Long-Tang. The traditional mixed herbal formula has been used for hundreds of years in Asian countries to treat cold-pattern allergic rhinitis in the large airways. Clinical trials have shown the formula to improve hay fever symptoms in selected patients presenting with cold symptoms.[51]

Schisandra chinensis	½ ounce (15 g)
Pinellia ternata	½ ounce (15 g)
Glycyrrhiza uralensis	⅓ ounce (10 g)
Paeonia lactiflora	⅙ ounce (5 g)
Zingiber officinale	⅙ ounce (5 g)
Cinnamomum cassia	1/30 ounce (1 g)
Ephedra sinica	1/30 ounce (1 g)

Combine the dry herbs and gently simmer 1 heaping tablespoon of the blend per cup of hot water for 10 minutes. Remove from the heat and let sit in a covered pan for 10 minutes more. Strain and drink 3 or more cups each day for acute episodes or as a preventive in those with a history of mold sensitivity, seasonal allergies, or other triggers.

Formulas for Acute and Chronic Bronchitis

Bronchitis is an inflammation of the bronchial passages and may be acute or chronic. Acute bronchitis is usually due to an underlying infection, typically a virus initially, although as with rhinitis, the abundant secretions promoted by viruses may lead to secondary bacterial infections. This progression may be extremely rapid in some cases, without accompanying viral symptoms, and the onset of bacterial infections in lungs seems sudden. Rhinoviruses, influenza, coronaviruses, and respiratory syncytial viruses (RSV) are the most common infectious agents associated with acute bronchitis. Acute RSV infections are a common cause of hospitalization in infants and young children. Several potential treatments and vaccines are in development, but there is no specific treatment for RSV lower respiratory tract infection, and herbal and alternative therapies may be extremely useful. *Mycoplasma* and *Chlamydia* strains are also potential causative pathogens, as is *Bordetella pertussis* in unimmunized populations. The symptoms of acute bronchitis include tightness or burning in chest, a productive cough, and possibly fever and general malaise. Particularly high fevers may signal that bronchitis is progressing to pneumonia. Hemoptysis is not common but may occur in those with compromised airways, such as smokers or those with COPD and damaged vasculature. In atopic individuals, infectious bronchitis may trigger wheezing and dyspnea.

In immunocompromised individuals, the elderly, or those with cardiovascular weakness or cystic fibrosis, simple infectious bronchitis may develop into pneumonia, requiring more vigilant monitoring and aggressive therapy. Dyspnea, malaise, and especially hypoxia are worrisome findings in such populations. Chest X-rays may be appropriate to rule out pneumonia when symptoms are severe or rapidly progressing, but are usually not warranted.

Phytoconstituents are useful in deterring viral infections such as the common cold rhinovirus and the H1N1 and other strains of influenza viruses. Antimicrobial herbs with an affinity for the lungs may help deter viral infections and they include *Sambucus* spp., *Allium*, *Lomatium*, *Ligusticum* spp., *Inula helenium*, *Eucalyptus*, or *Thymus*. At times, the treatment protocol can be made fairly aggressive to match the efficacy of antibiotics with the use of tinctures plus teas and encapsulations plus steam inhalation and topical applications. Due to the effort this requires, it is tempting for some to simply use antibiotics; however, antibiotics harm beneficial intestinal microbes and can foster resistant organisms if used repeatedly.

When coughing is pronounced, the use of antitussive herbs in bronchitis formulas is appropriate. Bronchial antispasmodics and anti-inflammatories include *Petasites*, *Prunus serotina*, *Lobelia*, or *Tussilago*. These herbs may improve the efficacy of and offer palliative comfort in herbal formulas. Expectorating and anti-inflammatory additions would be *Grindelia*, *Zingiber*, or *Cinnamomum* spp. Once the airways are inflamed, it can take a month for them to recover, and it is important that patients understand this prognosis. After the acute infection has been resolved, the use of convalescent formulas is appropriate, but the medical community often overlooks or omits this step. Convalescent restorative formulas are particularly useful for those with underlying vascular or respiratory diseases, for the elderly, or for those with any other concerns or risk factors. Many of the herbs used in convalescent formulas have not been researched to any great extent, such that details and mechanisms of action are not yet available. However, they are folkloric standards for coughing and lung infections, and I have found them useful empirically.

Many herbs recommended for acute respiratory infections folklorically are now being studied for their antiviral effects. *Terminalia chebula* binds to host cell surface glycosaminoglycans, inhibiting viruses from binding at the same location, and thereby blocking viral entry into cells and preventing cell to cell spread.[52] *Terminalia* may deter influenza and RSV in this manner. *Scutellaria baicalensis* has antiviral activity because baicalein promotes interferon-driven virucidal effects, as well as impedes viral attachment to host cells.[53] *Pelargonium sidoides* (African geranium) root extracts inhibit viral attachment to cell surfaces and have demonstrated activity against influenza viruses.[54] *Taraxacum officinale* inhibits influenza due to inhibition of viral nucleoprotein RNA levels and polymerase activity.[55] *Glycyrrhiza inflata* chalcones, *Polygala karensium* xanthones, and *Caesalpinia sappan* (also known as *Biancaea sappan*) homoisoflavonoids are all active against influenza A via effects on glycoprotein receptors and binding.[56] *Lonicera* spp. (honeysuckle) are commonly used as a traditional remedy for viral infections in China, and modern research has demonstrated activity against influenza viruses, including the H1N1 flu strain.[57] Indirubin is a compound found in *Isatis tinctoria* (also known as *I. indigotica*) and *Persicaria tinctoria* (also known as *Polygonum tinctorium*) that has antiviral activity, including effects against influenza strains. Indirubin has poor solubility and bioavailability, which may be overcome when combined with other herbs, as is traditional in Chinese formulas for sore throat and flu. Research is underway to prepare semisynthetic forms of indirubin with immunomodulating activity.[58] Agents that may deter RSV include *Actaea cimicifuga* (also known as *Cimicifuga foetida*),[59] *Selaginella uncinata*, and *Lophatherum gracile*.[60] Readers with a good background in Western herbalism will notice the striking lack of Western herbs from these investigations. That is because the Western herbs have not been given any research attention for their antiviral effects, but herbs traditionally used for bronchitis and pneumonia can reasonably be expected to also have a variety of antiviral actions, such as *Lomatium* or *Ligusticum*. *Sambucus nigra* (elderberry) has indeed been shown to interfere with viral attachment to respiratory mucous membranes and is briefly detailed in a formula in this section. *Sambucus* may help prevent and shorten the duration of the common cold[61] and is one of the few Western influenza herbs to receive research attention.

Chronic bronchitis is not infectious in nature, but rather related to long-standing irritation of the airways, such as occurs with smokers or possibly chronic lung irritation from fumes and chemical solvents. Chronic bronchitis often involves fairly constant low-grade symptoms, such as the morning hacking cough of smokers, accentuated by episodic exacerbations and remissions, with the occasional superimposed bacterial infection. Those with chronic lung irritation often have thick sticky or rubbery mucus of scant quantity that will respond best to stimulating expectorants and lung tonics. Eosinophilic bronchitis is another type of chronic bronchitis and is a hypersensitivity phenomenon. Lung cancer may also first present as a chronic cough, and excessive watery expectoration is a sign of alveolar cell carcinoma. All unexplained chronic coughs, wheezing, rhonchi, chest pain, hemoptysis, or unusual expectoration should be evaluated with a chest X-ray. Bronchoscopy and additional imaging techniques should also be performed should any tumors be suspected.

The treatments for chronic bronchitis may be more varied than those for acute bronchitis and should, of course, address the underlying cause. Cough suppressants may supply palliative comfort, but using them is

---

Stimulating Expectorants for Bronchitis

Stimulating expectorants are indicated for chronic problems in which respiratory mucus is scanty, rubbery, or difficult to expectorate. Smokers will often benefit from stimulating expectorants, as will those with emphysema, dry burning sensations, or other chronic inflammatory conditions of the airways.

Asclepias tuberosa	*Ligusticum porteri*
Aspidosperma quebracho	*Marrubium vulgare*
Eriodictyon californicum	*Prunus serotina*
Grindelia spp.	*Sanguinaria canadensis*
Inula helenium	*Sticta pulmonaria*

illogical without also addressing the cause of the cough as directly as possible. Compliance can be a significant issue, but patients must be educated as to the urgency of reducing smoking, chemical exposures, or other direct causes of chronic bronchitis, because any lasting cures may be impossible without this all-important step. While many patients are aware of the association between smoking and lung cancer, it is worth explaining the gradual demise of chronic bronchitis into COPD, for which there is no cure or even significant therapy other than relying on constant use of supplemental oxygen. While certainly not cures, lung tonics

and counterirritant herbs may help to expel old mucus and stimulate circulation in the lungs, and anti-inflammatories such as *Astragalus membranaceus, Curcuma, Centella,* and other herbs may slow or even prevent fibrotic changes in the airways. Antimicrobial herbs are also indicated when frequent colds and episodes of infectious bronchitis complicate recovery, or especially when recurrent episodes of pneumonia and dyspnea occur. As chronic bronchitis devolves into COPD, the sequelae of pulmonary hypertension, bronchiectasis, and other complications may ensue. (See the "Formulas for COPD" section on page 147.)

Tincture for Acute Infectious Bronchitis

Thyme has strong antimicrobial effects,[62] and thymol is among the strongest antimicrobial volatile oils in the Lamiaceae family.[63] *Sambucus* offers significant effects against respiratory viruses. *Sambucus nigra, S. racemosa,* and *S. ebulus* species have antiviral activity credited to the lectins they contain, which are capable of binding to virions and agglutinating virus particles, preventing their ability to penetrate host cells. *Sambucus* lectins also bind to glycoprotein receptors on host cells and interfere with viral docking and viral replication.[64] *Sambucus* is palatable as well, making it an attractive herb to include in cold and flu formulas for children. Garlic inhibits a variety of secondary lung bacterial microbes. In addition to using garlic in tinctures, it can also be taken in capsule form.

Sambucus nigra	15 ml
Ligusticum porteri	15 ml
Thymus vulgaris	15 ml
Allium sativum	15 ml

Take 1 teaspoon of the combined tincture hourly for several days, reducing frequency as symptoms improve.

Encapsulations for Acute Bronchitis

Garlic has demonstrated antimicrobial effects against respiratory pathogens including against *Streptococcus pneumoniae, Bacillus cereus, Pseudomonas aeruginosa,* and *Klebsiella pneumoniae,* as well as a wide variety of other bacteria and microbes.[65] Much of the antimicrobial activity is credited to ajoene and related sulfides, such as allicin.[66]

Bromelain is a well-known cysteine protease from *Ananas comosus* (pineapples), and although most research centers around vascular and musculoskeletal effects, bromelain also reduces mucous production and

---

The Use of Counterirritants in Respiratory Formulas

For long-standing respiratory issues, a philosophical concept in folkloric herbalism is to include herbs considered to be *counterirritants* in the formula. A counterirritant is an herb that has a slightly irritating quality, but when given in a small, physiologically appropriate dose will stimulate immune, circulatory, and tissue responses for stuck, deficient, atonic conditions. Herbs that are warming and stimulating enough to be considered counterirritants to respiratory tissues include *Grindelia* spp, *Lobelia, Stillingia sylvatica, Zanthoxylum clava-herculis, Euphorbia* spp., *Capsicum,* and *Armoracia.* Combine one such counterirritant with nourishing base herbs in formulas for scant expectoration of a tough, tenacious, or rubbery quality. Counterirritants will often provoke new, thin mucous flow, helping to move out old, sticky mucus as well as stimulate the mucociliary "elevator" that helps move mucus upward so it can be expectorated. Counterirritants may be useful in formulas for long-term smokers, chronic bronchitis, emphysema, cystic fibrosis, and other conditions.

inflammatory responses in animal models of asthma and cytokine hyperreactivity.[67] Consider complementing herbal tinctures and teas for allergic congestion of the respiratory passages with bromelain capsules and others, as follows:

Allium (garlic) capsules
Bromelain capsules
N-acetylcysteine (NAC) capsules

Take 2 *Allium* capsules 4 times daily. Take 2 500 to 1,000 milligram bromelain capsules 2 to 4 times daily. Take 1 or 2 NAC capsules 2 to 4 times per day. Note that *Allium* capsules are available in different strengths and preparations such as 500 milligram enteric-coated tablets, or aged garlic for those who experience flatulence or bloating from other products.

Tea for Acute Bronchitis

I keep this basic respiratory tea on hand at all times. I mix it in a large plastic bag, using 1 pound quantities of each herb. I then weigh out the mixtures into 4- and 8-ounce sizes. This formula covers the bases to address coughing and bronchial irritation due to a variety of causes.

Thymus vulgaris
Eucalyptus globulus
Mentha piperita
Glycyrrhiza glabra
Achillea millefolium
Lobelia inflata
Tussilago farfara

Combine the dry herbs in equal parts; you can adjust quantities of *Mentha* and *Glycyrrhiza* to taste. Steep 1 tablespoon of the herb mixture per cup of hot water and then strain. Drink as much as possible through the day.

Acute Bronchitis Tincture from Bill Mitchell, ND

Bill Mitchell, who was a close friend and teacher of mine, was one of the founders of Bastyr University in Seattle, Washington, and professor of botanical medicine. This is his acute bronchitis formula.

Grindelia spp.	20 ml
Marrubium vulgare	20 ml
Sanguinaria canadensis	10 ml
Lobelia inflata	10 ml

Take 1 dropperful of the combined tincture as often as every 15 to 30 minutes, reducing as symptoms improve.

Drying Agents for Over-Copious Mucus Production

Drying agents are useful for new onset of acute infections, or for allergic people and asthmatics who produce abundant free-flowing mucus. In many cases, excessive mucous production triggered by allergic reactivity or recent viral infections creates a fertile breeding ground for bacteria, and secondary bacterial infection ensues. The drying antimicrobial herbs listed here are excellent choices to include in herbal formulas in such cases.

Hydrastis canadensis	*Myrica cerifera*
Hyssopus officinalis (also known as *H. vulgaris*)	*Rosmarinus officinalis* *Salvia officinalis* *Thymus vulgaris*

Acute Bronchitis Convalescent Tincture

Inula helenium (elecampane) is a traditional remedy for lung complaints and lung infections. Research suggests that the plant is active against methicillin-resistant *Staphylococcus aureus* (MRSA), various gram-positive bacteria, yeasts, parasites, and *Mycobacterium tuberculosis*.[68] Much of the medicinal activity is credited to alantolactone and isoalantolactone. *Inula* is also noted to protect the lungs against sepsis-induced and inflammatory injuries. *Allium* has antibacterial effects and helps keep mucus thin, and *Glycyrrhiza* has anti-inflammatory and immunomodulating effects. Both *Panax* and *Glycyrrhiza* offer adrenal supportive effects, which may enhance recovery following illness and help a person regain energy and vitality.

Panax ginseng	15 ml
Inula helenium	15 ml
Allium sativum	15 ml
Glycyrrhiza glabra	15 ml

Take 1 teaspoon of the combined tincture 3 to 6 times daily for several weeks, reducing to 2 or 3 times daily as recovery ensues. Elderly or weakened individuals may use this formula for several months following acute illnesses.

Tincture for Chronic Bronchitis in Smokers

Because smoking damages the airways and the cilia on the mucosal cells and tends to reduce mucous quality and quantity, a formula for smokers should include some of the warmer, stimulating expectorants such as *Grindelia* and *Capsicum*, both also considered to be "counterirritants." Antimicrobial agents are not needed in this particular formula, which instead features many anti-inflammatory and immunomodulating herbs.

Curcuma longa	15 ml
Ginkgo biloba	10 ml
Scutellaria baicalensis	10 ml
Petasites hybridus	10 ml
Grindelia squarrosa	10 ml
Capsicum spp.	5 ml

Take 1 dropperful of the combined tincture as often as hourly, reducing as symptoms improve.

Tincture for Chronic Eosinophilic or Allergic Bronchitis

Those with reactive airways can develop bronchospasms and excessive mucous production following exposure to irritants, chemicals, fumes, and smoke. All of the herbs in this formula are noted to help reduce allergic reactivity, with the fennel providing antispasmodic effects.

Curcuma longa	20 ml
Ginkgo biloba	15 ml
Foeniculum vulgare	15 ml
Zingiber officinale	10 ml

Take 1 dropperful of the combined tincture as often as hourly, reducing as symptoms improve.

Formulas for Cystic Fibrosis

Cystic fibrosis is a congenital inherited disease affecting primarily Caucasians. It results from a defect in the body's exocrine glands, affecting digestive and respiratory secretions and leading to lung disease and other pathologies; the disease is associated with a high mortality rate. Research in recent decades has identified that the cystic fibrosis transmembrane conductance regulator (CFTR) is altered by a single gene defect,[69] and ongoing investigation is seeking methods of attenuating dysregulated sodium, chloride, and bicarbonate ions, which leads to dehydration of the airway surface liquid and unusually thick, tenacious respiratory secretions. Resveratrol and other flavonols found in *Vitis* species (grapes) may block excessive CFTR activity and inhibit excessive chloride flux.[70] Gastrointestinal pancreatic secretions are also insufficient with cystic fibrosis and can lead to liver disease, as well as promote general digestive issues. *Rhodiola* species may attenuate secretory diarrhea via effects on CFTR and may improve digestive symptoms in cystic fibrosis.[71] The pancreatic and biliary insufficiency of cystic fibrosis can lead to fibrotic changes in the liver and cirrhosis, and impaired liver function can contribute to the development of diabetes. Inflammation of the liver can lead to portal hypertension. The sweat glands are also affected, and sampling of sweat may show characteristically altered electrolytes and is a diagnostic marker for cystic fibrosis.

The chronic obstruction of the lungs by thick, viscous mucus increases the susceptibility to *Staphylococcus aureus* infections and, as the disease progresses, to more serious pathogens such as *Pseudomonas aeruginosa* and possibly *Burkholderia cepacia*, which is associated with a poor prognosis. Disease progression involves a vicious cycle of airway obstruction triggering inflammation and excessive secretions that are difficult to expectorate. This in turn invites pathogens that lead to epithelial and lung tissue damage. As the airways are damaged, bronchial hypersensitivity may occur, and chronic inflammation leads to bronchiolitis and bronchiectasis. As alveoli become damaged, hypoxemia may ensue, and increased airway resistance causes pulmonary hypertension, putting great strain on the heart and leading to heart disease.

Herbal therapies for cystic fibrosis can have multiple goals: thinning respiratory secretions, preventing airway changes, reducing the risk of lung infections, protecting the liver from fibrosis, guarding against hypertension and diabetes, and optimizing digestive secretions. Although difficult expectoration is the most obvious symptom to address, herbal formulas should be aimed at preventing all long-term issues and risks of cystic fibrosis.

Many people who have cystic fibrosis may respond to bronchodilators, from inhalers to herbal teas and tinctures. Garlic, N-acetylcysteine (NAC), and other mucolytic agents and expectorants are indicated to palliate daily symptoms, as well as reduce the risk of lung infections. Among the herbs most noted to deter opportunistic lung infections are *Ligusticum*, *Lomatium*, *Terminalia* spp., *Allium*, *Thymus*, *Origanum vulgare*, and oil of oregano and other essential oils. Essential oil inhalation may be added to the daily routine during the morning shower, at midday, and before bed.

The use of sputum cultures helps to assess the resident microbial flora of the lungs, and more aggressive lung infection therapies may be implemented as warranted. Airway anti-inflammatories and vascular protectants such as *Angelica sinensis, Ginkgo, Curcuma, Petasites, Crataegus* spp., *Astragalus,* and *Scutellaria baicalensis* may help prevent vascular and airway damage. Similarly, lifelong liver support with *Pueraria montana* var. *lobata, Silybum marianum, Curcuma, Berberis,* and lipotropic and berberine supplements may offer hepatoprotection. (For more details on liver support formulas, refer to chapter 4 of Volume I.)

Croton lechleri (sangre de drago) is an Amazonian tree that exudes large quantities of blood red resin when the bark is sliced. Products containing the resin are increasingly becoming available in the United States. *Croton lechleri* contains a proanthocyanidin named crofelemer, shown to affect epithelium transport proteins including CFTR. Inhibition of CFTR by crofelemer results in stabilization of chloride-gated ion channels,[72] which may decrease secretions and thereby benefit cystic fibrosis patients. Crofelemer and whole *Croton lechleri* resin products are also well-established remedies for secretory diarrhea.

Croton lechlerii Tincture for Cystic Fibrosis

In addition to *Croton lechleri,* this formula contains an Asian species, *Rhodiola kirilowii,* which is also shown to stabilize CFTR fluid transport in a manner that attenuates secretory diarrhea. It can be included in protocols for the intestinal effects on cystic fibrosis.[73] Both herbs may require a bit of searching, but would be worth the effort.

Croton lechlerii	60 ml
Rhodiola kirilowii	60 ml

Take ½ teaspoon of the combined tincture 3 or 4 times daily, using more or less as needed as respiratory symptoms wax and wane.

Cystic Fibrosis Expectoration Tincture

These herbs are basic expectorants and are not unique to treating cystic fibrosis. Due to the nature of the condition, expectorants are an important part of the daily routine. These herbs offer antimicrobial effects as well, reducing the risk of lung infections.

Allium sativum	10 ml
Petasites hybridus	10 ml
Eriodictyon californicum	10 ml
Lobelia inflata	10 ml
Grindelia squarrosa, G. robusta	10 ml
Armoracia rusticana	10 ml

Take 1 teaspoon of the combined tincture as often as every ½ hour when thick tenacious mucus is present; reduce the frequency on days when the lungs are relatively clear.

Pulmonary Protection Tincture for Cystic Fibrosis

All the herbs in this formula have been researched to provide anti-inflammatory and antifibrotic effects on the lungs and are safe and gentle to use on a daily basis, long term. There is no research on herbs specifically for cystic fibrosis, but because these herbs have been shown helpful in other conditions and in research models of fibrotic degeneration, they will certainly do no harm.

Ginkgo biloba	10 ml
Astragalus membranaceus	10 ml
Scutellaria baicalensis	10 ml
Curcuma longa	10 ml
Crataegus spp.	10 ml
Zingiber officinale	10 ml

Take 1 teaspoon of the combined tincture 3 to 6 times daily.

Hepatoprotective Tincture for Cystic Fibrosis

These are among the most studied hepatoprotective herbs, and this formula is also appropriate for hepatitis or other liver assault.

Silybum marianum	15 ml
Curcuma longa	15 ml
Berberis aquifolium	15 ml
Glycyrrhiza glabra	15 ml

Take 1 teaspoon of the combined tincture 3 to 6 times a day. The lower dosages are appropriate for general liver support and prevention, while higher doses are indicated for actual liver signs or symptoms.

Formulas for Emphysema

Emphysema is a type of obstructive pulmonary disease that involves the gradual loss of functional alveolar cells as they enlarge and lose their tightly clustered form, which reduces the surface area for oxygen assimilation and causes the lungs to look empty in chest X-rays. The elastic quality of lung parenchyma is lost, and the structural integrity of the lungs is greatly impaired due to massive loss of parenchyma and alveoli. This can lead to collapse of the airways, causing bronchiectasis and the formation of large bullae. Spontaneous pneumothorax may occur. Due to the loss of the natural elastic properties of the lungs, people with emphysema must instead use chest muscles to support inspiration and expiration, leading to caloric expenditure, weight loss, and exhaustion. A buildup of the small intercostal accessory muscles of the chest often gives emphysema patients a barrel-chested appearance. Mucous secretion can become excessive in response to the inflammation, but be difficult to expectorate due to hyperinflation of the lungs and loss of elastic recoil. This can lead to formation of solid mucous plugs and occurrence of superimposed microbial infections. Pneumonia can be fatal in those with emphysema.

A genetic deficiency in α-1 antitrypsin results in decreased protease inhibitor activity in the serum and predisposes to emphysema, as well as to liver disease. α-1 antitrypsin is produced in the liver and protects elastic tissue in lungs, among other functions. A deficiency in this protein may not result in serious disease in non-smokers, but when smoking or other respiratory assault creates inflammation and toxic stress in the lungs, those genetically predisposed may develop lung damage at a young age. Liver disease may also result when the α-1 antitrypsin enzyme is abnormally formed and secreted, accumulates in the liver, and damages hepatic cells.[74] Long-term exposure to respiratory irritants including smoking, asbestos, volatile fumes, mineral dust, and other paints and chemicals can contribute to emphysema. Because there is no cure for emphysema, people should be educated as to the consequences of such exposure. Therapy for emphysema is mainly palliative, although herbal medicine may offer connective tissue tonics with the goal of preserving what functional lung tissues exists. Oxygen therapy is often needed, and needed more constantly as the disease progresses. Oxygen therapy may also slow deterioration in the lungs because chronic hypoxia contributes to the development of pulmonary hypertension and puts a strain on the heart due to increased vascular resistance; this leads to cor pulmonale. (See "Formulas for Cor Pulmonale" on page 154.)

Expectorants, connective tissue tonics, breathing therapies, hydrotherapies, immune modulation, and respiratory antimicrobials may all be offered to patients with emphysema to improve comfort and oxygen delivery and to thwart opportunistic infections. High doses of green tea catechins and other flavonoids such as curcumin, quercetin, and resveratrol may reduce inflammation, and several have been noted to normalize elastase enzymes typically elevated in emphysema. Thiol molecules from *Allium sativum* (garlic), *Allium cepa* (onions), *Armoracia rusticana* (horseradish), and *Raphanus* (radishes) may offer mucolytic effects as well as reduce infection risk and may be included in the diet as well as in herbal medicines. Serine protease inhibitors may reduce elastase activity and protect elastic fibers, preventing the loss of elastic recoil and attenuating emphysema.

Daily Emesis and Hydrotherapy Routine for Emphysema

Those with emphysema should be encouraged to breathe steam through the mouth for 10 to 15 minutes upon arising each morning, while taking the Expectoration Tincture (page 137) every few minutes, nearly to the point of nausea. The use of emetics such as mustard powder or *Lobelia* are folkloric methods of clearing the lungs of mucous plugs, because the vomiting reflex will strongly stimulate lung expectoration as well, but the therapy may be too harsh and debilitating for weak or elderly patients or simply unappealing to others. However, for those who are willing, the use of emetics and vomiting is sometimes helpful in clearing stagnant mucus and deterring opportunistic infections. At dosages below the nausea and vomiting threshold, *Lobelia* and *Capsicum* can help stimulate saliva flow and the mucociliary apparatus of the lungs and facilitate expectoration without actually vomiting. Gentle but vigorous pounding of the back and chest by a family member may also help loosen settled and stuck mucous plugs and enhance the therapy. This whole procedure is best carried out in the bathroom so that the person can spit or even vomit into the toilet as secretions are freed and able to be expectorated. Carry a big pot of steaming water into the bathroom, set in the bathroom sink, and allow a person to breathe deeply over the steam. The person might also take a hot shower immediately upon arising to help start the process and get the bathroom humid and steamy, then follow with steam inhalation. Commercial facial and sinus steamers are available and worth the minimal investment to make this morning routine convenient and easy to comply with.

Expectoration Tincture for Emphysema

No research on herbs for emphysema has been carried out, and thus this formula is based on folkloric notions. It combines some of the most powerful expectorating agents. This tincture could be used throughout the day, but may be especially beneficial to clear the night's accumulation of mucus in the lungs. Upon waking, drink several large glasses of water, shower, inhale moist steamy vapor for another 5 to 10 minutes, while taking frequent doses of this tincture in an effort to expectorate as much mucus from the airways as possible.

Grindelia spp.	15 ml
Armoracia rusticana	15 ml
Lobelia inflata	15 ml
Iris versicolor	10 ml
Capsicum annuum	5 ml

Combine the herbal tinctures and take an entire dropperful of the combined tincture every 3 to 5 minutes for a half an hour while breathing in steam, coughing, and spitting into the toilet. Reduce the frequency should nausea occur; however, those with enough constitutional and muscular strength to be physically able to vomit may more easily rid themselves of mucus.

Using Essential Oils in Nebulized Form

Biofilms are bacterial colonies that adhere to the surfaces of tissues or even to nonliving surfaces such as metals and plastics and can be a constant source of infection. Biofilms can be difficult to eradicate with antibiotics because the biofilm acts as a physical barrier, preventing drug absorption, and can quickly mutate and become resistant to antibiotics. Patients with endotracheal tube ventilation are subject to bacterial biofilm colonization of the airway device, leading to ventilator-associated pneumonia, with *Klebsiella pneumoniae* and *Acinetobacter baumannii* being the most frequently isolated bacteria. *Pseudomonas aeruginosa* may also form biofilms in the airways and may result in pneumonia.

Nebulized medications may help deliver antimicrobial agents into the airways where they may have more direct effects against biofilm-related bacterial infections and reduce the incidence of pneumonia. Essential oils are excellent agents to use daily in nebulized form such as via inhalation, steam, and hydrotherapy due to numerous anti-inflammatory and antimicrobial effects. Furthermore, essential oils may permeate tissues readily due to their volatile nature and fat-soluble molecules. Nebulized eucalyptus oil is shown to reduce pathogenic bacterial populations and help reduce infections.[75] For COPD patients, essential oil inhalation may reduce cytokine release and production of reactive oxygen species.[76]

Essential oils can cleanse cellular receptor sites of medications, petrochemicals, and other disruptors of intercellular communication and can chelate heavy metals and other toxins, helping to remove and flush them through the kidneys, lungs, skin, colon, and liver. In fact, essential oils increase the body's ability to absorb nutrients and vitamins.[77] Essential oils most useful for the respiratory system include *Eucalyptus globuls* (eucalyptus), *Mentha piperita* (peppermint), *Mentha spicata* (spearmint), *Thymus vulgaris* (thyme), *Pimpinella anisum* (anise), and *Foeniculum vulgare* (fennel), which offer bronchodilating, decongestant, expectorant, anti-inflammatory, and antimicrobial effects. Inhalation of spearmint volatile oil is reported to increase lung capacity in healthy subjects,[78] as shown by spirometry assessment. Sandalwood, geranium, and *Ravensara aromatica* essential oil may alleviate the symptoms of allergic rhinitis in adults, improving total nasal symptom score, and are especially useful in alleviating nasal obstruction and fatigue in patients with perennial allergic rhinitis.[79]

Long-Term Lung Support Tincture

This formula is aimed at reducing inflammation in the lungs with *Ginkgo* and *Curcuma*, while supporting the connective tissue of the existing lung parenchyma with *Equisetum* and *Centella*. The lungs of those suffering from emphysema need all the connective tissue support possible, and this tincture could be used in tandem with the Tea for Emphysema that follows.

Ginkgo biloba	15 ml
Panax ginseng	15 ml
Curcuma longa	10 ml
Equisetum spp.	10 ml
Centella asiatica	10 ml

Take 1 dropperful of the combined tincture, 3 or more times daily.

Tea for Emphysema

Centella, *Equisetum*, and *Rosa canina* are connective-tissue tonics, combined in this tea with anti-inflammatories *Camellia*, *Glycyrrhiza*, *Petasites*, and *Zingiber*. *Zingiber*, *Glycyrrhiza*, and *Thymus* may also help to protect against opportunistic infections. This is just one example of a broad-spectrum tea for emphysema; many other options exist.

Camellia sinensis
Centella asiatica
Equisetum arvense
Thymus vulgaris
Rosa canina
Glycyrrhiza glabra
Petasites hybridus
Zingiber officinale
Eucalyptus globulus

Combine the dry herbs in equal parts and steep 1 heaping tablespoon per cup of hot water. Steep 10 minutes, strain, and drink 3 or more cups per day.

Encapsulations for Emphysema

As muscle fibers are lost from the airways, airspaces enlarge, and it becomes increasingly difficult to clear stagnant mucus from the alveoli. The following encapsulations can help thin the mucus and protect against further inflammation and secondary infection in emphysema patients.

Bromelain
NAC
Resveratrol
Curcumin
Immune support

Take at least 1 capsule of each, 3 times per day. Add any of the bronchitis or pneumonia antimicrobial formulas in this chapter as needed, immediately at the onset of the slightest cold.

Formulas for Pleurisy

Pleurisy is an inflammation of the pleural membranes of the lungs, causing characteristic pain with inspiration or the least respiratory excursion. Pleurisy is most commonly due to viral infections such as coxsackieviruses and, less commonly, echoviruses. Patients with pleuritis attempt to limit the motion of the lungs when coughing, bracing the ribcage with their hands, and physicians can hear friction rubs (high-pitched squeaky sounds) through a stethoscope with each inspiration. Pleurisy may involve other viral infection symptoms such as fever and sore throat, but these may just as often be absent. Chest X-rays can help rule out pulmonary embolism or distinguish between pneumonia and other causes of chest pain when necessary.

Herbal therapies for pleurisy are antiviral agents, anodyne options, and anti-inflammatory herbs with an affinity for the lungs. *Asclepias tuberosa*, aptly named pleurisy root, is a folkloric herb emphasized for the symptoms of pleurisy, as well as bronchitis and consolidation in the lungs. *Lomatium*, *Ligusticum*, *Thymus*, and *Allium* are all appropriate antimicrobial options to include in herbal formulas, and *Bryonia* is specific for pain that is worse with motion, typical of pleurisy. *Hypericum perforatum* (St. Johnswort) can help allay acute neuralgic pain, as well as offer anti-inflammatory and antiviral support. Topical approaches, even simple heat packs, are extremely helpful for treating the discomfort of pleurisy. Hot applications may be complemented by the use of *Hypericum* oil, castor oil, or essential oils such as choices rich in menthol or camphor, placed over the affected area, and exemplified in the following formulas.

Tincture for Pleurisy

Asclepias and *Bryonia* are considered specific for the symptoms of pleurisy. This tincture combines them with general lung infection choices *Lomatium* and *Allium*. *Lobelia* may help quiet painful coughing, and *Hypericum* may help with pain management as well.

Asclepias tuberosa	15 ml
Lomatium dissectum	10 ml
Allium sativum	10 ml
Hypericum perforatum	10 ml
Lobelia inflata	10 ml
Bryonia alba	5 ml

Take 1 teaspoon of the combined tincture every ½ to 1 hour, reducing frequency as symptoms improve.

Topical Herbal Oil for Pleurisy

A topical application can help reduce pain and inflammation of the pleural membranes. This example uses *Ricinus* (castor) oil as part of the base, which may help to drive the essential oils into tissue. *Lobelia* oil would be another valuable herb to include in this formula, but must be made for one's self as there are no commercially available products. A recipe follows.

Ricinus communis oil	20 ml
Hypericum perforatum oil	20 ml
Mentha piperita essential oil	10 ml
Eucalytpus globulus essential oil	10 ml

Combine the oils in a 2-ounce (60 ml) bottle and label. Shake well prior to each use to homogenize, because the essential oils will separate out to the top. Gently rub liberal amounts of the combined oils into the entire torso—chest, back, and both sides of the rib cage, and cover with heat, as detailed in the following formula. Repeat 4 or more times daily, reducing as symptoms improve.

Lobelia Oil for the Lungs

Lobelia is an effective herb to allay tightness in the lungs and spasmodic coughing, and *Lobelia* leaves are included tea and tincture formulas throughout this chapter. *Lobelia* oil is not commercially available, but when topically applied to the chest and back, is an excellent complement to teas, tinctures, and syrups used to treat chest pain and severe coughing in cases of bronchitis, COPD, pleurisy,

and other conditions. I have seen *Lobelia* oil improve dyspnea immediately, albeit temporarily, in those with chronic pulmonary disease.

Lobelia inflata	8 ounces (240 g)
Olive oil	3 cups (720 ml), or more as needed
Castor oil	3 cups (720 ml), or more as needed

Place the dry *Lobelia* leaves in a high-speed blender and add 3 cups each of olive and castor oils. Blend on low speed, and then increase to high speed to achieve the smallest particulate size possible. Transfer to a gallon-size glass jar and allow to sit undisturbed overnight or for 12 hours. The plant material should slowly sink to the bottom of the jar and be completely covered with oil—molding will occur if the plant material is exposed to air. Add an additional cup or two of oil, if necessary, to ensure the plant material is fully submersed. Shake the jar each day, allowing the *Lobelia* to settle out. Strain out the plant material after 6 to 8 weeks, and transfer the resulting *Lobelia* oil to individual 8-ounce stock bottles. Use *Lobelia* oil topically on the back and chest to allay spastic cough or to improve respiration in those with COPD, or include in chest oil formulas such as the Topical Herbal Oil for Pleurisy.

Hot *Lobelia* Pack

Lobelia is effective to quiet the cough reflex and can also help when topically applied. Vinegar helps draw out the active alkaloids, such as lobeline.

Water	2 cups (480 ml)
Vinegar	2 cups (480 ml)
Lobelia inflata leaves, dry	1 cup (240 g)

Combine the water and vinegar and bring to a simmer. Turn off the heat and add the *Lobelia* leaves, allowing them to steep for 15 minutes. Strain out the herbs. Use the fluid to prepare the moist hot pack by soaking an old small towel in the infusion. Place the wet towel over the most painful area of the chest and lungs, cover with plastic wrap, and add heat, such as a heating pad, another hot moist towel, or hydrocollator pack. Leave the pack in place for 20 to 30 minutes. The *Lobelia* pack combines well with the Topical Herbal Oil for Pleurisy. Oil the person's skin first, then apply the hot *Lobelia* pack, then cover with heat. Repeat every few hours during the acute phases of pleurisy.

Treating Hyperventilation Syndrome

Hyperventilation usually occurs as part of a panic attack or anxiety syndrome, but may also result from the use of stimulants or amphetamine abuse. Hyperventilation is more common in women, particularly younger ones, but may affect anyone. The condition may come on suddenly—people experience air hunger and feel as though they are suffocating, causing gasping and rapid breathing. The abnormal breathing itself dysregulates blood gases, and paresthesias and tetany of the fingers may result as hypoxia ensues, along with a dizzy sensation, faintness, or actual fainting. The classic remedy of breathing into a paper bag increases the partial pressure of carbon dioxide and usually corrects the blood gas imbalance in a matter of minutes. Less dramatic cases may also occur when respiratory alkalosis deranges blood calcium and phosphorus levels and causes people to sigh deeply and often, as with Kussmaul breathing. Such cases require investigation into systemic acid-base balance and metabolic functions, as well as renal, hepatic, and vascular function.

When hyperventilation is associated with a general anxiety disorder or acute stress reactions such as adrenaline surges, nervine herbs and lifestyle approaches are indicated. Encouraging and supporting patients to implement daily meditation, exercise, and other relaxation techniques help to address the underlying anxiety state. Adrenal supportive herbs may regulate cortisol activation and feedback pathways, and *Panax*, *Glycyrrhiza*, or *Eleutherococcus senticosus* may serve as useful base herbs in formulas. Nervine herbs such as *Actaea*, *Scutellaria lateriflora*, *Melissa officinalis*, or *Avena sativa* may also prove useful to calm stress reactions.

Nervine Tincture for Those Prone to Hyperventilation

This general adaptogen and nervine formula may improve hyperventilation associated with panic disorders and general anxiety disorders.

Eleutherococcus senticosus	20 ml
Melissa officinalis	15 ml
Scutellaria lateriflora	15 ml
Glycyrrhiza glabra	10 ml

Take 1 teaspoon of the combined tincture 3 to 6 times per day at times of stress-induced breathing disorders.

Formulas for Pneumonia

Pneumonia is an acute lung infection that can be caused by viruses, bacteria, mycobacteria, fungi, and even parasites. Prior to the advent of antibiotics, pneumonia was a leading cause of death worldwide. The disease remains a serious cause of mortality, especially in developing countries, and in patients with preexisting heart and lung diseases, as discussed throughout this chapter and in chapter 2. The underlying pathogens that cause pneumonia are closely related to a person's age and situation. Bacterial pneumonia is most common in adults, and viral pneumonia is most common in infants and toddlers. *Streptococcus pneumoniae* is the most common bacterial culprit, followed by *Haemophilus influenzae* and *Chlamydia pneumoniae* (also known as *Chlamydophila pneumoniae*). *Mycoplasma pneumoniae* is another possible pathogen and occurs most commonly in otherwise healthy adolescents and young adults. It is sometimes referred to as "walking pneumonia" because, although somewhat ill with a cough and lung symptoms, those afflicted may feel well enough to be active and walking around. In infants and young children, the most common viral pathogens are cytomegalovirus, respiratory syncytial virus (RSV), and parainfluenza. In many cases, the original pathogen cannot be clearly identified, and diverse secondary opportunistic bacterial infections become superimposed in the inflamed and congested airways.

Those predisposed to pneumonia, such as individuals with COPD, cor pulmonale, emphysema, or immunodeficiency, might be helped by herbal immune support, as well as by having a more specific respiratory antimicrobial protocol at the ready. The causative microbes are endemic in the general environment, especially for those in care facilities and hospitals. Strengthening the immune system to help confer resistance to common pathogens is advised for the elderly and at-risk patient populations who may develop pneumonia following simple colds. Cigarette smoking, diabetes and high sugar intake, and poor nutritional status may also predispose to pneumonia, even among those who do not have reactive airway disease and underlying heart and lung disease.

Pneumonia involves consolidation of mucus in large areas of the lungs and is thereby associated with more chest pain than bronchitis is. It typically involves severe coughing, fever, and malaise. Antibiotic therapy is appropriate in many cases, especially for the at-risk populations mentioned, but herbal therapies can still be complementary and used in tandem. Infected lungs, even in otherwise healthy children and adults, can take a full month or two to repair and recover. As with bronchitis, convalescent therapies are useful but frequently overlooked to help shorten the recovery time and support a person's energy, quality of life, and immune status while recovering. Hydrotherapy can complement the herbs, and simple steam inhalation deters microbes and helps loosen and expectorate the consolidated fluid in the lungs. Essential oils may be added to steams, heat packs, chest rubs, and *Lobelia* packs, as described in the section on "Creating Topical Applications for Lung Complaints" on page 109. Such topical applications are highly recommended, both for comfort and to add additional therapy to speed recovery. The indicated herbs are the same as those discussed for bronchitis and rhinitis.

Herbal Tincture for Pneumonia

Similar to the formulas exemplified in the bronchitis section, all the herbs in this formula have antimicrobial effects. *Allium* is antibacterial and expectorating, *Ligusticum* is antiviral, *Phytolacca* is immune-stimulating, and *Thymus* is antibacterial, drying to excess mucus, and antispasmodic to help allay vigorous coughing. Complement this formula with essential oils and steam inhalation, a tea, and a topical therapy.

Thymus vulgaris	15 ml
Allium sativum	15 ml
Ligusticum porteri	15 ml
Phytolacca decandra	15 ml

Take 1 teaspoon of the combined tincture every ½ to 1 hour, reducing as symptoms improve.

Pneumonia Simple from Bill Mitchell, ND

Dr. Bill Mitchell was one of the founders of Bastyr University where he taught botanical medicine courses for many decades. He recommended *Bryonia* at the onset of pneumonia, reporting that the herb can halt the exudative process and promote the absorption of accumulated fluid. This style of using potent herbs diluted in water and consumed by the spoonful was commonly employed by the Eclectic physicians of the late 1800s through the early 1900s.

Bryonia alba	
Water	¼ cup (60 ml)

Place 10 drops of *Bryonia* in the water and take ½ teaspoon every hour at the onset of acute respiratory infections. Typically, the dosage could be reduced to every 2 or 3 hours for several days, and then to 5 or 6 times a day until the acute nature of the infection has passed.

Pneumonia Tea from Cascade Anderson Geller

Cascade was another one of my teachers, known for her wise-woman ways and gentle, lifestyle approaches to healing. Her herbal formulas were often aimed at nourishing the body, and her pneumonia formula would be a useful preventive approach for those at risk of pneumonia or as a convalescent therapy.

Usnea spp.	1 ounce (30 g)
Sticta pulmonaria	1 ounce (30 g)
Verbascum thapsus	1 ounce (30 g)
Tussilago farfara	1 ounce (30 g)
Glycyrrhiza glabra	1 ounce (30 g)
Lomatium dissectum	1 ounce (30 g)

Combine the dry herbs, and steep 1 heaping tablespoon per cup of hot water. Strain and drink freely, 3 or more cups per day until the infection subsides.

Steam Inhalation for Chest Pain Due to Pneumonia

This combination of essential oils can be mixed when needed, or preblended into a 1-ounce or ½-ounce bottle.

Eucalyptus essential oil	10 ml
Mentha essential oil	10 ml
Lavendula essential oil	10 ml

Dispense 10 to 20 drops of the essential oil blend onto the surface of a large pot full of freshly boiled, steaming water, and set the pot in an accessible location, such as on a tabletop, so the person may sit comfortably for 15 minutes with their head over the pot. Inhale the vapors through the mouth using a towel over the head to keep the local area as humid as possible. Keep the eyes closed, because the evaporating essential oils can sting and irritate the eyes. Repeat hourly if possible throughout the day during acute phases of lung infections, reducing to 2 or 3 times per day as recovery ensues. This therapy may be complemented by placing the feet in a bucket of hot water to encourage free sweating and promote general circulation—a type of hydrotherapy.

Herbal Convalescing Lung Recovery Tincture

This formula is an example of how to support the lungs and general vitality following recovery from an acute infection. These herbs are less powerful antimicrobial agents than those in the Herbal Tincture for Pneumonia, and instead offer anti-inflammatory, circulatory-enhancing, and immunomodulating effects to support

repair and regeneration of the lungs, as well as to support general energy and strength for those who make have been weakened or hospitalized.

Astragalus membranaceus	15 ml
Panax ginseng	15 ml
Tanacetum parthenium	15 ml
Ginkgo biloba	8 ml
Allium sativum	7 ml

Take 1 teaspoon of the combined tincture 3 to 6 times daily, reducing as recovery progresses. Some individuals may find it helpful to continue to take this tincture twice a day for 2 to 3 months following recovery.

Formulas for Asthma

Asthma is a chronic inflammatory disease—described by Hippocrates and other early medical authors—that tends to be a lifelong condition. Asthma prevalence ranges from 1 to 18 percent in various countries and is considered a significant global health issue. Asthma has increased in both prevalence and incidence in recent decades, in part due to pollution and environmental influences. The worldwide number of asthma deaths is about 180,000 per year,[80] with a wide variation among continents, regions, and age and economic groups. Interestingly, asthma incidence may be increased among higher economic tiers, and some studies have shown that lack of exposure to basic environmental microbes in early life is associated with an increased incidence of asthma.[81] While exposure to basic microbes appears beneficial, heavy exposure to allergens and environmental pollutants may induce hyperreactivity via upregulation of proinflammatory genes. It is estimated that more than 60 percent of asthmatics also have allergic rhinitis and that at least 10 percent have chronic sinusitis, hinting at underlying atopic constitutional tendencies. While some children "outgrow" asthma, other cases may worsen over time and necessitate frequent emergency-room visits for nebulized steroids when asthmatics fail to respond to their medications and inhalers.

Typical features of asthma include shortness of breath, wheezing, airway hyperreactivity, abundant airway mucous production, and structural changes in the airways. The typical symptoms of asthma are episodes of wheezing, dyspnea, coughing, and a sense of tightness in the chest, often with identifiable individual triggers, such as exposure to dust, pollen, or other allergens, as

well as hypersensitivity to stress, cold air, and exercise. Other allergic and atopic phenomena such as eczema, hay fever, or chemical sensitivity frequently accompany asthma. Chronic asthmatic inflammation can trigger angiogenesis, and inhibition of new vascular proliferation can have therapeutic effects. For example, agents that inhibit vascular endothelial growth factor (VEGF) may be beneficial to prevent lung changes over time.[82] Tumor necrosis factor, the TNF superfamily member, has numerous immunomodulating effects and promotes airway remodeling in asthma. Human mast cells possess TNF receptors that play a role in immune and allergy response, triggering the release of immunoglobulins from T helper cells.[83] Many herbs and their individual molecular constituents are shown to reduce airway inflammation via mechanisms involving VEGF, TNF, and many other inflammatory mediators.

Among the main pharmaceutical bronchodilators used to open the airways for acute asthma are β-adrenergic agonists such as albuterol, methylxanthines such as theophylline, and corticosteroids such as prednisolone, all associated with various side effects, sometimes significant.[84] Newer antileukotriene medicines include montelukast. Many people have demonstrated allergies to aspirin and other nonsteroidal anti-inflammatory drugs (NSAIDs), developing acute bronchospasms or asthmatic exacerbations,[85] and I have seen one case of a salicylate-containing herb inducing wheezing in an asthmatic patient.

A large number of herbs are used traditionally for cough, wheezing, and asthma, and many of them have

Alternatives to Pharmaceuticals for Asthma

Bronchodilators and anti-inflammatories are the main broad categories of drugs used to manage asthma pharmaceutically, and all have herbal alternatives. Note the mechanisms of action of respiratory herbs in this chapter that match each of the following drug categories. Thoughtful herbal formulas for cough, wheezing, and dyspnea will mix herbs having different such mechanisms. Their use can create natural approaches to wean from steroidal inhalers, treat acute asthma attacks, and reduce chronic hyperreactivity of the airways.

BRONCHODILATORS	HERBAL ALTERNATIVES
β-adrenergic agonists: e.g., metaproterenol, albuterol	Lobeline
Anticholinergics: e.g., ipratropium bromide, tiotropium bromide	Tropane alkaloids
Methylxanthines: e.g., theophylline, aminophylline	Ephedrine, caffeine
ANTI-INFLAMMATORY AGENTS	
Corticosteroids: e.g., prednisolone, dexamethasone	*Glycyrrhiza*
Antileukotrienes: e.g., probilukast, montelukast	Many anti-inflammatory herbs
Mast cell stabilizers: e.g., cromolyn sodium, nedocromil sodium	*Ammi visnaga*, *Angelica*, *Petasites*, and many other herbs

Sulfur Gasotransmitters and Lung Function

Some sulfur compounds are shown to have regulatory actions in mucous membranes, where they are produced endogenously as hydrogen sulfides with numerous physiological functions. These sulfur compounds are referred to as gasotransmitters because they have neurotransmitter-like activities, but exist in a gaseous, rather than solid, state. Epithelial cells in mucous membranes both protect themselves from exogenous sulfides and use sulfides as signaling molecules.[86]

Many sulfur-containing molecules in plants that are in a solid form become a gas, following ingestion by humans, when those substances travel in the blood to the alveoli and volatilize there. The amino acids methionine and cysteine can provide sulfur, which volatilizes in the lungs. The supplement N-acetylcysteine is used in alternative medicine to break apart thick, stuck mucus from the lungs and promote expectoration, and this is accomplished, in part, via sulfur's effects. Other natural donors of sulfur are garlic and onions. These *Allium* species are rich in the sulfur-containing S-allylcysteine and allicin and may react with intracellular thiols like glutathione to produce hydrogen sulfide gas.

Antioxidant deficiencies have been strongly associated with poor asthma control and accelerated lung function decline. The ingestion of sulforaphane compounds—which are found in high amounts in cruciferous vegetables such as broccoli, radishes, and cabbage—offer protective antioxidant effects in the airways, increasing small airway diameter, reducing airway resistance, and reducing hyperresponsive bronchoconstriction.[87] The glucosinolates are one type of sulforaphane found in cruciferous vegetables.

been shown to exert anti-inflammatory, mast cell–stabilizing, and bronchodialating effects, via animal or molecular studies. Very few human clinical trials on such plants have been done, however. *Bacopa monnieri* is traditionally used in India for asthma and other allergic phenomena, and is a mast cell stabilizer. *Cnidium monnieri*

is a traditional medicine in China used for allergic activity, with many medicinal effects credited to osthol, the coumarin it contains. *Eclipta prostrata* (also known as *E. alba*) is a traditional antianaphylactic and antihistamine agent included in various prescriptions for asthma and allergic airway disorders. *Euphorbia hirta*, popularly

Herbal Mast Cell Stabilizers for Allergic Airway Disorders

Mast cells are widely distributed throughout the body, residing in close proximity to blood vessels, smooth muscle cells, and mucus-producing glands. In airway disorders, mast cells contribute to excessive mucous production and bronchoconstriction by releasing histamine excessively or inappropriately. Agents that prevent the release of histamine and other inflammatory mediators from mast cells are referred to as mast cell stabilizers. Numerous individual flavonoids have been found to be mast cell stabilizers; quercetin, luteolin, disometin, and apigenin are among the most powerful.[88] Ellagic acid—which is found in pomegranates, strawberries, and other common fruits—is a mast cell stabilizer. Legume family isoflavones, including genistein, are also credited with mast cell–stabilizing activity. Mast cell stabilizers include the following herbs and herbal constituents:

HERB	HERBAL CONSTITUENT
Albizia lebbeck	Saponins
Allium cepa	Thiosulphinates
Ammi visnaga	Khellin
Angelica dahurica	Furanocoumarin
Artemisia keiskeana	Artekeiskeanol
Azadirachta indica	Nimbin, nimbinine, nimbandiol, quercetin
Bacopa monnieri	Bacosides, alkaloids, glycosides
Camellia sinensis	Epigallocatechin gallate (EGCG)
Cnidium monnieri	Osthol
Plectranthus baratus, also known as *Coleus forskohlii*	Forskolin (diterpenoid)
Curcuma longa	Tumerones, curcuminoids
Eupatorium chinense	Sesquiterpene lactones
Ginkgo biloba	Ginkgetin
Hydrangea macrophylla	Thunberginol B

HERB	HERBAL CONSTITUENT
Inula racemosa	Inulolide, sesquiterpene lactone
Isatis tinctoria	Indoline
Magnolia obovata	Honokiol, magnolol
Magnolia officinalis	Honokiol, magnolol
Mentha piperita	Flavonoid glycosides
Ocimum tenuiflorum	Myrcenol, nerol, eugenol
Picrorhiza kurroa	Picrosides, androsin
Punica granatum	Ellagic acid
Senna alata	Anthraquinones, flavonoids
Silybum marianum	Silymarin
Tanacetum parthenium	Parthenolide
Tephrosia purpurea	Flavonoids, tephrosin
Terminalia chebula	Ellagic acid, tannins, chebulagic acid
Tinospora cordifolia	Tinosporin
Tylophora asthmatica	Tylophorine
Vitis vinifera	Resveratrol
Xanthium cavanillesii	Xanthatin lactone

known as asthma weed, has noted antihistamine and anti-allergy activity. *Hemidesmus indicus* is a twining shrub commonly used in India against asthma, and *Lepidium meyenii* (maca), a crucifer from the high Andes, is noted to inhibit bronchospasm. *Mentha spicata* (spearmint) has antihistamine effects. Many of the herbs used in the formulas in this section have not been researched for asthma per se, but have been shown to reduce histamine, mast cell hyperreactivity, cytokine response, and other contributors to reactive airways. Many formulas are simply inspired by folkloric traditions and my own clinical experience.

Diet is also an important component of all antiallergy protocols, because certain dietary changes can exponentially increase consumption of antioxidant nutrients. And one might reduce exposure to numerous chemicals, pesticides, preservatives, and other allergens by adopting a clean and chemical-free, whole-foods–based diet. In fact, poor maternal nutrition and lack of breastfeeding are both risk factors for the development of childhood asthma. Particularly notable in asthma epidemiology are deficiencies of essential fatty acids and vitamins C and E. Herbs, foods, and supplements supplying these nutrients are all appropriate.

Tincture for Allergic Asthma

This all-purpose antiallergy formula is appropriate for daily long-term use. It works systemically to reduce inflammatory and allergic phenomenon at the cellular level. The formula has no harsh ingredients or energetically specific herb choices and may be nourishing and restorative when used long term to reduce the severity and frequency of asthmatic episodes. This formula may be a starting place to help people wean themselves off steroidal inhalers once their immune hyper-reactivity has been improved. This formula is not as powerful a bronchodilator as the Tincture for Acute Asthmatic Wheezing, but may be more effective in reducing airway reactivity over time than more powerful bronchodilators such as *Lobelia* and *Thymus*.

Ginkgo biloba
Tanacetum parthenium
Angelica sinensis
Scutellaria baicalensis

Combine the tinctures in equal parts. Take one dropperful of the combined tincture 3 to 4 times daily for many months. The use of antioxidant nutrients and essential fatty acids and following an anti-inflammatory diet can be supportive therapies.

Tincture for Chronic Asthma

Once the allergic phenomena have been improved and the frequency and severity of asthmatic episodes has been reduced, patients may be able to switch from immune-suppressive steroids to other pharmaceuticals, and even no pharmaceuticals at all, depending on the situation. This tincture can be helpful in the process of weaning off steroid medications. *Panax*, *Glycyrrhiza*, and *Lepidium* may support adrenal gland function and the body's own cortisol output, typically suppressed by long-term steroid use. A separate formula to address acute symptoms, such as the Tincture for Acute Asthmatic Wheezing, will be necessary to use as needed.

Ginkgo biloba
Glycyrrhiza glabra
Lepidium meyenii
Panax ginseng

Combine the tinctures in equal parts. Take 1 dropperful 3 to 6 times daily, for at least 3 months.

Tincture for Acute Asthmatic Wheezing

This is an example of a formula that may be strong enough to relax bronchospasms during moments of acute asthmatic wheezing. For best results, the dose should be large and repeated often, reducing as symptoms improve or in the event that *Lobelia* promotes nausea. This formula should be the palliative prong of an asthma protocol. It will alleviate acute symptoms, while other components of the protocol, such as dietary, nutritional and other herbal support, simultaneously aim to prevent future episodes. *Lobelia* is a powerful bronchial relaxant, but can promote nausea and even have emetic effects at high and repetitive dosages. *Angelica sinensis* and *Ammi visnaga* are gentle antiallergy herbs, which are used in many formulas for antiallergy effects on the airways. Thyme is drying to the abundant mucus that often accompanies acute asthma and contributes to airway distress.

Lobelia inflata	15 ml
Angelica sinesis	15 ml
Ammi visnaga	15 ml
Thymus vulgaris	15 ml
Foeniculum vulgare essential oil	5 drops

Combine the herb extracts and essential oil in a 2-ounce (60 milliliter) container. Shake well prior to each use to distribute the fennel essential oil, which will separate

Ammi visnaga for Allergic Airways

Ammi visnaga (khella) is indigenous to the Middle East and Mediterranean but has become naturalized in other parts of the world. Khella has been traditionally used for kidney stones, asthma, and cough, with many medicinal effects credited to the coumarin derivatives khellin and visnagin. Several pharmaceutical drugs have been developed from *Ammi visnaga* including amiodarone, nifedipine, and cromolyn,[89] the latter having been developed into an inhalant for asthma. Khellin has bronchial antispasmodic effects via calcium channel inhibition.[90] Khellin is also a mast cell stabilizer, useful to reduce allergic and histamine-driven airway inflammation.

Cromolyn sodium (or sodium cromoglycate) is a pharmaceutical derivative of the furanochromone khellin used to reduce chronic asthma due to mast cell–stabilizing effects and inhibitory effects on other inflammatory cells including macrophages and eosinophils, also involved with allergic inflammation. Cromolyn sodium inhalers have few side effects compared to other types of inhalers. Both khella as a whole plant and cromolyn sodium need to be taken regularly to reduce airway reactivity gradually.

Amiodarone is used in atrial and ventricular fibrillation therapy due to the ability to block sodium and L-type calcium currents, but amiodarone can have many adverse effects, so herbalists often prefer whole plant herbal remedies over isolated molecules or pharmaceutical derivatives. In the lungs, amiodarone may inhibit the degradation of lung surfactant protein A.[91]

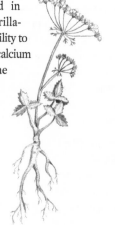

Ammi visnaga, khella

out and float on the top when the tincture is left sitting. Take 30 drops every 15 minutes, reducing the frequency as normal breathing is restored.

Tea for Asthma

These herbs offer support for asthma, but will not alleviate acute wheezing on their own. This tea complements the Tincture for Allergic Asthma on page 145 and the Tincture for Chronic Asthma on page 145.

Lobelia inflata	4 ounces (120 g)
Thymus vulgaris	4 ounces (120 g)
Eucalyptus globulus	4 ounces (120 g)
Mentha piperita	4 ounces (120 g)
Glycyrrhiza glabra	4 ounces (120 g)
Pimpinella anisum	4 ounces (120 g)
Verbascum thapsus	4 ounces (120 g)
Ginkgo biloba	4 ounces (120 g)

This recipe yields 2 pounds of tea and is intended for long-term use. Store the tea in several large airtight containers such as large canning jars in a cool dark cupboard. Steep 1 tablespoon per cup of hot water for 10 minutes, and then strain. Drink 3 or more cups per day when asthmatic reactivity has flared, or as a preventive at any time when a person is exposed to known triggers such as hay fever season, stress, cleaning the garage, and so on.

Tincture for Cough-Variant Asthma

Some people who have asthma develop a spastic cough instead of the more typical wheezing. These plants are more specific for coughing and are combined here with *Ammi visnaga* for general support in reducing airway reactivity. This formula would be beneficial in allaying acute coughing episodes, but may not act as deeply as formulas intended for long-term use.

Ammi visnaga	30 ml
Tussilago farfara	30 ml
Prunus serotina	30 ml
Thymus vulgaris	15 ml
Drosera rotundifolia	15 ml

Take 1 dropperful of the combined tincture every 15 minutes for several hours, reducing as symptoms improve.

Tincture for "Humid Asthma"

Humid asthma is a traditional term for the condition in those who produce abundant mucus that triggers bronchospasm or those whose asthmatic reactivity involves hypersecretory responses. This situation can lead some asthmatics to develop secondary bacterial infections such as bronchitis or pneumonia. *Allium* and *Thymus* both offer antimicrobial effects: *Allium* thins mucous secretions, which helps them to be better expectorated; *Thymus* dries excessive secretions. *Euphorbia* and *Grindelia* are both high-resin plants to stimulate expectoration. This formula would be most effective for short-term use when profuse mucus is present, switching to one of the long-term formulas above for daily maintenance. *Tylophora* should be used in small quantities, short term, and can be removed from the formula if nausea occurs and once effective expectoration has been accomplished.

Allium sativum
Thymus vulgaris
Euphorbia hirta
Grindelia spp.
Tylophora asthmatica

Combine the tinctures in equal parts and take 1 or 2 dropperfuls of the combined tincture every hour, reducing as mucus abates and wheezing subsides.

Tincture for Asthma as Part of Atopic Tendency

A formula for those whose asthma is accompanied by hay fever, eczema, or hives might feature herbs especially noted for atopy, to stabilize mast cells and reduce allergic reactivity. *Bacopa monnieri* is a mast cell stabilizer and *Hemidesmus indicus* has immunomodulating properties.

Petasites is a traditional herb for allergies with an affinity for the lungs, and *Ammi visnaga* makes a useful base for nearly any allergy formula. The use of *Petasites* should be limited to the short term due to possible liver toxicity.

Bacopa monnieri
Hemidesmus indicus
Petasites hybridus
Ammi visnaga

Combine the tinctures in equal parts and take 1 or 2 dropperfuls of the combined tincture, 3 times per day for several months, reducing after that time as symptoms improve.

Encapsulations for Asthma

These individual capsules are all readily available options, each anti-inflammatory to the airways via a variety of mechanisms. Use of these capsules to complement a tea and tincture creates an aggressive protocol for asthma.

Ginkgo biloba capsules
Curcuma longa capsules
Zingiber officinale capsules
Allium sativum capsules
Bromelain capsules
Resveratrol capsules

Choose 2 or more of the above, and take 2 of each of the chosen capsules at a time, 2 to 3 times a day, reducing as airway reactivity is improved.

Formulas for COPD

Chronic obstructive pulmonary disorder (COPD) is a disease of progressive airflow obstruction that can result from vascular disease, where the heart fails to deliver oxygenated blood to the airways, as well as from destruction of the airways due to inflammatory processes such as emphysema and chronic lung inflammation. The major risk factor for developing COPD is cigarette smoke exposure, with 90 percent of COPD deaths attributable to smoking. Those in the early stages of the pathology can prolong life by quitting smoking. Exposure to smoke in areas where people cook over wood fires and burn trash is another cause of respiratory disease. Exposure to fungal spores and mycotoxins may cause sick-building syndrome and also induce respiratory illness, as can

exposure to a long list of airborne fumes and particulates from asbestos to heavy metals to silica.[92] Approximately 10 percent of the world's population suffers from COPD, and it is currently the third leading cause of death.[93]

The main pharmacological therapy for treatment of COPD is long-acting bronchodilators, which mainly provide symptomatic relief, because COPD progression is typically refractory to steroid treatment. Long-term oxygen therapy improves survival in patients with COPD or other cause of severe hypoxia when used for at least 15 hours a day. The natural course of COPD involves progressive worsening of pulmonary hypertension, hypoxemia, and right heart failure, and oxygen therapy has been shown to delay mortality in all such situations.[94] Metabolic syndrome, diabetes, and a high inflammatory load in the body may contribute to pulmonary fibrosis when not due to smoking or other identifiable cause, a condition known as idiopathic pulmonary fibrosis. Advanced glycation end products (AGEs) result from elevated lipids and glucose. AGEs contribute to accelerated aging, poor wound healing, and abnormal epithelial-mesenchymal restorative capacity in idiopathic pulmonary fibrosis.[95] Therefore, therapies that optimise blood sugar and lipids are also important for slowing the progression of the disease.

Herbal therapy for COPD may be aimed at halting ongoing inflammation, preserving lung function, preventing pulmonary hypertension, and prolonging life. As some of the tissue destruction involves active inflammation and release of enzymes from cells, herbs noted to be general anti-inflammatories and immunomodulators might be appropriate. Herbal medicines may improve pulmonary function and diaphragm muscular tension, lessening symptoms, increasing exercise capacity, and reducing exacerbations in COPD patients. Thus, herbal therapies may be complementary to the sparse Western medical therapies.

Due to limited airflow, acute exacerbations of COPD may require hospitalizations, and each new episode may contribute to demise of the airways and speed the progression to end-stage consequences. While human studies on herbs for COPD are mostly lacking, there has been research on botanical agents that may slow the progression of the disease, help protect the airways from smoke and irritants, and help preserve lung tissue elasticity. Antioxidant nutrients and phytochemicals such as flavonoids are among the most studied tissue protectants. Such natural compounds include quercetin, resveratrol, curcumin, and other flavonoids, all of which should be supplemented and consumed liberally in the diet to help protect lung parenchyma, reduce systemic inflammation, and prolong life. Luteolin, quercetin, kaempferol, and likely other flavonoids can reduce pulmonary hemorrhage and neutrophilic inflammation, as well as interstitial edema. *Camellia sinensis* (green tea) catechins are all-purpose antioxidant and anti-inflammatory compounds that can be used in capsules to supplement teas and tinctures for COPD, or the whole dry herb can be included in tea recipes. *Curcuma, Zingiber, Astragalus,* and *Scutellaria baicalensis* are all-purpose anti-inflammatory herbs that are useful in medicinal foods, as well as teas and tinctures. *Cordyceps sinensis* protects bronchial epithelia from the damaging

effects of cigarette smoke,[96] but smoking cessation is essential, because nothing will compensate or protect the tissue from such an assault as to make smoking acceptable for heart or lung disease patients.

Additionally, herbs believed to strengthen connective tissue, collagen, and elastin, such as *Centella, Urtica,* and *Equisetum,* may be useful to strengthen the alveoli and lung structure. As mucous hypersecretion is typical, herbs that help dry this excess, such as *Achillea, Origanum,* and *Thymus,* may be both palliative and helpful in reducing the inflammatory processes. Stagnant fluid in the lungs makes a hospitable environment for microbes such as *Haemophilus influenzae* and *Pseudomonas aeruginosa,* and so the use of antimicrobial herbs with immunomodulating properties may be useful for preventing devastating superimposed infectious bronchitis and pneumonia. Regular use of *Thymus, Ligusticum, Lomatium,* and *Allium* as lung antimicrobials is warranted, and essential oil inhalation may also deter these opportunistic infections.

Many herbs have been studied for limiting excessive neutrophil and macrophage activation, which are involved in the inflammatory processes of COPD, and herbs have been noted to reduce vascular endothelial growth factor (VEGF), which drives some of the destruction of lung tissue. Such herbs include *Ginkgo, Panax, Curcuma,* and *Zingiber,* all of which are available in encapsulated form and may complement herbal teas and tinctures. Airway epithelial cells are also important to immune system defenses because they secrete antimicrobial peptides and proteases, as well as cytokines to activate neutrophils and macrophages. Thus, immunomodulating herbs and preventive measures may be warranted, especially during the peak cold and flu season. Stimulating endogenous growth factors may promote tissue repair and limit the fibrotic demise of the lungs, but studies on antifibrotic herbs for the lungs are lacking.

Tincture for Long-Term Lung Support

This is an example of a lung formula that features herbs shown to deter fibrosis and prevent lung infections.

Panax ginseng	10 ml
Picrorhiza kurroa	10 ml
Eriodictyon spp.	10 ml
Scutellaria baicalensis	10 ml
Lomatium dissectum	10 ml
Origanum vulgare	10 ml

Take 1 teaspoon of the combined tincture 3 to 6 times daily.

Examples of Current Research on Herbs for COPD

Many studies have been done on herbs that may help reduce inflammatory and fibrotic processes in the lungs. Here is a summary of that research.

Any of the herbs listed below may be included in herbal protocols to deter fibrosis of the lungs and prolong life.

HERB	RESEARCH FINDINGS
Allium sativum	Garlic inhibits the pro-inflammatory cytokine interleukin 17 (IL-17), which is often upregulated in inflammatory airway diseases.[97]
Andrographis paniculata	Andrographolide activates nuclear factors in a manner that may help protect the lungs from smoke-induced oxidative injury.[98]
Eriodictyon californicum	Eriodictyol, a flavonoid found in *Eriodictyon* and also in the Chinese herb *Dracocephalum rupestre*, has been shown to modulate many inflammatory cytokines in the lungs,[99] and although no clinical trials have been done, the plants have long been used for lung diseases.
Lobelia inflata	Veterinary medicine is exploring lobeline for COPD in horses.
Neopicrorhiza scrophulariiflora (also known as *Picrorhiza scrophulariiflora*)	Picrosides have immunomodulating and anti-inflammatory activities and may reduce airway inflammation.
Paeonia lactiflora, *P. emodi*	Peony's paeoniflorin and related compounds are noted to reduce oxidative stress and fibrosis in several tissues and organs, and peony is an ingredient in several traditional Chinese formulas for COPD and acute respiratory distress syndrome (ARDS).
Phyllanthus emblica (also known as *Emblica officinalis*)	The leaves contain polyphenols credited with numerous antioxidant and anti-inflammatory effects able to reduce alveolar damage and macrophage infiltration and may help repair pulmonary damage and attenuate fibrotic processes.[100]
Raphanus sativus var. *niger*	This plant has been used in the folk medicine of China, Iran, Turkey, and other regions in the treatment of asthma, bronchitis, and other chest complaints. It may help reduce pulmonary fibrosis via effects on transforming growth factor $\beta 1$(TGF-$\beta 1$).[101]
Syzygium aromaticum (also known as *Eugenia caryophyllata*)	The essential oil eugenol found in cloves reduces proinflammatory cytokines and may be included in inhalants. Clove powder may be included in medicinal teas and *Syzygium* tincture in a variety of formulas for COPD and ARDS.
Uncaria tomentosa	This South American vine is traditionally used to treat cancer, infection, and inflammation. It contains the alkaloid mitraphylline, which is noted to inhibit pro-inflammatory cytokines.

Scutellaria baicalensis for Respiratory Formulas

Scutellaria baicalensis is traditionally used for allergic and inflammatory issues in Traditional Chinese Medicine (TCM), and baicalin and baicalein are *Scutellaria* flavonoids credited with many beneficial effects on the airways. Numerous anti-inflammatory effects on cytokines have been identified, helping to explain the plant's long-standing use for airway infection, inflammation, and allergic reactivity.[102] Fibroblasts play a crucial role in chronic respiratory inflammation, secreting inflammatory cytokines and chemokines, such as interleukins, that drive upper airway inflammation. Baicalin may also have stabilizing effects on respiratory extracellular matrix molecules and attenuating excessive collagen production and interleukin release from activated fibroblasts. These actions make *S. baicalensis* flavonoids useful in the treatment of asthma, COPD, and chronic rhinosinusitis.[103] Baicalin from *S. baicalensis* inhibits the synthesis and accumulation of collagen in animal models of pulmonary hypertension[104] and end-stage sequelae of COPD and heart failure.

Scutellaria baicalensis, scute

Tea for COPD Based on Yu Ping Feng San

Yu Ping Feng San is widely used in China for treatment of diseases in respiratory system, including COPD. Modern investigations of Yu Ping Feng San have shown beneficial effects on transforming growth factor β1 (TGF-β1) and inflammatory cytokines, which are both involved in airway destruction characteristic of COPD. The formula protects the lungs from cigarette smoke extract in animal models of COPD and also mitigates inflammation in human bronchial epithelial cell cultures.[105] Other versions of the formula may contain other herbs, with this trio considered a base combination.

Atractylodes macrocephala dried root	8 ounces (240 g)
Astragalus membranaceus dried root	4 ounces (120 g)
Saposhnikovia divaricata dried root	4 ounces (120 g)

This formula yields a 1-pound blend that can be stored in a large glass jar. Decoct 1 heaping tablespoon in 5 cups of water; simmer gently down to 4 cups, and then strain. Drink daily.

Encapsulated Herbs for a COPD Protocol

Choose a combination of supplements to complement teas and tinctures and create an aggressive protocol for those suffering from COPD and related situations of pulmonary fibrosis to slow the fibrotic process and prolong life.

Curcuma longa (turmeric) capsules
Ginkgo biloba (ginkgo) capsules
Allium sativum (garlic) capsules
Zingiber officinale (ginger) capsules
Camellia sinensis (green tea) capsules
Andrographis paniculata (andrographis) capsules
Resveratrol capsules
N-acetylcysteine (NAC) capsules

For any of the herbal capsules, take 2 capsules, 3 times per day. For the resveratrol or the NAC capsules, take 1 capsule 3 times per day.

Connective Tissue Tonic Tea for COPD

The herbs in this tea support the immune system, offer anti-inflammatory effects, and act as a connective tissue tonic. The tea may be amended for simplicity's sake, or different herbs may be used in rotation, as therapy for COPD is lifelong.

Astragalus membranaceus
Equisetum spp.
Medicago sativa

Centella asiatica
Urtica spp.
Thymus vulgaris
Camellia sinensis
Glycyrrhiza glabra

Combine the dry herbs in equal parts, or use more mint and licorice as desired for flavor. Steep 1 tablespoon per cup of hot water. Prepare a pot every morning and drink as much as possible, 3 or more cups each day. For long-term usage, take weekends off, or skip 1 or 2 days per week.

Tincture for Lung Infection Prevention

Those at risk of serious complications from bronchitis and pneumonia—such as those with COPD or heart failure or the frail elderly—should take this antimicrobial formula at the first sign of a cold or during the entire winter.

Ligusticum porteri	15 ml
Andrographis paniculata	15 ml
Allium sativum	10 ml
Thymus vulgaris	10 ml
Echinacea spp.	10 ml

Take 1 teaspoon of the combined tincture hourly at the onset of any new cold, reducing as symptoms improve.

Inhalant Therapy for Those with COPD

Patients with compromised airways should be encouraged to adopt a daily practice of inhaling antimicrobial essential oils to deter opportunistic infections such as *Pseudomonas*. Breathing in the volatilized vapors of essential oils can be useful in steam inhalation practices, and this formula places the essential oils in a fixed oil carrier to inhale off the palms of the hands or to rub over the throat and upper chest prior to showering.

Eucalyptus essential oil	50 drops
Citrus aurantia (orange) essential oil	50 drops
Mentha (mint) essential oil	50 drops
Thymus (thyme) essential oil	20 drops
Sweet almond oil	2 ounces (60 ml)

Add the essential oils to the sweet almond carrier oil and shake well. Place 20 drops of the combined oil on the palms of the hands, cup the palms near the face, and breathe gently for 10 minutes, morning and evening. Apply ¼ to ½ teaspoon directly to the chest and/or back immediately before or after a hot shower; the moist heat will enhance vapor inhalation and absorption through the skin.

Opportunistic Infections and Pulmonary Disease

Pseudomonas aeruginosa is one opportunistic pathogen that may infect the lungs of those with emphysema, COPD, and bronchiectasis. It is involved in the vicious cycle of ongoing exacerbations, airway inflammation, dilatation, and damage in these chronic respiratory diseases, and the presence of this bacteria is generally associated with health demise and a poor prognosis.[106] Several herbs and their individual essential oils may deter *P. aeruginosa* lung infections when inhaled or otherwise assisted in reaching the airways. For example, *Thymus* species have activity against *P. aeruginosa*[107] and may help when used in nebulizers, or simply placed on the surface of a pot of steaming water, or even put on the palms of the hands, and then inhaled. Concentrated thyme essential oil is extremely irritating to the skin and mucosa, but is well tolerated when highly diluted. Just a single drop in a facial steamer may be an appropriate dose.

Topical Emodin Oil for COPD

Animal and molecular studies suggesting that emodin and rhein in *Rhamnus* and other plants will deter fibrosis, even when topically applied over abdominal organs. Although there are no published studies on its use for lung fibrosis, I believe that it could do no harm. It is included here in an effort to slow fibrosis of the lungs and prolong life, however possible.

Rhamnus purshiana or *Reynoutria japonica*
  (also known as *Polygonum cuspidatum*,
  Polygonum japonica, and *Fallopia japonica*)
  dried roots   1 cup
                                        (about 100 grams)

Castor oil

Place the *Rhamnus* or *Reynoutria* roots in a high-speed blender and cover with castor oil. Pulverize as finely as possible and transfer to a canning jar. Store in a dark cupboard and shake every day for 6 weeks. Then strain out the particulate and use the resulting oil topically over the lungs immediately after a morning shower. Prepare a new batch each week to be able to have an ample supply.

This oil could also be used as the base oil in other recipes for topical applications, such as those that include essential oils. This oil could also be mixed 50:50 with *Lobelia* oil, as described in "*Lobelia* Oil for the Lungs" on page 139.

Tropane Alkaloids for the Lungs

The parasympathetic innervation of airway smooth muscle stimulates the release of acetylcholine that may be received by muscarinic receptors or nicotinic receptors, each having several subtypes. There are at least five muscarinic receptor subtypes, and in the airways, the M3 type induces muscle contraction and promotes respiratory secretions when agonized by acetylcholine, while the dominant M2 type limits sympathetic-driven muscle contraction. Thus, blocking either the M3 or M2 type of muscarinic receptor can result in bronchodilation.[108]

The Solanaceae family is noted for its tropane alkaloids. *Atropa belladonna*, *Datura stramonium*, and *Hyoscyamus niger* are important herbal medicines to ancient Europe, and related plants are important throughout South America, where the Solanaceae family evolved. Tropane alkaloids include stramonium in *Datura*, hyoscine or scopolamine in *Hyoscyamus*, and atropine in *Atropa*. The tropane alkaloids are said to have antimuscarinic effects, by binding to muscarinic receptors in various tissues and preventing the endogenous ligand, acetylcholine, from binding. Tropane alkaloids are therefore credited with strong anticholinergic effects.

The tropane alkaloid atropine was isolated in the 1900s and quickly became the basis of a variety of medicines; it remains the standard mydriatic eye drop to the present day. These herbs and medications were used as bronchodilators, but when high and frequent doses are needed, tropane alkaloids will cause decreased salivation and perspiration, mydriasis and blurred vision,

urinary retention and decreased gut mobility, nausea, tachycardia, and hallucination and central nervous system effects. Due to these significant side effects, beta blockers became the preferred inhalers for treating asthma and emphysema upon their development in the 1920s.

Tropane alkaloids have been synthesized into anticholinergic medicines, including medicines for asthma and respiratory disease. Several such synthetic compounds, ipratropium and tiotropium bromide, can improve obstructive symptoms in the lungs due to the bronchodilating effect and were approved for the long-term treatment of COPD in the United States in 2004.[109] Such drugs are possible inhalant medications, are alternatives to beta blocking and steroidal inhalers, and have few side effects.[110]

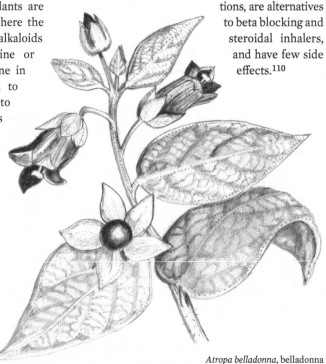

Atropa belladonna, belladonna

Traditional Chinese Formulas to Slow COPD Progression

Tongsai and Bufei Yishen granules, two traditional Chinese herbal formulas, are shown to reduce inflammatory response in the lungs, improve pulmonary function, and shorten the recovery time in acute exacerbations of COPD. Formulations such as these may inhibit inflammatory cytokine expression, help balance protease-antiprotease ratios, and reduce collagen deposition, which are all part of the fibrotic process of COPD. Additional research suggests that these formulas may have stabilizing effects on tight junctions and adherens molecules to help preserve structural integrity of lung tissue. Granules are produced commercially by preparing large volumes of a traditional decoction and gently simmering it down to a thick syrup that is then dried and processed into dry granules. The granules can be encapsulated and offer a more concentrated product than crude herbs or may be reconstituted into a tea, made as strong or as weak as desired.

Tongsai granule consists of the following:

Ardisia japonica	15 g
Lepidium apetalum	12 g
Fritillaria cirrhosa	12 g
Paeonia anomala	12 g
Ophiopogon japonicus	12 g
Ephedra sinica	9 g
Rheum officinale	6 g

This formula or something similar may be available from a vendor of Chinese herbs, or it may be prepared for oneself by searcing out individual granules. Take 5 grams of the blended formula at a time in hot water, or stirred into honey, nut butter, or apple sauce, once a day for long-term maintenance or 3 to 5 grams at a time 3 times a day to help limit acute exacerbations.

Bufei Yishen granule consists of the following:

Astragalus membranaceus	15 g
Lycium chinense	12 g
Cornus officinalis	12 g
Panax ginseng	9 g
Epimedium brevicornu (also known as *E. rotundatum*)	9 g
Schisandra chinensis	9 g
Ardisia japonica	9 g

Take 5 grams at a time once a day for long-term maintenance, or 3 to 5 grams at a time 3 times a day to help limit acute exacerbations.

Treatment for Acute Respiratory Distress Syndrome

Acute respiratory distress syndrome (ARDS) is a rapidly progressive lung disease that is often fatal (30 to 50 percent of cases). ARDS is associated with fibrotic inflammation of the lungs characterized by neutrophil infiltration, interstitial edema, and acute hypoxemia. Despite its severity, ARDS remains poorly understood, and there are few effective therapies other than heroic measures, such as being put on a ventilator. The condition is associated with acute elevation of cytokines in the lungs, and the elevation can be so severe that unless it is immediately controlled, lung damage can manifest within 72 hours; such damage is often permanent. The acute episodes are characterized by alveolar edema and uncontrolled neutrophil migration to the lungs, leading to tissue damage through the release of proteases, oxidants, and peptides. ARDS must be managed in an acute care setting, and the herbal research offered here is to provide ideas on herbs that may be appropriate to possibly prevent ARDS in patients with advanced lung disease. However, herbal therapy is not a substitute for expert care in a pulmonology unit.

Several small clinical trials have shown the traditional Xuanbai Chengqi decoction—which is described in the "Enemas for Respiratory Distress?" sidebar on page 154—to reduce mortality in those with acute exacerbations of COPD, ameliorating cytokine levels and oxidation/antioxidant index compared with a placebo.[111] Another study showed the formula to improve the efficacy of Western therapy for patients with COPD experiencing an acute deterioration associated with systemic inflammatory responses.[112] *Rheum officinale* contains emodin, discussed in the Topical Emodin Oil for COPD formula on page 151. Emodin from *Rheum* species may attenuate lipopolysaccharide components of endotoxins, helping to protect the lungs from acute injury and manifestation of ARDS.[113]

Systemic inflammation may be triggered by an infection and lead to ARDS. Systemic inflammatory

Enemas for Respiratory Distress?

Patients with advanced lung disease or acute respiratory distress episodes develop increased intra-abdominal pressure as the pulmonary vasculature become congested, simultaneous with an increase in elastic resistance of the lungs. The increased abdominal pressure causes intestinal mucosal hyperemia, and slowing peristalsis contributes to intestinal obstruction, which, in a vicious cycle, aggravates the progression of acute respiratory distress syndrome (ARDS). The connection between the lungs and the large intestine has been recognized in TCM since ancient times, and the use of intestinal herbs has been employed to "clear heat" from the lungs in many traditional formulas.

A series of cases studies from the Henan University of Traditional Chinese Medicine reported that the use of a traditional enema formula may help protect the lungs in ARDS. Xuanbai Chengqi decoction—which contains gypsum, *Rheum officinale*, *Prunus dulcis* (also known as *P. amygdalis*), and *Trichosanthes kirilowii*—may improve static and dynamic lung compliance as well as reduce the complication incidence and fatality rates[114] when administered as an enema.

Picrosides to Guard Against Acute Lung Injury

Neopicrorhiza scrophulariiflora (also known as *Picrorhiza scrophulariiflora*) has been used traditionally for relief of asthma. This herb contains picrosides, a group of iridoid compounds that exhibit numerous anti-inflammatory activities. Animal studies suggest an ability to control infection and reduce lung injury in situations of sepsis.[115] Picrosides may downregulate cytokine release and response and help treat inflammatory airway diseases.[116] A molecular compound derived from picrosides is currently in clinical trials for COPD. Picrosides may suppress neutrophilic inflammation in acute exacerbations of COPD and attenuate inflammatory cytokine release and the resulting cascades that may contribute to acute lung injury and distress.[117]

response to infectious processes, known as sepsis, is most apparent in the lungs, where circulating endotoxins initiate immunoreactivity. In those with advanced heart and airway disease, this can be fatal. Those who survive usually experience a significant decline in health and general prognosis following acute episodes. Any assault that induces increased vascular permeability can cause microvascular thrombosis and impair circulation in a manner that promotes the release of inflammatory cytokines and leads to acute respiratory distress. *Melilotus suaveolens* is a folkloric antithrombotic herb, and research shows it to downregulate vascular endothelial growth factor in septic situations, stabilizing the microvasculature against hyperpermeability.[118] Lacking more specific herbal research, these herbs and agents noted to reduce inflammatory cytokines in general (as cited with COPD research above) may be appropriate to prevent ARDS and would certainly do no harm.

Formulas for Cor Pulmonale

Cor pulmonale is a type of heart failure that results from lung diseases and primary COPD, as well as cystic fibrosis, obstructive bronchitis, scleroderma, or any other lung disorder that results in pulmonary hypertension, which places a high workload on the heart to move blood through vessels affected by increased vascular resistance. Severe thoracic skeletal deformities such as kyphosis or scoliosis may also predispose to cor pulmonale due to the mechanical limitations these conditions place on respiratory excursion, reducing lung capacity. The heart muscle enlarges in cor pulmonale, as with cardiomyopathy; circulation is impaired leading to edema, and the jugular veins distend. Liver disease may follow as pulmonary hypertension leads to portal congestion. Vascular stasis ensues, leaving patients susceptible to thrombi and embolism. Poor circulation and inflammation in the lungs increases the risk of pneumonia and ARDS. Cor pulmonale is considered an end stage of COPD, emphysema, cystic fibrosis, and other chronic lung disease. There is no cure and most people will succumb to disseminated sepsis and pneumonia, often following several episodes of acute crises, clots, or infections.

The only herbal therapies are those botanical therapies that may help preserve elasticity of the lungs, prevent opportunistic infections, and reduce the systemic inflammatory load on the body. Herbs to prevent the development of pulmonary hypertension might also prevent progression of lung diseases to cor pulmonale. *Crataegus, Ginkgo, Curcuma, Camellia, Allium,* and other herbs may help protect the pulmonary blood vessels, improve circulation in the lungs, and prevent hypertension. Herbs that inhibit platelet aggregation, such as *Allium, Capsicum,* and *Zingiber,* may help prevent clotting due to circulatory stasis. If edema is pronounced, diuretics and cardiac glycosides may become necessary. Herbal

diuretics include *Taraxacum* leaf, *Urtica* spp., *Equisetum,* and heart tonics including *Selenicereus, Digitalis,* and *Convallaria.* The primary therapy for cor pulmonale should be individualized and aimed at treating any underlying lung disease as vigilantly as possible. As with all congestive lung conditions, preventing or aggressively treating any opportunistic infections is also appropriate. (See "Formulas for Cystic Fibrosis" on page 134.) End-stage disease usually requires mechanical oxygen therapy and treatments focused on keeping patients comfortable.

Look back at the Tincture Tonic for Cor Pulmonale on page 85 in chapter 2 for an appropriate formula that aims to reduce vascular stress in general.

Heart and Diuretic Support Tincture

Selenicereus grandiflorus (night-blooming cactus) is traditionally used for an enlarged heart. It is supported here with the diuretic *Taraxacum* leaf to treat edema, *Andrographis* to deter opportunistic infections, and *Melilotus* to help reduce the risk of thrombi and microvascular derangements.

Selenicereus grandiflorus	15 ml
Taraxacum officinale leaf	15 ml
Andrographis paniculata	15 ml
Melilotus suaveolens	15 ml

Take 1 teaspoon of the combined tincture 3 to 6 times daily.

Formulas for Tuberculosis

Mycobacterium tuberculosis is a dangerous pathogen with a high human mortality and morbidity rate, especially in India, Africa, and Southeast Asia. The World Health Organization (WHO) has ranked tuberculosis (TB) the seventh most urgent public health concern and public health crisis, with roughly 2.2 million cases each year. Human immunodeficiency virus (HIV) infection increases the likelihood of developing TB, especially in African regions. Some of the main pharmaceuticals for treating tuberculosis include rifampicin, isoniazid, ethambutol, pyrazinamide, and second-line drugs include cycloserine, cephalosporin, viomycin, and others. However, many strains of TB are now resistant to these drugs. Although there are numerous reports of herbs and constituents that deter TB, few clinicians would opt to experiment with such a dangerous pathogen and rely on the advice of public health experts.

That being stated, however, when a person does not respond to drug therapy, or when the mycobacteria is resistant to all such drugs, herbal medications may be used in tandem with the pharmaceutical options and boost the response. Thus, there is an urgent need for novel agents to combat the global tuberculosis epidemic. Artemisinin is one such antimalarial agent derived from *Artemisia annua* and recommended by the WHO. Artesunate is a semisynthetic derivative of artemisinin that may also effectively modulate or inhibit tuberculosis mycobacteria. Artesunate demonstrated an antitubercular action with a daily dose of 3.5 milligrams/kilograms in an in vivo test for 4 weeks, without toxicity and side effects.[119]

Garlic and onions should be included in the diet, and garlic capsules and *Allium*-containing formulas can be included in complementary protocols for treating TB, because the antimicrobial sulfides in the plants deter the microbe. Because 25 percent of the world's cases occur in India, Ayurvedic medicine has proposed numerous herbal medicines to help treat the infection as well as to palliate symptoms and support overall health. I include one such formula in this section. *Withania somnifera* is often included as a single herb to use in combination with other formulas and approaches in India. *Juniperus communis* is another traditional remedy for TB, used as an inhalant and an oral medication. Modern research confirms its activity against many mycobacteria species, including drug-resistant strains, with terpenoids credited with antimycobacterial effects.[120] Specifically, activity is credited to the sesquiterpene identified as longifolene and to two diterpenes, totarol, and trans-communic acid.

Ayurvedic Tea for TB

This formula is said to be a *rasayana*, an Ayurvedic term for a rejuvenating tonic. This rasayana is thought to be rejuvenating to the lungs, helping to repair and regenerate tissue and preserve function. Such a formula may be powdered and encapsulated but is offered here as a tea. It is reported to decrease cough, dyspnea, and fever, reduce hemoptysis, and increase body weight.[121] Many rayasanas are traditionally prepared as teas and blended with milk to create chai-like beverages. Others are stirred into honey and taken by the spoonful.

Emblica officinalis	1 ounce (30 g)
Tinospora cordifolia	1 ounce (30 g)
Withania somnifera	1 ounce (30 g)
Glycyrrhiza glabra	1 ounce (30 g)
Piper longum	½ ounce (15 g)
Hemidesmus indicus	½ ounce (15 g)
Saussurea costus (also known as *S. lappa*)	½ ounce (15 g)
Curcuma longa	½ ounce (15 g)
Alpinia galanga	¼ ounce (8 g)

Combine and blend all herbs and store in a large glass jar. Decoct the herbs by gently simmering 1 teaspoon per cup of water for 10 minutes. Remove from the heat, cover the pan, and let stand 10 minutes more before straining. Drink 3 cups per day, long term. This tea may be used in tandem with pharmaceutical drugs.

Tuberculosis Support Protocol

Artemisinin capsules
Allium sativum capsules
Juniperus communis tincture

Inhalant of eucalyptus, juniper, and other essential oils is complementary to the use of these capsules.

Specific Indications: Herbs for Respiratory Conditions

Achillea millefolium • Yarrow

Teas and tinctures of yarrow flowers may be included in formulas for coughs and simple upper respiratory infections and in sinus and allergy formulas for its decongestant anti-inflammatory effects. *Achillea* is also specific for hemoptysis. Folkloric texts suggest yarrow for every organ system and an extremely broad list of medical issues including colds, coughs, laryngitis, bronchitis, hay fever, and influenza.

Aconitum napellus • Aconite

Caution: Aconite, also called fuzi, is a potentially dangerous herb due to the highly toxic alkaloid aconitine it contains. This herb is for use by skilled clinicians only. Very small doses of root medicines may be used just a drop or two at a time internally for the acute onset of bronchitis, laryngitis, dyspnea, and bronchospasms.

Aconitum has been especially recommended in the early phases of infection where a person feels chilly and restless, and pulse is elevated. Some early American herbal authors purported aconite to be one of the best remedies for acute tonsillitis and quinsy before pus forms, and for the initial stages of respiratory congestion due to the onset of viral infections. The Eclectic physicians commonly diluted aconite by placing 10 drops in a full glass of water, resulting in a liquid of nearly a homeopathic dilution, which was dosed a single teaspoon at a time, hourly, for acute inflammatory conditions. There is a long list of respiratory complaints for which a tiny dosage of aconite may be included in formulas, including acute colds, acute coryza, lobar pneumonia, broncho-pneumonia, pleurisy, croup, acute nasal and faucial catarrh, acute pharyngitis, acute bronchitis, acute laryngitis, and acute tracheitis.

Actaea cimicifuga • Shengma

Actein is a triterpene glycoside isolated from the rhizomes of the Chinese herb shengma, a common name used to refer to several species of *Actaea* (*Actaea* is also known as *Cimicifuga*).

The roots of several *Actaea* species have been used as antiviral agents in China indicated for respiratory infection with fever and pain. *A. cimicifuga* (also known as *Cimicifuga foetida*), *A. dahurica*, and *A. heracleifolia* are used for heat-clearing and detoxifying effects. Hundreds of triterpene glycosides, also known as cycloartane triterpenoids, have been identified in these species and credited with immunomodulating effects.[122]

Actaea racemosa • Black Cohosh

Also known as *Cimicifuga*, the roots of *Actaea* may be useful for lung pain and musculoskeletal inflammation of the chest due to viruses, autoimmune disorders, or stress. The roots are popular for menopause and anxiety; however, the plant may also be used as an antispasmodic in cases of asthma, cough, and pertussis. *Actaea* has tonifying effects on membranes, and while not a first-choice respiratory remedy, it may be included as an adjuvant in formulas for cough and influenza where myalgia, anxiety, tension, and headache accompanies. It has been included in formulas for febrile respiratory infections and for skin eruptions due to viral infections in children. It may be included in cold, cough, and flu formulas where there is muscle pain and conjunctival inflammation. Embrocations have been employed for musculoskeletal pain, including achiness accompanying viral lung infections. There are conflicting reports of potential liver toxicity from the use of *Actaea*. Monitoring liver enzymes may be prudent to assess any possible concern in those using the plant long term, as may be the case for fibromyalgia or menopause protocols, but unlikely necessary for short-term use in formulas to treat respiratory infections. Avoid black cohosh during pregnancy. Macrotys is another common name for this species.

Adhatoda • See Justicia

Albizia lebbeck • Siris

There are approximately 150 *Albizia* species, mostly trees and shrubs native to warm areas of Asia and Africa. Many are used medicinally for infectious diseases, epilepsy, anxiety, and for cough and bronchitis. All plant parts are used including the leaves, roots, flowers, and bark. The plant is traditionally used in Ayurvedic medicine for asthma, sinusitis, and tuberculosis. *A. lebbeck* is one of the most common species and goes by the name Siris (in Sanskrit), as well as lebbeck tree, flea tree, frywood, koko, and Laback. *A. lebbeck* contains bronchodilating, mast cell–stabilizing saponins. Other medicinal species include *A. julibrissin* (silk tree) and *A. odorata*.

Allium cepa • Onion

Onion bulbs are used as a traditional remedy to thin mucus and are a time-honored base for homemade cough syrup having expectorant effects for cough and chest colds. Onion thiosulfinates are mast cell stabilizers and antimicrobial agents that may help with infection and allergic disorders of the airways. *Allium* sulfides may also offer chemoprotective effects in the lungs and help reduce pulmonary hypertension and metabolic stress.

Allium sativum • Garlic

Due to broad antimicrobial and expectorating properties, garlic tinctures, homemade syrups made using garlic bulbs, or commercial encapsulations make excellent ingredients in formulas and protocols for simple URIs to more serious bronchitis, sinusitis, and pneumonia formulas. Because garlic thins mucus, it is particularly indicated for thick, rubbery, tenacious, or otherwise difficult-to-expectorate mucus. Garlic inhibits the pro-inflammatory cytokine IL-17 often upregulated in inflammatory airway diseases.

Alstonia scholaris • Dita Wood

The bark of this tree in the Apocynacea family is used for a variety of inflammatory illnesses, including malarial fevers. *Alstonia* is especially indicated for periodic fevers and for infectious illness where diarrhea and poor digestion accompany. Lung infections and hypersensitivity disorders may benefit from the bronchodilating and antiviral activity of a group of alstotides[123] (including ditamine, echitamine, and echitenine) along with anti-inflammatory triterpenes found in the plant. Other common names are blackboard tree, devil tree, ditabark, milkwood-pine, and white cheesewood.

Althaea officinalis • Marshmallow

Althaea roots can be a useful demulcent in formulas, especially teas, to help soothe sore throats, hoarse voice, and tickling or spastic coughs in cases of respiratory infections.

Ammi visnaga • Khella

Khella seeds are a traditional remedy for bronchospasms, asthma, dyspnea, spastic coughs, and respiratory allergies, with many medicinal effects credited to the coumarins and khellin, a mast cell–stabilizing flavonoid. Khella may also be included in "rescue" formulas for asthmatics to use first, before resorting to an inhaler. This can help reduce reliance on steroids while a broader protocol to treat underlying reactivity is being employed. The plant also goes by the common name bishop's weed.

Ananas comosus • Pineapple

Pineapple's fruits and stems are the source of bromelain, a useful encapsulated product with mucolytic and anti-inflammatory effects. Bromelain capsules can be started at the first sniffle for those with a tendency to bronchitis and pneumonia. Taking the capsules may also help speed recovery from the same when already present.

Andrographis paniculata • King of Bitters

The leaves and roots of this herbaceous annual have been used traditionally for infectious illness. This herb is now cultivated commercially to support its popularity. The leaves, which are extremely bitter, may also improve metabolic functioning in diabetes and cardiovascular disease. *Andrographis* may be included as a tonic herb in formulas for lung disorders related to oxidative stress and to fibrotic demise due to long-term inflammation. Andrographolide is a diterpene lactone in the plant shown to activate nuclear factors in a manner that may help protect the lungs from smoke-induced oxidative injury and may help deter opportunistic infections in those with respiratory disease. Clinical trials suggest efficacy for common upper respiratory infections.[124] All parts of the weedy herbaceous shrub have a bitter flavor; the plant is referred to as *Maha-tikta* in India, which translates as the King of Bitters. In Ayurvedic medicine the herb is referred to as Kalmegh or Kalamegha, meaning "dark cloud."

Anemopsis californica • Yerba Mansa

Yerba mansa is a resinous herb that has a warming and stimulating effect on the respiratory mucous membranes. The roots of the wetland plant can be included in formulas for bronchial and upper respiratory mucous congestion. Yerba mansa is also useful in throat sprays because its sticky nature helps formulas cling to the pharynx and tonsils, helps check secretions, and helps decongest swollen tissues.

Angelica dahurica • Bai Zhi

The roots of this nothern Asian species are used for allergic phenomena and lung inflammation. A compound in the plant, angelicotoxin, has an excitatory effect on the respiratory system, acting as an expectorant and sialagogue. Like other *Angelica* species, bai zhi is said to be a blood mover and may help allay the pain of congestive ailments such as headaches, sinusitis, and toothaches. Bai zhi is used in formulas for allergic airway disorders, rhinitis, and headaches and is noted to contain furanocoumarins with mast cell–stabilizing effects. Also known as *Dahurian angelica*, this plant also goes by the common names Chinese angelica and root of the holy ghost.

Angelica sinensis • Dong Quai

This and several other species of *Angelica*, are widely researched as having antiallergy and anti-inflammatory effects; the roots are useful in formulas for hay fever, airway reactivity, and allergies. *Angelica sinensis* is specifically indicated for allergic individuals and those with atopic constitutions in cough, cold, sinus, dyspnea, and other respiratory formulas. The antispasmodic compound butylidenephthalide has been identified. The common name dong quai is an anglicized version of the Chinese pinyin *dānggūi*.

Apium graveolens • Celery

Celery seeds and extracts are a traditional antiasthmatic and brochodilating remedy. Butylphthalide contributes to celery's flavor and aromatic odor and, along with apigenin, acts as a calcium channel blocker and contributes to the bronchodilating effects.[125]

Ardisia japonica • Marlberry

Ardisia species are native to Asia and considered one of the 50 fundamental medicines of TCM. Root juice and medicinal preparations have been traditionally used to treat cough, fever, chest pain, and a variety of inflammatory complaints. *Ardisia* species are being explored as anticancer agents. Jamaican and West Indian species including *A. esculenta*, *A. tinifolia*, and *A. coriacea* (beefwood) produce edible fruits. Large or repetitive doses are to be avoided due to possible renal toxicity. *Ardisia japonica* also goes by the common name coralberry.

Armoracia rusticana • Horseradish

Horseradish is a folkloric remedy for bacterial infections of the respiratory tract and is one of the best remedies to move stagnant mucus out of the sinuses. Its use can lessen sinus pain and pressure, and it is a valuable ingredient to include in sinusitis and head cold formulas. Fresh horseradish root, when tolerated, is also very useful to eat simply spread on crackers, or it can be prepared into a tea with honey and lemon. For those prone to sinus infections from hay fever, respiratory allergies, or simple colds, using horseradish at the initial onset of symptoms may help prevent progression into sinus congestion and infection. Sinigrin is a selenium-containing glucosinolate found in horseradish and credited with anticancer and detoxifying effects. Sinigrin breaks down into allyl isothiocyanate with many beneficial physiologic effects, including antimicrobial activity. Horseradish is credited with anti-inflammatory effect due to regulating effects on cyclooxygenase and lipoxygenase.[126]

Artemisia annua • Sweet Annie

The aerial parts of *Artemisia annua* are a traditional medicine for febrile illness. *A. annua* is recommended by the World Health Organization for tuberculosis, with numerous antimicrobial effects credited to artemisinin. Artesunate is a semisynthetic derivative of artemisinin, which may also effectively modulate or inhibit the tuberculosis mycobacteria with a daily dose of 3.5 milligrams/ kilogram. No toxicity and side effects are reported. Other common names of *A. annua* include sweet wormwood and sweet mugwort, as well as annual wormwood and annual mugwort. Other species of *Artemisia* have bronchodilating and mast cell–stabilizing effects, including *A. caerulescens* and *A. keiskeana*.

Asclepias tuberosa • Pleurisy Root

Asclepias roots are a classic herbal remedy for pleurisy, as the common name implies, and are specifically indicated for pain in the lungs and pleural membranes. *Asclepias* is a relaxing antispasmodic and a gentle and well-tolerated expectorant, even for considerable consolidation of the lungs, as with pneumonia. *Asclepias* is appropriate for all respiratory infection symptoms including fever, cough, soreness in the chest, and runny nose. Taking a dropperful of tincture in hot water can have a gentle diaphoretic effect, and the Eclectic physicians purported *Asclepias* to enhance elimination via the skin, helping to reduce inflammation and congestion in the lungs. Gardeners know this plant by the name butterfly weed as the flowering plant attracts numerous butterflies.

Aspidosperma quebracho • Quebracho

Aspidosperma is a South American plant used as a respiratory tonic and expectorant. The bark of the tree is high in alkaloids such as aspidospermidine. Due to a stimulating effect on cardiac and pulmonary nerve plexi, the plant may help to decongest the respiratory mucosa, especially in cases where circulation and general vitality are weak. Include *Aspidosperma* in formulas to move stagnant mucus out of the lungs for asthma, emphysema, and bronchitis, especially when associated with underlying pulmonary congestion. Specific indications for *Aspidosperma* include dyspnea in those with lung and cardiac disease, including heart failure, emphysema, and COPD with pulmonary congestion and feeble circulation. A pale face, weak or irregular pulse, cardiac cough, and orthopnea—the symptoms of congestive heart failure— are additional specific indications for *Aspidosperma*.

Astragalus membranaceus • Milk Vetch

Astragalus roots have extensive immunomodulating effects and are useful for improving immune deficiency and the tendency to infections, as well as for reducing immune hyperreactivity, respiratory allergies, and asthma. *Astragalus* is most effective when used for 3 to 6 months to improve chronic infections or reduce hyperreactivity states and is less appropriate in acute formulas for asthma or infections. Due to the mild flavor, I encourage patients with chronic respiratory disease to use *Astragalus* powder in smoothies and the dried root slices prepared with bone broths and vegetable stock to consume as a food.

Atropa belladonna • Belladonna

The aerial parts of *A. belladonna* are a traditional remedy for wheezing and chest tightness. *Atropa* contains atropine, which, as an anticholinergic agent, may help relax bronchospasms when topically applied or included in small amounts in oral formulas. **Caution:** Due to possible central nervous system (CNS) effects, belladonna should be formulated only by skilled clinicians familiar with the plant. An excessive dose can result in dry mouth, visual disturbance, and hallucinations. Tropane alkaloids such as atropine can be altered into semisynthetic compounds, several of which are approved by the FDA for dyspnea and available in inhalant forms including tiotropium bromide.

Azadirachta indica • Neem

Native to India, Africa, and other tropical parts of the world, *Azadirachta indica* is traditionally used for the treatment of conditions including cancer, hypertension, heart diseases, and skin disorders. The bitter leaves have also been used as a general bitter tonic beverage, and several oil preparations are obtained from the plant in India where they are used medicinally. Medicinal preparations of *Azadirachta* are used for allergies, as an antiviral herb, and against tuberculosis. Nimbin, nimbinine, nimbandiol, and quercetin in the plant are credited with mast cell–stabilizing effects, and glycoproteins in neem leaves have immunomodulating and anticancer effects.[127] The oral consumption of neem oil must be limited to a small amount and for a limited duration due to toxicity concerns. The plant is also known as *Melia azadirachta*. Other common names include bean tree, China tree, false sycamore, and pride of India.

Bacopa monnieri • Brahmi

The leaves of this herbaceous plant have been listed in ancient Ayurvedic texts as a brain tonic and its cholinesterase-inhibiting effects have been demonstrated. *Bacopa* is traditionally used in India for asthma and other allergic phenomena and is a mast cell stabilizer. Bacopasides, alkaloids, and glycosides in the plant are all credited with immunomodulating effects, including an ability to reduce mast cell activation and histamine release. Nicotine and related alkaloids have been identified, explaining some of the plant's traditional usage for acute and chronic asthma. The bacopasides are a group of triterpenoid saponins credited with many anti-inflammatory and immunomodulating effects.

Berberis aquifolium • Oregon Grape

Oregon grape is also known as *Mahonia aquifolium*, and its roots can be employed as an alterative ingredient in formulas for chronic infections in mucous membranes. By supporting elimination of wastes, enhancing digestion, and improving glucose and lipid metabolism, *Berberis* can improve the tendency to colds, sore throats, and lung infections. It can be employed in teas or tinctures taken consistently for 3 months or more to support liver and metabolic function and reduce chronic respiratory infections. Berberine, one of the active alkaloids credited with antimicrobial and metabolic effects, is also available in concentrated capsules and may also be part of a protocol to reduce chronic colds and lung and sinus infections.

Brassica nigra • Black Mustard

Mustard poultices and plasters are traditional remedies to treat lung pain due to pneumonia and to help thin and move consolidated mucus in the lungs. Seeds from both *Brassica nigra* (also known as *Sinapis nigra*) and *Sinapsis alba* (white mustard) are ground to prepare topical applications that have expectorating, pain-relieving, and circulation-enhancing actions when applied to the chest. Mustard is too irritating to put directly on the skin and must be cut with flour. Use 1 part mustard to 4 parts flour, or even 1 part to 10 parts or more of flour for children or those with sensitive skin.

Bryonia dioica • Bryony

Bryonia roots are specifically indicated for respiratory infections associated with chest pain, especially sharp pain that is worse with motion, producing cutting, shooting, or lancinating sensations. *Bryonia* may be used for acute respiratory infections and is often used in small dosages as a simple or in small proportions in combination herbal tincture formulas. Some of the herbal physicians of the early 1900s recommended *Bryonia* for those who are weak, apathetic, and devitalized and experiencing hyperesthesia in the skin or in the lungs. *B. alba* is also used medicinally in the same manner. *B. dioica* is also known as *B. cretica*.

Caesalpinia sappan • Su Mu

Also known as *Biancaea sappan*, this leguminous tree, also called brezel wood, is a dye plant. It is used therapeutically to support circulation and treat infections. Homoisoflavonoids have been found in the bark and wood that are active against influenza A via effects on glycoprotein receptors and binding.[128]

Camellia sinensis • Green Tea

The catechins in green tea are well-studied all-purpose antioxidant and anti-inflammatory compounds that can be used in capsules to supplement teas and tinctures. Green tea can also be included in herbal tea recipes to treat allergies and lung inflammation. Due to the level of anticancer and tissue-protectant effects that green tea offers, the herb may also be considered in protocols for lung cancer and in chemotherapy support.

Capsicum annuum • Cayenne

Also known as *C. frutescens*, cayenne may be added as a tincture in small amounts to cough formulas and throat sprays for its anodyne effects on the throat. Many folkloric texts, especially those by Jethro Kloss, recommend cayenne to potentiate the effects of *Lobelia*, and the pair are commonly seen together in respiratory formulas.

Carum carvi • Caraway

Caraway seeds have been used as a general carminative, and like other Apiaceae family plants, *Carum* has anti-inflammatory and bronchodilating effects.

Cassia • See Senna

Centella asiatica • Gotu Kola

The leaves of gotu kola are edible and have a long history of use as nourishing and tissue restorative herbal medicine. Little research exists for this purpose, but gotu kola tea and other preparations are historically used to heal ulcers, broken bones, injured nerves, and skin lesions and may be considered in formulas for respiratory diseases such as emphysema and bronchiectasis due to its tissue-regenerating and protecting properties. Gotu kola may be included in formulas to protect elasticity of the lungs and is safe to use long term in chronic inflammatory conditions of the airways. Gotu kola is also sometimes known as *Hydrocotyle asiatica*, and also goes by the common name pennywort.

Cimicifuga • See Actaea

Cinnamomum camphora • Camphor Tree

This species of cinnamon contains camphor, an aromatic compound in the tree wood that is purified and commercially available as a pure essential oil. Camphor is classically used in chest rubs to help treat chest pain, congestion, and cough. Camphor is specifically indicated for abundant mucus in the airways that causes a suffocating sensation and dyspnea. Camphor essential oil is also appropriate to use in throat sprays for laryngitis and tonsillitis and to help decongest the throat in cases of problematic hacking cough and postnasal drip.

Cinnamomum verum • Ceylon Cinnamon

Cinnamon bark from several different species of *Cinnamomum* are traditionally used for bronchitis and may be used to help control bleeding from the lungs. *C. verum* is also known as *C. zeylanicum*. *C. cassia* (cassia cinnamon) is another medicinal species. These species are high in cinnamaldehyde and eugenol instead of camphor that is found in *Cinnamomum camphora*.

Cissampelos sympodialis • Cissampelos

The bark of this vining plant has been traditionally used to treat cough, bronchitis, and asthma. A bronchodilating isoquinoline alkaloid called warifteine has been identified in *C. sympodialis*. Current research is reporting antiallergy and immunomodulating effects via eosinophil and lymphocyte regulation. A related species, *C. pareira*, goes by the name abuta in South America and laghu patha in Ayurvedic practice, where it is used to treat viral and febrile illnesses.

Citrus spp. • Citrus Fruits

The rinds of oranges, lemons, limes, and grapefruit contain limonene, credited with many immunomodulating and medicinal actions. In TCM, citrus rinds are said to "open the orifices," helping medicines enter the digestive and respiratory passages. Citrus peels are common ingredients in TCM tea formulas for cough and respiratory infection and disease and are also commonly combined with Western herbs, both for flavor and medicinal purposes. Bitter orange (*Citrus* × *aurantium*) essential oil and other citrus essential oils can be used in steam inhalations to treat respiratory infections. Cell culture studies show orange essential oil and purified limonene found in the oil to deter lung cancer proliferation.

Clerodendron serratum • Glorybower

The *Clerodendrons* are tropical species. *C. serratum* has bronchodilating effects and may be used in formulas for respiratory disorders. The leaves have been traditionally used in Ayurvedic medicine for *shwasa*, which is translated as breathlessness. The leaves are said to regulate water and resolve dampness in TCM. Apigenin, phenolic glycosides, and a wide variety of medicinal compounds have been identified in *Clerodendron* species. The plant is also called bagflower, bleeding-heart, and bharangi.

Cnidium monnieri • Snow Parsley

Snow parsley seeds are high in coumarins such as osthol, imperatorin, and xanthotoxol, which are noted to reduce allergic reactivity of the airways. *Cnidium* may be used for asthma and allergic airway disease and to reduce oxidative stress in the airways in a variety of chronic pulmonary conditions. *C. monnieri* is a traditional medicine in China used for allergic activity, with many medicinal effects credited to osthol.

Coleus forskohlii • See *Plectranthus forskohlii*

Collinsonia canadensis • Stoneroot

All parts of the plant are used medicinally, but the roots are considered to be the most powerful. Primarily an herbal remedy for vascular congestion, *Collinsonia* root medications are specifically indicated for coughing when due to a tickle in the throat or for the sensation of swelling, congestion, or a foreign body in the throat or a sensation of constriction in the throat. Hoarse voice, laryngitis, respiratory catarrh, and congested tonsils and pharyngeal tissues are traditional indications for *Collinsonia*. Small doses are warming and stimulating and may be diaphoretic, but large doses can become emetic and are to be avoided. Root decoctions are indicated for simple colds and for spastic cough, even whooping cough (pertussis), and are traditionally said to be most effective as a hot tea.

Commiphora myrrha • Myrrh

Resin from the *Commiphora myrrha* tree is collected by incising the living tree bark and allowing the sticky sap to harden. Myrrh resin was powdered and mixed in wine in ancient times to use as an antimicrobial medicine and as an ingredient in antiseptic mouthwashes, throat gargles, and various remedies for cough and lung afflictions. Myrrh stimulates mucous secretion in swollen, congested, boggy tissues. It is especially effective in treating tonsillitis that accompanies respiratory infections. Myrrh is also useful for chronic bronchitis when the tissues are lax and chronically congested or for chronic postnasal drip associated with atonic tissues, subject to swelling and stagnation. *C. molmol* is considered to be synonomous with *C. myrrha*.

Cornus officinalis • Japanese cornel

This dogwood relative, also called shan zhu yu, is used in TCM, where the bark may act as a quinine substitute in treating malarial fevers. It can be included in complex formulas for treating respiratory infection and inflammation. *Cornus* has been used in TCM for at least 2,000 years to invigorate circulation and treat chronic inflammatory diseases. The edible, nutritious fruits are referred to as Cornelian cherries and are also used medicinally in formulas for vascular inflammation and vasoprotective effects. Japanese cornel contains iridoid glycosides, credited with numerous immunomodulating activities.

Corydalis cava • Corydalis

Corydalis is a poppy family plant whose underground tuber contains opiate-like alkaloids traditionally used for pain and for calmative effects. This and other species of *Corydalis* contain bulbocapnine, an acetylcholinesterase inhibitor that blocks dopamine in the brain. It causes fatal poisoning in cattle if they are allowed to overfeed on the plant. However, in small doses, *Corydalis* may help wean from opiate medications, improve sleep in painful or infectious conditions, and help soothe distressed respiration. Cholinesterase inhibitors can also support motor function, and *Corydalis* may help to allay musculoskeletal irritation and spasms of intercostal muscles associated with severe coughing. Cholinesterase inhibitors are also shown to reduce inflammation

in situations of acute lung injury, explaining the use of *Corydalis* in TCM cough and lung formulas.[129]

Crataegus spp. • Hawthorn

The berries of the hawthorn tree are high in anti-inflammatory flavonoids and are most studied for their stabilizing and protective effects on the heart and vasculature. *Crataegus* may also be a supportive herb in formulas for allergic airway diseases, for mast cell activation in asthma, and for those with heart disease aiming to prevent pneumonia or other lung infection. *Crataegus* may be one ingredient in formulas for difficulty breathing due to underlying circulatory and heart issues. *Crataegus monogyna*, *C. oxyacantha*, and *C. laevigata* are all used medicinally.

Croton lechleri • Sangre de Drago

Euphorbiaceae (spurge) family plants are often rich in latex and resin, and sangre de drago, meaning blood of the dragon, exudes a blood red resin traditionally used for wound healing, digestive ulcers, and repairing internal tissues. *Croton lechleri* resin stores well and may be included in formulas for cystic fibrosis due to effects on the cystic fibrosis transmembrane conductance regulator (CFTR) chloride channel found on the apical surfaces of epithelial cells lining the airways. Because cystic fibrosis involves defects in the CFTR ion channel, salt and water transport in epithelial cells is affected resulting in abundant, tenacious mucus in the airways. Sangre de drago can be mixed into tincture formulas, and resin extracts can be mixed into tinctures or used in drinking water long term.

Curcuma longa • Turmeric

Curcuma rhizomes are highly popular and robustly researched, credited with anti-inflammatory, tissue-protective, antimicrobial, and anticancer properties. *Curcuma* may be included in formulas for asthma, hay fever, and airway reactivity due to underlying chemical and allergen hypersensitivity. *Curcuma* is also appropriate in formulas to reduce chronic respiratory infections and as a supportive herb or capsule to treat sinusitis, bronchitis, and pneumonia. *Curcuma* may also protect connective tissue in chronic inflammatory and degenerative diseases of the lungs, and it can be used as a medicinal food, prepared into beverages, or taken in teas, capsules, and tinctures long term to help slow fibrotic processes. *Curcuma* has been used to reduce allergic reactivity with reported success, and both the turmerones and curcuminoids have mast cell–stabilizing effects.

Datura stramonium • Jimson Weed

Caution: *Datura* is an anticholinergic agent that may help relax bronchospasms; however, its use is limited due to drying, and potentially hallucinogenic, effects. *Datura* is a folkloric plant used for asthma, where the dried leaves were used in smoking blends or simply crumbled over glowing coals and the vapors inhaled to alleviate acute dyspnea. The plant is high in tropane alkaloids, which can be drying to mucous membranes, and therefore is best used for copious moist lung secretions; it is best avoided in cases where there is thick, sticky mucous congestion in the lungs. *Datura* can also have psychoactive effects and even be fatal if inappropriately used.

Digitalis purpurea • Foxglove

Most well known for treating heart failure, *Digitalis* may also be appropriate in formulas for febrile diseases, acute lung inflammations, asthma, hemoptysis, and whooping cough. *Digitalis* is specific for a "cardiac cough," referring to coughing, dyspnea, and fluid in the lungs due to congestive heart failure or heart valve disease. See "Cardiac Glycosides for Cardiomyopathy" on page 49 for further details.

Dracocephalum rupestre • Dragon's Head

Dracocephalum rupestre is native to western China and is used in folk medicine for treating respiratory infections, including colds, coughs, and laryngitis. The plant contains dracocephins, which are composed of the well-studied anti-inflammatory compound naringenin and which are also credited with numerous anti-inflammatory and cytoprotective effects. The plant also contains the flavonoid eriodictyol, found to limit lipopolysaccharide-induced acute lung injury. *Dracocephalum* has been shown to modulate many such inflammatory cytokines in the lungs, and although no clinical trials have been done, the plants have long been used to treat lung diseases. These actions may help protect the lungs in cases of acute respiratory distress syndrome and limit permanent alveolar damage and fibrotic degeneration of the lungs.

Drosera rotundifolia • Sundew

Drosera is specifically indicated for croup, whooping cough, or similar tight spastic coughs associated with constriction in the throat. The entire plant is tinctured and used for choking sensations and laryngeal spasm.

Echinacea angustifolia • Coneflower

Echinacea is indicated for chronic infections with slow recovery time and especially for destruction and decay of the tissues, such as foul-smelling breath, ulcers on the tonsils, large quantities of tonsillar exudate, or a thick purulent coating on the tongue. When colds or lower respiratory infections are chronic, *Echinacea* may be useful to reduce toxicity, improve immune resistance, and treat new infections. It may be used in both acute and chronic formulas when the preceding symptoms accompany. *Echinacea* is also indicated for serious septicemia, lung abcesses, malignant lesions, and acute tissue destruction.

Eclipta prostrata • Eclipta

Eclipta prostrata (and other species) is a weedy herb of the Asteraceae family, traditionally used in Ayurvedic medicine for anti-inflammatory effects, including for allergic reactivity of the airways and asthma. *Eclipta* species are also traditional for anaphylactic reactivity, and numerous antimicrobial, anti-inflammatory, and antioxidant mechanisms have been reported. *Eclipta prostrata* is also known as *E. alba*. *E. prostrata* is a cosmopolitan species and goes by a number of names throughout the world including yerba de tago, False Daisy, karisilanganni, and bhringraj.

Elaeocarpus serratus • Bhadrasey

Several species of *Elaeocarpus* bear olive-like fruits that are pickled and used to make chutneys in India. The seeds are used in meditation beads, known as Bodhi beads (*Rudraksha* in Hindi), and in ceremonial garlands. Traditional medicinal uses of the plant are as a hypotensive sedative, to calm agitation, and to soothe bronchial and cardiac excitation. Leaves and fruits of the plant have been used folklorically in formulas for asthma. Bronchodilating glycosides, steroids, alkaloid, and flavonoids have been identified in *Elaeocarpus*. Due to its many uses, the plant has become threatened, and ecologically grown products are an important consideration. This herb is also known as *E. ganitrus*.

Emblica • See *Phyllanthus*

Ephedra sinica • Ma Huang

Ephedra sinica is native to China and the aerial parts are used traditionally for asthma. Several *Ephedra* species contain the bronchodilating alkaloid ephedrine. The North American species, *E. nevadensis* (Mormon tea), was used by Native Americans and Mormon pioneers as a mildly stimulating beverage. Ephedrine is a caffeine-like alkaloid useful for dyspnea and allergic airway disorders, but *Ephedra* products were removed from the market by the FDA in 2004 when their abuse led to cases of acute hypertension and tachycardia, including several fatal cardiac arrhythmia cases. Such side effects are extremely unlikely with the consumption of *Ephedra* tea, but pills containing ephedrine were used as stimulants, party drugs, and even as raw material for making methamphetamines. *Ephedra* teas are the only medicines currently available, prepared from the photosynthetic twigs.

Epimedium brevicornu • Horny Goatweed

This species is also known as *E. rotundatum*, and several species of *Epimedium* have been used for a wide variety of medicinal effects. Other common names are yin yang huo or bishop's hat. Icariin is a phosphodiesterase inhibitor found in the plant; it contributes to the herb's cardiac and respiratory effects. *Epimedium* has been used in TCM for formulas to invigorate chi when energy is deficient and the voice is weak, as well as in cases of chronic lung disease. Other symptoms of chi (qi) deficiency include cough with a large amount of sputum, loss of appetite, fatigue and weakness, pale and bulgy tongue, spontaneous sweating, loose stools, and a weak thin pulse.

Equisetum spp. • Horsetail

Also called scouring rush, *Equisetum* species are primarily known as urinary and connective tissue tonics. The aerial parts are high in minerals and silica, and low levels of silica in connective tissue are seen with aging and loss of strength and integrity in collagen and other extracellular matrix proteins. Dietary insufficiency of silicon leads to impaired bone and connective tissue repair and regeneration. I include *Equisetum* in formulas for teas intended for long-term use to help protect elasticity of the lung connective tissue in cases of emphysema, COPD, and other situations of fibrosis. The dried stalks can be used in teas and the powder can be encapsulated for oral consumption or used in smoothies and medicinal foods. Commonly used North American species include *E. arvense* and *E. hyemale*.

Eriodictyon californicum • Yerba Santa

Yerba santa or mountain balm is a high-resin plant whose aerial parts are used for chronic respiratory congestion of the laryngeal mucosa, as well as for bronchial and pulmonary congestion. Eriodictyol, a flavonoid found in the leaves, has been shown to modulate many inflammatory cytokines in the lungs, and although no clinical trials have been carried out, the plants have long been used for lung diseases. This traditional Native American remedy is specifically indicated for "humid asthma" and coughs, meaning when there is abundant watery or free-flowing mucus in the airways. Yerba santa resin has a warming, drying, and antimicrobial effect in the lungs and upper respiratory passages and can be included in formulas for simple colds as well as for coughs, bronchitis, and rattling of mucus in the airways. Yerba santa can also mask bitter flavors, having the ability to affect bitter receptors and change bitter flavors to sweet, a phenomenon also credited to eriodictyol. Modern research shows that bitter receptors occur throughout the digestive system, not just on the tongue, and are distributed in the respiratory and urinary passages as well. Eriodictyol may bind bitter receptors in the airway, promote the mucocillary apparatus on airway epithelia, and relax bronchial smooth muscle, promoting expectoration.

Eschscholzia californica • California Poppy

This poppy is native to the Pacific Coast and is the state flower of California. The entire plant is used in pain and anxiety formulas. Although this plant contains only a small fraction of the opiate compounds found in opium poppies, *Eschscholzia* can be used in protocols for opiate withdrawal and to allay pain and improve sleep in chronic ailments. Research suggests that the alkaloids do not bind to opiate receptors, but rather to γ-aminobutyric acid (GABA) and possibly other neurotransmitters, offering gentle sedative effects. Large and repetitive doses of the tincture can act as a respiratory sedative for problematic coughs and nervous cough. *Eschscholzia* may also be included in nighttime cough remedies intended for use before bed and when coughs are interfering with sleep.

Eucalyptus globulus • Eucalyptus

Eucalyptus leaves are one of the key respiratory herbs used in saunas, inhalants, chest rubs, tinctures, and teas. *Eucalyptus* has a warming and stimulating quality, and while all-purpose for respiratory infections, it is specifically indicated for a sensation of chill. *Eucalyptus* has general antimicrobial effects due to the abundance of volatile oils and is an effective respiratory expectorant when applied topically, inhaled, or ingested orally in tinctures and teas. *Eucalyptus* may be especially effective when there is consolidation in the lungs and mucus is difficult to expectorate, or when colds and bronchitis linger because of abundant mucous production. Eucalyptol (cineol) in the volatile oil fraction of *Eucalyptus* promotes expectoration in COPD patients, improving pulmonary function tests and reducing the frequency and severity of acute exacerbations. Other conditions associated with excessive mucous production such as rhinosinusitis and asthma may also respond to *Eucalyptus*, with steams and inhalations being particularly effective. *Eucalyptus* also has significant antimicrobial effects and can be included in formulas for infectious bronchitis.

Eupatorium perfoliatum • Boneset

Eupatorium is a traditional North American herb whose aerial parts are used for intermittent fevers (especially fevers worse in the morning) and flu. It is specifically indicated for severe and intense aching muscles, from which the common name boneset derives, in reference to viral myalgia that is so intense as to be likened to the pain of a broken bone. This species of *Eupatorium* is also specific for chilliness, nausea that is worse with motion, sore and sensitive eyes, sore chest, and laryngitis. Boneset has diaphoretic and gently stimulating actions, used by Native Americans for febrile respiratory infections and influenza with myalgia. Native Americans used the plants to promote sweating and fevers and speed the resolution of febrile respiratory illnesses. The gentle expectorating effects can become emetic if taken hot and in repeated dosage, so small doses of cold tea are best. Many early authors and sources report that the plant is also specific for febrile respiratory illness that results from living in damp marshy locations. Eupafolin found in the plant has antioxidant, anti-inflammatory, and antiproliferative activities, as well as antiviral activity. The plant also contains small amounts of pyrrolizidine alkaloids, compounds with known hepatoxicty with large and continuous exposure. For this reason, it is best to use boneset for acute respiratory infections and discontinue in a week or two. Other common names for *Eupatorium purpureum* include agueweed, feverwort, and sweating-plant. *E. chinense* is a related species used in China for acute and chronic inflammatory conditions of the lungs. In several species, the sequiterpene lactones are credited with mast cell–stabilizing effects.

Euphorbia spp. • Spurge, Asthma Weed

Several species of *Euphorbia* go by the common name of spurge, and the thick root bark is used to treat lung congestion. *E. hirta* (asthma plant) is used as an expectorant at low dosages and an emetic at high dosages. *E. parviflora* (also known as *E. pilulifera*) is also known as asthma weed or pill-bearing spurge, due to its folkloric usage for wheezing and dyspnea, and research has noted antihistamine and antiallergy activity. These and other species are useful expectorants, but can act as harsh emetics and cathartics if large and frequent doses are consumed, giving rise to another common name, wild ipecac. Small amounts of spurge can be included in formulas for hay fever, laryngeal spasms, emphysema, coughs including whooping cough, bronchitis, asthma, and febrile respiratory illness with lung congestion. It is usually advised to blend *Euphorbia* with more nourishing, antispasmodic, and soothing plants. Concentrated latex is a harsh emetic, and even poisonous, but in small amounts it contains medicinally valuable diterpenes and triterpenes. The Asian species *E. trigona* and *E. pekinensis* are used in traditional Chinese medicine and included in the 50 fundamental herbs, a list of the most commonly used herbal medicines. Broad antimicrobial effects are also reported from *Euphorbia* species, including activity against antibiotic-resistant microbes.[130] Toxicity studies have not revealed any issues with acute use—no hepatic, renal, or hematologic toxicity has been revealed with long-term useage.

Euphrasia officinalis • Eyebright

The common name eyebright is well deserved as this is one of the most reliable herbal medicines for treating allergic reactivity in the eyes. It is also a general hay fever remedy for the upper airways. *Euphrasia* is specifically indicated for watery discharges from the eyes and nose, due to either early stages of a viral infection or allergic reactivity. It is specifically indicated for allergic conjunctivitis, itchy eyes with lacrimation, sneezing and an itchy nose, and itchy ears, all of which may then progress to more purulent mucous accumulation, conjunctivitis, rhinitis, and otitis media related to swollen congested eustachian tubes. The aerial parts may be dried for teas and eyewashes, or the fresh plant may be tinctured. Modern research has demonstrated *Euphrasia* to have anti-inflammatory, antimicrobial, or immunomodulatory effects, including direct effects on the cornea, credited to flavonoids, phenolic acids, and the iridoid alkaloid aucubin.

Foeniculum vulgare • Fennel

The seeds may be used to prepare eyewashes in cases of eye symptoms associated with upper respiratory infections. Fennel teas, tinctures, and concentrated essential oils are also useful remedies in formulas for asthma, wheezing, and spastic coughing. Due to their pleasing flavor and gentle nature, fennel teas and topical chest compresses prepared from essential oils are especially good medicines for infants and toddlers suffering from chest colds or allergic airway disease. Modern investigations have employed fennel essential oil in nebulizers and various inhalants to treat acute bronchospasms. Fennel seeds are credited with many anti-inflammatory actions in the lungs, many of which are credited to the volatile oil anethole. Modern research suggests that fennel can reduce permanent damage to the lungs in acute inflammatory disease, effectively blocking inflammatory molecules. Fennel essential oil may be included in respiratory therapies, from simple steam inhalation to nebulized medications.

Fritillaria cirrhosa • Yellow Himalayan Fritillaria

The bulbs of some *Fritillaria* species are edible, and *F. cirrhosa* is the main species used medicinally in China, Japan, and the Tibetan plateau. One of the earliest Chinese herbals, the *Shen Nong Ben Cao Jing* (*The Divine Farmer's Materia Medica Classic*) details six species of *Fritillaria*. At the present time, there are more than 200 different products or formulas that contain *Fritillaria* on the market in China, with the majority being indicated for respiratory conditions including asthma and bronchitis. Significant antitussive, expectorant, and antiasthmatic effects are reported. Due to its popularity and limited habitat, the bulbs are threatened. Botanists have now separated what was three varieties of *Fritillaria cirrhosa* into three distinct species: *F. yuzhongensis*, *F. sichuanica*, and *F. taipaiensis*. All have been referred to as chuan bei mu in China.

Galphimia glauca • Calderona Amarilla

Galphimia glauca, also called flor estrella, has been used in Ayurvedic medicine as an antiasthmatic, muscle-relaxing anxiolytic, and anticonvulsant. In Mexico, the leaves are a traditional anxiolytic, and research has indentified galphimines contributing to the sedative effects. Research has also shown anti-inflammatory, antimalarical, antiallergic and antiasthmatic activities. *Galphimia* may be included in formulas for spastic cough, wheezing, and panic-related dyspnea. Broncho-dilating compounds have been identified, including tetragalloylquinic acid and quercetin. Related species are grown in other warm regions of the world where they are similarly used for anxiety, tension, and dyspnea.

Gardenia latifolia • Gardenia

Gardenia species are important ornamental and aromatic plants that are also used in TCM for "heat" symptoms.

G. latifolia is a long-lived tree known as papra in Hindi and is also referred to as Indian boxwood or Ceylon boxwood. The stem bark, leaves, roots, and fruits are all used medicinally in various traditional remedies as a hemostatic and to "drain fire" associated with heat and pain in joints, lungs, heart, or other tissues. Iridoid glycosides contribute to the pigments of the flowers as well as to the medicinal activity of *Gardenia* fruits. *Gardenia* can be included in formulas for lung inflammation and for cardiovascular disease contributing to pulmonary hypertension and loss of pulmonary elasticity. Bronchodilating saponins have been identified, and the iridoid glycoside geniposide and its metabolite genipin have numerous immunomodulatory effects and can limit respiratory inflammation by having direct effects on lipoxygenase and cyclooxygenase. The related species *Gardenia jasminoides* (cape jasmine) is also used medicinally to drain heat symptoms in TCM, and has antimicrobial and anti-inflammatory effects.

Ginkgo biloba • Maidenhair Tree

Ginkgo is most often regarded as a heart and vascular herb, for which it excels, but it is sometimes overlooked for its value in treating allergic airway diseases. Due to its many mast cell–stabilizing, antiallergy, and anti-inflammatory effects, *Ginkgo* is also useful for reducing hyperreactivity in the airways and improving circulation to the lungs in cases for pulmonary fibrosis and hypertension. Ginkgetin is a mast cell stabilizer found in the leaves, that along with the bronchodilating ginkgolides explain the traditional use of *Ginkgo* for hypersensitivity disorders in the lungs. Include *Ginkgo* in formulas for atopic individuals with asthma, hay fever, and hypersensitivity disorders of the airways.

Glycyrrhiza glabra • Licorice

Licorice roots are often used in teas to help improve the flavor of bland herbs, but offers its own antiviral, anti-inflammatory, and mucous membrane–healing effects. *Glycyrrhiza* triterpenoid saponins and flavonoids have broad anti-inflammatory effects, which together with mucilage in the roots offer a soothing demulcent action on respiratory mucous membranes. The major saponins, including glycyrrhizin and its aglycone glycyrrhetinic acid, are anti-inflammatory and credited with antiviral, antiulcer, antiallergy, and other effects. There have been more than a thousand papers published on the beneficial effects of licorice on the lungs, against lung cancer, mechanisms of action, and traditional formulas using licorice respiratory complaints. Other species used medicinally in China include *G. inflata* and *G. uralensis*. Glycyrrhizin has mineralocorticoid effects that may occasionally elevate the blood pressure in genetically susceptible individuals, so blood pressure should be monitored when using licorice regularly or long term.

Grindelia robusta, G. squarrosa, G. hirsutula • Gumweed

Gumweed is a high-resin plant with warming, stimulating, and expectorant effects. Its leaves have been used traditionally in cough and respiratory formulas where there is stuck and stagnant mucus, dry coughs, and asthma with scant expectoration and a sense of constriction in the chest. Native Americans used gumweed species to treat asthma and bronchitis, sometimes in smoking mixtures, and the plant is emphasized folklorically for harsh dry coughs in those with a pale or cyanosed countenance, and where there is painful hacking and spastic coughing. *Grindelia* is an expectorant and sialagogue, and while increasing new respiratory secretions, *Grindelia* medicines are capable of assisting with the removal of thick or stagnant mucus and reducing stagnant cough and asthmatic dyspnea. *Grindelia, Lobelia*, and *Datura* comprise a classic smoking mixture for the relief of acute asthma. *Grindelia* is also recommended for elderly patients with COPD or emphysema to aid in efficient expectoration, to help reduce paroxsysmal coughing, and to alleviate laryngeal spasm and mucosal congestion. The flavonoid hispidulin found in some *Grindelia* species is credited with antiasthmatic activity. Hispidulin has antioxidant, antifungal, anti-inflammatory, antimutagenic, and antineoplastic properties and is found in a handful of other plants also used for asthma and respiratory disorders. Hispidulin also binds GABA receptors in the brain and has an anticonvulsant action. Hispidulin may be 100-fold more potent than theophylline in its property of inhibiting platelet aggregation, contributing to the antihistamine and anti-inflammatory effects. *G. hirsutula* is also known as *G. cuneifolia*.

Hemidesmus indicus • Hemidesmus

Hemidesmus indicus is a twining shrub whose roots are commonly used in India against asthma, lung cancer, and autoimmune hyperreactivity in the lungs including asthma and bronchitis. This traditional Ayurvedic herb is also used for inflammation, ulcers, fever, and digestive inflammation. The root is said to have sweet, bitter, and cooling properties and to be specific for burning sensations. It is also used to neutralize poisons or recover from toxicity states and may be helpful for respiratory inflammation and allergic reactivity. Specific indications include loss of taste, dyspnea, and chronic cough. Modern research shows antiangiogenic and apoptotic effects, explaining its traditional use in treating lung tumors. *Hemidesmus* may also help treat obesity and hyperlipidemia, which can contribute to systemic oxidative stress and heart and lung disease. *Hemidesmus* is traditionally decocted with *Nigella* and *Smilax* in Ayurvedic formulas to treat chronic inflammatory diseases, cancer, and liver ailments. The plant is sometimes referred to as East Indian sarsasparilla or false sarsasparilla.

Hydrangea macrophylla • Hydrangea

Hydrangea may be most well known as a perennial landscaping plant, but the leaves are also fermented into a beverage known as amacha in Japan and used in folkloric herbal medicine for kidney stones and cough. Hydrangeae Dulcis Folium is an herbal medicine prepared by fermenting the dried leaves of *H. macrophylla* and credited with antiallergic and antimicrobial properties.[131] Thunberginol compounds in the leaves are noted to be mast cell stabilizers and to reduce histamine release. *H. paniculata* is widely distributed in China where it is used for malaria, fevers, and other ailments.

Hydrastis canadensis • Goldenseal

Goldenseal is a popular antimicrobial agent, specifically indicated for excessive mucous production and purulence in mucous membranes. Goldenseal powder has been used as a snuff for rhinitis and nasal polyps and is a useful ingredient in neti pot rinses. *Hydrastis* is very effective to astringe and disinfect pharyngitis and tonsillitis and may be included in cough, cold, flu, and bronchitis remedies when runny nose and enlarged mucous-filled tonsils accompany.

Hyoscyamus niger • Henbane

Hyoscyamus leaves have been used in ancient times for pain, spasm, cough, renal colic, and urinary retention. Hyoscyamine along with atropine and scopolamine are among the plant's tropane alkaloids long noted to have anticholinergic effect that may help relax bronchospasms, and the herb can be included in small amounts in formulas for cough. Because of its serious possible side effects, *Hyoscyamus* is to be used by skilled clinicians only.

Hyoscyamus is best for hacking dry cough that interferes with sleep or is worse lying down. It is also helpful for tense, nervous individuals, but should be avoided for those in exhausted or weakened states. The Eclectic authors state that *Hyoscyamus* is not effective for whooping cough but is effective for bronchitis and pneumonia. **Caution:** *Hyoscyamus* has many possible side effects including mydriasis, tachycardia, arrhythmia, agitation, convulsion and coma, dry mouth, thirst, slurred speech, difficulty speaking, dysphagia, warm flushed skin, pyrexia, nausea, vomiting, headache, blurred vision and photophobia, urinary retention, distension of the bladder, drowsiness, hyperreflexia, hallucinations, confusion, disorientation, delirium, aggressiveness, and combative behavior.

Hypericum perforatum • St. Johnswort

St. Johnswort is most well known in the present era for its mood-modulating effects, but it also has broad anti-inflammatory and some antiviral and anticancer properties. Both oral and topical preparations of St. Johnswort leaves are used traditionally as anodynes, especially for nerve injury or severe strains and sprains where there is extensive bruising and inflammation. The vulnerary and anti-inflammatory properties of St. Johnswort also make it valuable in formulas for severe intercostal pain and possibly pleuritic inflammation. Where there is pulmonary inflammation, the herb's highly regarded ability to heal vascular and connective tissue trauma may be a reason to include it in topical oils and chest liniments, as well as in teas and tinctures. Its antiviral effects are photo mediated, and hypericin is shown to accumulate in cancer cells and contribute to anticancer effects. *Hypericum* speeds up the liver's metabolism of many drugs, so care must be taken if using the plant in patients who are also tenuously balanced on a lifesaving medication, such as a heart antiarrhythmic or anticoagulant.

Hyssopus officinalis • Hyssop

Hyssop leaves have drying and antimicrobial effects and can be included in teas and tincture formulas to treat congestion in the mucous membranes and to help treat coughs, colds, and bronchitis. *Hyssopus officinalis* is also known as *H. vulgaris*.

Inula helenium • Elecampane

The roots of elecampane are a traditional remedy for lung diseases including asthma, bronchitis, and whooping cough. *Inula helenium* occurs in both China and North America and is often used in cases of lung infection for its diaphoretic and expectorant effects; it has been used in the past for tuberculosis. Elecampane has expectorating action and may be active against some respiratory viruses. Sequiterpene lactones, including inulolide, are credited with mast cell–stabilizing effects, explaining some of the traditional recommendations for allergic and inflammatory disorders of the airways.

Iris versicolor • Wild Iris

Iris roots yield a potentially irritating medicine, esteemed by Native Americans who used the plant as a digestive and secretory stimulant. The plant strongly promotes various glandular secretions that one may hypothesize muscarinic effects, but little modern research has been done. The plant is such a reliable sialagogue that it was referred to as "vegetable mercury," in reference to the use of heavy metals such as mercury to promote secretions. *Iris* was highly respected by the Eclectic physicians, who included iris in formulas for respiratory congestion for old, dry, rubbery mucus, especially when associated with chest pain associated with digestive discomfort, biliary insufficiency, chronic constipation, and chronic complexion issues. Hence, *Iris* may be more of an adjuvant herb than a lead herb in formulas for respiratory conditions. *Iris* should be used only in small, diluted dosages, such as placing five drops in 4 ounces of water and taking a teaspoon of that mixture hourly. Combining *Iris* with ginger, black pepper, or camphor may reduce oral and digestive irritation.

Isatis tinctoria • Woad

Isatis tinctoria (also known as *I. indigotica*) is an important traditional Chinese medicine mainly used for the treatment of influenza and infectious diseases, and antiviral and immunomodulating activites are reported. The roots possess sulfur-containing alkaloids credited with antiviral and anti-inflammatory activity.[132] The species name *tinctoria* indicates the history as a dye plant and woad is traditionally used as a blue dye. It is irritating and possibly harmful to consume internally in large or prolonged doses. However, indoline, a constituent of this species, is noted to have anti-inflammatory effects by stabilizing mast cells.[133] *I. tinctoria* goes by the name ban lan gen in TCM.

Juniperus communis • Juniper

Sometimes said to be the only spice derived from conifers, juniper berries are used in only small amounts in cooking and in herbal formulas due to their strong and potentially irritating volatile oils, most notably pinene. Due to antimicrobial effects, juniper essential oil can be included in inhalant steams for those with pneumonia or as a preventive in those with cardiopulmonary weakness at the onset of simple colds. Juniper oil inhalants have been shown to deter the establishment of biofilms, as well as to penetrate antimicrobial-resistant respiratory biofilms of *Staphylococcus aureus* and *Pseudomonas aeruginosa*.[134] Juniper-based medications are also traditional for tuberculosis, and activity against mycobacteria, including drug-resistant strains, has been confirmed. Juniper is also a traditional remedy for asthma and hypersensitivity of the airways and gut; calcium channel-blocking effects have been noted, explaining the traditional use in smooth muscle spasm.[135] Juniper berry may be used as a synergist in formulas for hay fever with a tendency to nasal ulcers. Related species include *J. phoenicea, J. oxycedrus,* and *J. excelsa.*

Justicia adhatoda • Malabar Nut

This plant is traditionally used in India as an expectorant and antispasmodic, especially in the treatment of asthma and chronic bronchitis. The species is also known as *Adhatoda vasica.* The leaves contain bitter bronchodilating and antianaphylactic alkaloids including vasicine and vasicinone, and a commercial cough syrup containing *Justicia* (called Adulsa) is available in some Southeast Asian countries. The leaves have been smoked in the form of cigarettes, as well as prepared in oral medications to alleviate wheezing and spastic coughing. The drug is said to be nonpoisonous to mammals but is toxic to frogs and able to kill fish, insects, and other organisms. Larger doses are irritating to the digestive system and can cause vomiting and diarrhea in humans. The plant is included in the World Health Organization's manual for health workers in Southeast Asia, owing to its recurrent usage and safety profile. Other common names are adulsa, adhatoda, vasa, and vasaka.

Lepidium meyenii • Maca

Maca tubers are a traditional food in the Andes, where consumption is said to improve energy and stamina at high altitudes. Powders, encapsulations, and medicinal food are used at the present time as tonics and cardiovascular remedies. Maca may improve lung inflammation and spasm and be used for those with pulmonary congestion due to heart disease. *Lepidium* medications may inhibit bronchospasm, and bronchodilating flavonoids have been identified in the plant. *Lepidium meyenii* is also known as *L. peruvianum. L. apetalum,* and *L. sativum* are related species.

Ligusticum porteri • Osha

The roots of this North American species of *Ligusticum* were used by Native Americans to treat heart and lung disease and to remedy influenza, bronchitis, pneumonia, dyspnea, and sore throat. The plants are popular in lung tonics to improve hoarseness of the voice, treat viral infections of the lungs and upper airways, clear sinuses, and treat headaches. Osha may stimulate the appetite, improve digestion, and allay nausea and vomiting in cases of influenza. Butylidenephthalide in the roots is shown to have antispasmodic and vasodilating effects, and ligustilide has broad antimicrobial effects.

Ligusticum striatum • Szechuan Lovage

Also called chuanxiong, this Asian species of *Ligusticum* is considered to be one of the 50 fundamental herbs in TCM. The roots are traditionally used for lung infections, allergies, and inflammatory and hyperreactivity conditions of the airways. TCM recommends the medicine for pain due to blood and qi stagnation, such as stabbing pains in the chest. *Ligusticum striatum* is also specific for headaches and head pain that may accompany influenza and respiratory infection where there is a cough, sore throat, and red eyes, and coumarin-based phytoestrogens, such as a group of ligustilides, are credited with some of the medicinal effects. *Ligusticum striatum* is also known as *L. chuanxiong. Ligusticum* is sometimes decocted in wine in medicinal formulas.

Linum usitatissimum • Flax

Flaxseeds can be processed to concentrate the quality oils, which are rich in omega-3 fatty acids including α-linolenic acid (ALA). Flaxseed oil also contains short chain polyunsaturated fatty acids (PUFA), and the regular consumption of flax oil essential fatty acids can help reduce chronic inflammation and allergic reactivity and can be part of protocols for asthma, allergic rhinitis, allergic sinusitis, and other atopic conditions. Because essential fatty acids help regulate prostaglandins, flaxseed oil can improve allergic reactivity, including airway hypersensitivity disorders.

Although flaxseed oil is naturally high in antioxidants like tocopherols and beta-carotene, it is easily oxidized after being extracted and must be stored in cool dark locations to preserve its medicinal virtues. Quality products are expressed in small batches, packaged in opaque bottles, and stored in the refrigerator.

Lobelia inflata • Pukeweed

Lobelia is a traditional herb for asthma, cough, and respiratory spasms. It is specifically indicated for dyspnea and pressure in the chest, croupy and spastic coughing, and shooting pains in the chest. The aerial parts of *Lobelia*, especially the fresh young seedpods, have muscle-relaxing effects and will improve tight muscles in the chest and back. The herb has a brief stimulating effect on respiratory smooth muscle, prior to a more prolonged bronchorelaxing effect. *Lobelia* can be prepared into oil, plasters, and poultices to treat chest pain and spasm or taken orally in tinctures and teas. Large and repetitive dosages can cause nausea, and *Lobelia* has been historically used as an emetic to stimulate the mucociliary apparatus and support expectoration. Because *Lobelia* has strong and potentially harsh effects, it is not usually used for children, the elderly, or feeble patients. *Lobelia* oil and plasters will need to be made for one's self, because there are few commercial products available.

Lomatium dissectum • Biscuitroot

Lomatium is similar to *Ligusticum*; both belong to the Apiaceae family. *Lomatium* roots are especially indicated for lung congestion of a viral origin, and this herb was used to treat the 1918 influenza pandemic. *Lomatium* is a traditional remedy for cough, bronchitis, colds, and lung viruses, and it may be used to manage influenza. *Lomatium* may have activity against a number of viruses including hepatitis-C, HIV, AIDS, Epstein-Barr, and common rhinoviruses. *Lomatium* may be included in formulas for pneumonia, bronchitis, herpes simplex, sinusitis, and common colds. There are anecdotal reports of high or repetitive dosing of *Lomatium* resulting in a skin rash. I have hypothesized that photosensitizing coumarins in *Lomatium* may play a role both in subduing viruses and occasionally inducing a rash.

Lonicera japonica • Honeysuckle

Also called ren dong teng, *Lonicera* fruits are widely used in TCM as a cooling agent to help treat respiratory infections, viral epidemics, cough, and bronchitis. *Lonicera* helps to clear heat and toxicity from the body and to alleviate pain of sore throats and burning in the eyes that occurs with influenza and viral infections. *Lonicera* is specifically indicated for redness, fever, tender swelling, and burning sensations in the lungs. Modern research has demonstrated *Lonicera* to have activity against influenza viruses, including the H1N1 strain, as well as possible activity against lung cancer.[136] Some of *Lonicera*'s medicinal effects are credited to

chlorogenic acid. The North American species *Lonicera caprifolium* is also used for respiratory viruses. *L. xylosteum* (woodbine) has been used in asthma formulas. Some sources warn that the expectorating effects of *Lonicera* species can be overdone and like *Lobelia*, can have emetic and cathartic effects with high dosages.

Lophatherum gracile • Dan Zhu Ye Gen

The leaves of this grass family plant have been used traditionally in cold, flu, cough, and bronchitis formulas. The plant may also improve oral lesions and may be included in syrups and mouth rinses for viral lesions of the oral mucosa. Modern research has reported antiviral activity including against the respiratory syncytial virus (RSV).[137] *Lophatherum* has been included in traditional tonic soup blends in China and in traditional formulas for heat and dryness, such as pharyngitis with a dry sensation and a dry cough.

Lycium barbarum, L. chinense • Goji Berry

Also called wolfberry, the fruit of this Solanaceae family vine can be dried and eaten out of hand. The dried berries are commonly included in soups and decoctions for general nourishing effects. *Lycium* has been used as a tonic food in China for thousands of years. I consider *Lycium* to be a nonspecfic tonic herb that can be utilized in respiratory formulas as a nutritive base, to recover from chronic infections, and to support healing in convalescent formulas following debilitating illnesses. Goji berry powder is also available to use in smoothies and medicinal foods.

Lycopus virginicus • Bugleweed

The astringent properties of the leaves may be used to help reduce excessive secretions in respiratory disease. The plant is also diaphoretic and a sedative tonic and may allay irritative coughs in pneumonia respiratory disease with copious secretion of mucus or mucopus. *Lycopus* was included in formulas for simple colds with irritating or debilitating coughing. The Eclectic authors considered *Lycopus* to be one of the best remedies for hemoptysis, and they included it in formulas for tuberculosis and chronic pneumonia. In cases of pulmonary hemorrhage, *Lycopus* may be combined with cinnamon and ipecac to treat hemoptysis and support expectoration. *Lycopus* was combined with *Eupatorium* for respiratory viruses. Due to gentle nervine actions, *Lycopus* is also specific when tension and insomnia accompany, or for hyperthyroid-like presentations with tachycardia, vascular excitation, and elevated temperature.

Magnolia officinalis • Magnolia

Mature flowers and tree bark are used in TCM to treat respiratory infections and inflammatory conditions. The flower buds are used to treat sinus congestion and sinus headaches and may be taken orally or applied topically. Honokiol and magnolol are noted to be mast cell stabilizers contributing to the antiallergy effects. One specific indication is a "plum pit sensation" in the throat. In TCM, the medicine is used for coughing, wheezing, and dyspnea, with a sense of congestion and excess phlegm. *Magnolia* may be combined with *Rheum* in TCM to slow fibrosis in the lungs in cases of COPD and other chronic inflammatory diseases of the lungs. *Magnolia* may be combined with *Astragalus* and *Glycyrrhiza* to treat allergic rhinitis and hypersensitivity disorders of the lungs. *M. obovata* is a related species also found to contain honokiol and magnolol. Magnolia bark goes by the name chuan houpu in China, while the leaves are referred to as houpu, and the flowers buds as houpohua.

Mahonia • See Berberis

Marrubium vulgare • Horehound

The leaves and flowering tops of *Marrubium* are traditionally used as an expectorant and were boiled with sugar to prepare horehound candy. Also called hoarhound, *Marrubium* is specifically indicated for throat and laryngeal irritation, hoarseness, chronic bronchitis, cough, and asthma with abundant mucus. *Marrubium* has gentle stimulating effects, including expectorant, diuretic, and diaphoretic actions. *Marrubium* may be included in formulas for irritation of laryngeal and bronchial mucous membranes and prepared into cough syrups for simple colds, chronic catarrh, asthma, and all pulmonary infections.

Matricaria chamomilla • Chamomile

Chamomile is such an all-purpose carminative, nervine, and anti-inflammatory that the dried flowers can be included in teas for cough, cold, and respiratory infection, both for a pleasant flavor and to add its own medicinal actions. The antioxidant properties of *Matricaria* are shown to protect the lungs in animal models of toxin-induced lung damage.[138] *Matricaria* can protect the lungs by increasing levels of important antioxidants including superoxide dismutase and glutathione. A component of the essential oil fraction of the flowers, α-bisabolol can significantly reduce hypersensitivity reactions in the airways and help limit damage to the lungs and acute inflammatory disorders such as acute respiratory distress syndrome (ARDS). Because the flowers contain pollen, rarely handling the dried plant material can cause itchy eyes or trigger hay fever-like symptoms. Such rare individuals can usually still benefit from chamomile tea if someone else prepares it for them. *C. recutita* can be used interchangeably.

Medicago sativa • Alfalfa

In addition to its general nourishing, foodlike properties, alfalfa leaves have been used traditionally in asthma formulas and to help repair and rejuvenate injured connective tissues, as well as to recover energy and vitality following illnesses. Specific indications include weakness, poor digestion and assimilation, anemia, and weight loss. *Medicago* teas and powders prepared into medicinal foods may improve blood sugar regulation and thereby reduce oxidative stress in the lungs in those with poor circulation due to long-term metabolic disease.

Melaleuca alternifolia • Tea Tree

The essential oil may be included in steam inhalations to treat respiratory infections. Tea tree essential oil may help to break biofilms in the lungs and improve the efficacy of other medications in treating lung infections, or reduce antibiotic resistance. Consider daily steam inhalation with tea tree oil to deter *Pseudomonas aeruginosa* from becoming established in cystic fibrosis patients or others with chronic lung congestion. Castor oil and heat may help drive the volatile oil into the lungs when topically applied. Terpinen-4-ol, a component of the essential oil, has been shown to induce apoptosis in lung cancer cells.

Melilotus suaveolens, M. officinalis • Sweet Clover

Melilotus leaves and young stems are especially used for pain, headaches, neuralgia, and fever but is also noted for effects on blood congestion. *Melilotus* is a folkloric antithrombotic herb, and research shows it to downregulate vascular endothelial growth factor in septic situations, stabilizing the microvasculature against hyperpermeability. *M. suaveolens* may reduce the risk of thrombi and microvascular derangements. Even though *Melilotus* coumarins may reduce excessive clotting, the plant may also reduce capillary leakage in the lungs through effects on vascular endothelial cells,[139] explaining the traditional use of the plant for chronic epistaxis and lung diseases with hemoptysis. Sepsis involves a massive release of inflammatory mediators in response to the presence of bacterial toxins, and

Melilotus can help prevent damage to the lungs in such situations. Unfortunately, the need for such medicines becomes apparent only after a crisis has occurred, but *Melilotus* might be indicated in those with a history of lung disease, episodes of serious of pneumonia, diabetes and other metabolic stresses, long-term smoking with chronic lung infections, and especially a history of blood clots or emboli. Consider *Melilotus* in bronchitis and pneumonia formulas for patients with such conditions or to add into the therapy for COPD patients experiencing acute exacerbations.

Melissa officinalis • Lemon Balm

Melissa leaves are most well known as a nervine, anxiolytic, and antidepressant in the present era, but it has also been included in various tonics, cordials, and antiviral medications throughout folklore. *Melissa* may be included in antiviral teas for respiratory infections, and the essential oil is used as an inhalant against respiratory viruses and topically on cold sores as a deterrent to the herpes simplex virus. The essential oils in the highly aromatic plant have broad antimicrobial activity and are effective when consumed as a tea, as well as when the purified essential oil is inhaled as part of steam inhalation therapy for lung infections. *Melissa* may also have activity against influenza viruses including the epidemic avian (bird) flu strains and may also have a synergistic effect with antibiotics in the treatment of bacterial infections of the lungs, helping to break biofilms and reducing antibiotic resistance. *Melissa* contains rosmarinic acid credited with anti-inflammatory and immunomodulating activity. Human clinical studies have shown the plant to have an excellent safety profile.

Mentha piperita, M. spicata • Mint

Mentha piperita (peppermint) and *M. spicata* (spearmint) leaves can be included in teas and tinctures to help open the sinuses and airways and to improve the flavor of teas and medicines for children who have colds and cough. Mint essential oil can be used in throat sprays to allay pain and can be used topically on the chest to quiet cough and treat pain. Mint can alleviate bronchospasms and allergic reactivity, tightness in the throat, and congestion of the upper respiratory passages. Flavonoid glycosides in mint are shown to be mast cell stabilizers.

Mikania glomerata • Huaco

Also called guaco, this tropical South American vining aster is used folklorically for cough, asthma, and respiratory pain, and bronchodilating coumarins have been identified in the plant. The leaves are prepared into teas for respiratory inflammation, especially to help make bronchial and tracheal mucus more fluid, supporting expectoration.[140] Modern research suggests huaco to also relax bronchial smooth muscle, which together with improving mucous quality quiets the cough reflex.

Morella cerifera • Bayberry

Morella was written about under the name *Myrica cerifera* by the Eclectic physicians, with the plant being a useful astringent and circulatory stimulant for colds, stomatitis, sore throat, and upper respiratory catarrh. The berries are coated with wax, abundant enough to purify and use for making candles, hence the common names of candleberry, waxberry, and wax myrtle. As an herbal medicine, *Morella* acts as a respiratory astringent and is specifically indicated for boggy, swollen, and atonic mucous membranes. Where chronic mucosal stagnation and congestion leads to ulceration or dysbiotic changes, *Morella* may be used as a mouthwash, tonic, and astringent gargle or lavage. The combination of a small amount of *Sanguinaria* and/or *Hydrastis* with *Morella* may be used as a nasal lavage for polyps. Combine *Morella* with *Asclepias* in convalescent formulas to help recover from consolidation in the lungs. Myricetin, a flavonoid structurally similar to quercitin, is found in many *Morella*, *Myrica*, and other species and is credited with anti-inflammatory activity.

Morinda citrifolia • Noni

The fruit of this small tropical tree is a traditional medicine for inflammation and pulmonary diseases. Noni has been used in Polynesian folk medicine for more than 2,000 years, including as a remedy against fever, respiratory disorders and infections, and tuberculosis. Numerous anti-inflammatory and immunomodulating mechanisms are reported, as well as an ability to exert an antioxidant effect in the lungs of endurance athletes and protect against oxidative damage in heavy smokers.

Neopicrorhiza scrophulariiflora • Picrorhiza

Neopicrorhiza is a high-altitude Himalayan plant whose leaves, stems, and roots are used primarily as hepatoprotective agents, but may also be used for asthma and to reduce inflammation in the lungs. Picrosides, a group of iridoid compounds found in the plant, exhibit numerous anti-inflammatory activities. *Neopicrorhiza*, also known as *Picrorhiza*, also contains curcubitacins (a group of terpenes), betulinic acid, and the flavonoid apocynin, all credited with immunomodulating action. *Neopicrorhiza* may be included in formulas for dyspnea, tight and spastic cough, asthma, and COPD. *Picrorhiza kurroa* was originally used in Ayurvedic medicine, but due to severe overharvesting, it is now in danger and *Neopicrorhiza* is considered a substitute, also known in TCM where it is referred to as hu huang lian. Cell cultures are also being explored to produce picrosides. *Picrorhiza kurroa* was referred to as kutki in Ayurvedic medicine and the picrosides are sometimes collectively referred to as kutkin.

Nepeta cataria • Catnip

The leaves have a long history of use for cranky children, simple colds, and respiratory infections, and catnip was one of the most common household teas in early American homes. One of the traditional specific indications included using *Nepeta* for irritable children who are restless, crying, and inconsolable. Although most known for its profound effect on the nervous system of cats, *Nepeta* actually has calming effects on the human nervous system, improving sleep and reducing nightmares in children. Calmative terpenes and iridoid glycosides are credited with anxiolytic actions. The leaves contain rosmarinic acid, noted to have powerful antioxidant and anti-inflammatory effects. The plant also has antiviral and general antimicrobial effects, supporting its traditional use for children's respiratory infections. Use the tea as a gentle, mild-flavored ingredient in simple teas for upper respiratory infections, and the tincture in cough, cold, and flu formulas, especially for irritable toddlers. Catnip is also called catnep.

Ocimum tenuiflorum • Holy Basil

Also known as tulsi, the leaves of this relative of culinary basil have been used in India for spiritual and practical purposes, from religious rites and creating sacred space, to preparing the mind and body for meditative and spiritual practices—hence the name "holy basil." The leaves are used as a culinary spice in India and referred to as *kaphroa*, and the leaf powder is sometimes prepared in ghee (clarified butter) to use medicinally. Because of its many calming effects on the nervous system, holy basil can be included in formulas for infections, colds, flu, and fever, used topically to treat snake and insect bites, and used in wound and eye washes. The plant is highly aromatic due to high volatile oil content including eugenol, carvacrol, linalool, and β-caryophyllene. Modern molecular research has identified bronchodilating and mast cell–stabilizing compounds including the aromatic oils myrcenol,

nerol, and eugenol. The plant also contains organic acids including oleanolic acid, ursolic acid, and rosmarinic acid, each credited with powerful antioxidant and anti-inflammatory effects. Include holy basil in tinctures for allergic and inflammatory airway diseases. Use the dried leaves or powders in preventive or therapeutic teas for simple colds or serious airway illness or in the infusions in neti pot rinses for rhinitis, catarrh, nasal polyps, or other lesions. This species is also known as *O. sanctum*.

Oenothera biennis • Evening Primrose

Indigenous to North America, evening primrose was used as a food and medicine by the Ojibwe, Potawatomi, Iroquois, Cherokee, and other First Nations people for thousands of years. Many Native American communities used evening primose for a wide variety of ailments and "swellings in the body," and the plant began to be exported to England in the seventeenth century where it was sold under the name of "king's cure-all." While the roots, leaves, and seeds of the plant are all edible, the seeds are particularly prized for medicinal purposes. Evening primrose seeds contain quality oils, rich in essential fatty acids including gamma-linolenic acid (GLA), and regular consumption of essential fatty acids can help reduce chronic inflammation and allergic reactivity. *Oenothera* can be part of protocols for asthma, allergic rhinitis, allergic sinusitis, and other atopic conditions. GLA is a polyunsaturated omega-3 fatty acid and has been the subject of many studies showing its role in forming immunomodulating prostaglandins and its ability to inhibit the synthesis of pro-inflammatory leukotrienes. Evening primrose and other seed oils can be taken by the teaspoon, several times a day, to help optimize eicosanoid ratios in the body—immunomodulating compounds formed from dietary fatty acids.

Olea europaea • Olive

Olive fruits are used to prepare a culinary and medicinal oil, and the dried fruit and leaves are traditionally used to treat respiratory tract infections. Olive oil may be used as a vehicle to deliver upper and lower respiratory tinctures or prepared in an oral medicine without alcohol to provide a soothing, demulcent base for more irritating herbs. Many phenolic compounds, especially secoiridoids and iridoids, are pharmacologically active, and one major constituent of olives, oleuropein, is shown to have numerous physiologic benefits. Olive's anti-inflammatory, tissue-protective, and metabolic-supportive actions make it appropriate to include in formulas for chronic

Herbs for Respiratory Conditions

respiratory issues secondary to diabetes, metabolic syndrome, and hyperlipidemia.

Ophiopogon japonicus • Mai Men Dong

Ophiopogon japonicus is a grasslike lily family plant, sometimes used ornamentally due to the striking blue berries the plant yields. The tubers go by the name of mai men dong and are a traditional medicine for yin deficiency in TCM, moistening, nourishing, and tonifying the heart, lungs, and digestive system and helping to clear heat. In the case of respiratory infection, *O. japonicus* is specifically indicated for coughs with thick phlegm and "heat in the lungs," and for vascular diseases, especially those involving ischemia and thrombosis. Shin'iseihaito (xin yi qing fei tang) is a formula of traditional Japanese Kampo medicine and TCM that contains *Ophiopogon* roots and is used for the treatment of upper respiratory tract disease, including infectious and allergic complaints, especially rhinitis and sinusitis. Modern investigation on the entire formula shows activity against *Streptococcus pneumoniae*, including an ability to penetrate and deter biofilms.[141] Endophytic fungi in the root masses contribute to the broad antimicrobial effects. Steroidal saponins in the plant are collectively referred to as ophiogonons, and one such compound, ruscogenin, may protect against connective tissue damage and excessive capillary permeability in acute airway infection and inflammation. *Ophiopogon* may also be included in formulas for various pulmonary symptoms due to numerous perfusion-enhancing mechanisms of action. Horticulturists may refer to the plant as dwarf lilyturf, mondograss, fountainplant, or monkeygrass.

Origanum vulgare • Oregano

Oregano teas and tinctures, or possibly inhalation of the essential oil, may have protective effects on the lungs. *Origanum* has been shown to attenuate cyclophosphamide-induced oxidative stress in the lungs, helping the tissues to tolerate and benefit from this anticancer drug.[142] Consider oregano for patients undergoing chemotherapy for lung or other cancer with alkylating agents known to injure the lungs. Thymol, caryophyllene, and carvacrol are components of the essential oil fraction of oregano and credited with many anti-inflammatory, antioxidant, and antimicrobial effects. Oregano oil steam inhalations may also be included in protocols for pneumonia and as a preventive therapy for those with COPD or cystic fibrosis prone to opportunistic airway infections. These volatile oils are also present in oregano tinctures and teas and are shown to be taken up into cells of the pharynx and airway epithelia, contributing to the antioxidant and antimicrobial effects, as well as possible anticancer effects, inducing apoptosis.

Paeonia lactiflora • Peony

Several species of peonies are used in China for blood building and vascular effects. *Paeonia lactiflora* is the source of both white peony (bai shao yao or bai shao) and red peony (chi shao yao or chi shao) depending on how the roots are processed and cured, the red being the unpeeled roots and the white being the peeled roots. *Paeonia* is combined with a variety of different herbs to elicit specific effects on blood, pain, and inflammation and to best treat liver, heart, and lung diseases. Peony is indicated for blood-deficiency symptoms such as a pale complexion with dizziness, weakness, and ringing in the ears. Studies suggest paeoniflorin to reduce oxidative stress and help protect the lungs from cigarette smoke,[143] and to reduce fibrosis in several tissues and organs. Peony is an ingredient in several traditional Chinese formulas for COPD and ARDS. Other species used medicinally include *P. × suffruticosa* (moutan) and *P. anomala*.

Panax ginseng • Ginseng

Ginseng roots have long been used in TCM to improve energy and immune functions, and the plant may be used to improve resistance to colds and respiratory infections and to speed recovery for those who have slow recovery or lingering symptoms. Include ginseng in formulas for those with chronic infections and for people who are weak, cold, and deficient when simple colds progress into bronchitis, pneumonia, or sinus infections. Ginseng is also used for some elderly patients or patients with heart disease to take regularly to prevent potentially devastating lung infections. *Panax ginseng* is also traditional for improving strength, endurance, and physical stamina, and research suggests the plant optimizes nitric oxide in a manner that supports oxygen utilization in the lungs. Ginseng may also limit inflammatory responses in a manner that protects the lungs from septic processes. Saponins from the plant, referred to as ginsenosides, are responsible for some of these actions and are also shown to reduce allergic reactivity in the airways in animal models of asthma and atopic disease.

Passiflora incarnata • Passionflower

Passiflora is most well known as a nervine, and its leaves and flower buds have broad antispasmodic effects and may be included in formulas for cough or wheezing. Bronchodilating alkaloids in the plant have been identified. The antioxidant flavonoid chyrsin found in *Passiflora* has been studied for its broad antioxidant actions and may contribute to the plant's ability to reduce inflammatory processes and protect the lungs from chemical toxins and cigarette smoke. Isolated and concentrated chrysin may be more effective for such purposes, but *Passiflora* tea or tincture may, nonetheless, be helpful in lung-support formulas. Due to its calming nervine effects, *Passiflora* might be specifically indicated for those with stress or anxiety-induced asthma.

Pelargonium sidoides • African Geranium

The plant is active against respiratory viruses, and the essential oil is available to include in inhalants and steams to treat pneumonia. Root extracts inhibit viral attachment to cell surfaces and have demonstrated activity against influenza viruses. *Pelargonium* liquids and inhalants may be considered in therapies for bronchitis and rhinosinusitis and may reduce cough and congestion in asthmatics, as well as the frequency of attacks due to viral infections of the upper respiratory tissues. *Pelargonium* steams, syrups, and other oral medications may also improve tonsillitis and pharyngitis associated with colds and respiratory viral infections. The plant also goes by the name umckaloabo in proprietary products.

Perilla frutescens • Shiso

Perilla frutescens leaves are a popular seafood garnish in Asia and are a traditional antidote for allergic reactivity to fish and crab ingestion. The aromatic leaves are also high in antimicrobial and anti-inflammatory volatile oils and will change color from red to green depending on the pH of the soil. According to TCM, *Perilla* leaf has a "phlegm-dispersing action" useful in the treatment of common cold and upper respiratory infection symptoms. *Perilla* is also a general antiallergy herb in Japan and Korea, and modern research has found the plant to contain rosmarinic acid, well established to reduce hypersensitivity reactivity.[144] *Perilla* seeds are also used medicinally, being the source of anti-inflammatory omega-3 essential fatty acids.

Persicaria tinctoria • Chinese Indigo

Also known as *Polygonum tinctorium*, Chinese indigo is a blue dye plant and is also used to treat colds and flu. The dried stems and leaves contain indirubin credited with antiviral activity, including effects against influenza strains. Indirubin may help protect pulmonary vascular endothelial cells from the ravages of influenza-induced acute respiratory distress syndrome. Immunomodulating agents such as indirubin may reduce potentially fatal cytokine storm—a phenomenon of excessive production of proinflammatory cytokines, which can cause serious lung tissue injury and account for high mortality following influenza virus infection in susceptible individuals. Indirubin strongly inhibits enzymes involved with inflammatory reactions, particularly cyclin-dependent kinase enzymes. Indirubin has poor solubility and bioavailability, which may be overcome when combined with other herbs, as is traditional in TCM formulas for sore throat and flu. Research is underway to prepare semisynthetic forms of indirubin with immunomodulating activity.

Petasites hybridus • Butterbur

Petasites leaves and roots are traditional medicines for headache, cough, asthma, chest pain, and respiratory allergies. *Petasites* is best reserved for acute reactivity and for short duration due to the potentially toxic pyrollizidine alkaloids, less of which are found in the leaves. *Petasites* has antispasmodic, anti-inflammatory, and antiallergy effects on the airways, and can reduce hay fever, sinusitis, and hyperreactivity of the blood, vasculature, and respiratory passages. *Petasites* can be helpful in treating hay-fever symptoms as effectively as a prescription antihistamine, but without the drowsiness side effect. Due to potential hepatotoxicity, avoid in liver disease, alcoholism, and pregnant women.

Petroselinum crispum • Parsley

The leaves, seeds, and roots of *Petroselinum* have all been used medicinally, and traditional uses include as a renal and uterine stimulant, yet capable of relaxing the respiratory and digestive smooth muscle. As such, parsley-based medications may have bronchodilating and carminative effects. Flavonols (kaempferol and quercetin) and flavones (apigenin and luteolin) in the plant are credited with some of the anti-inflammatory and antiallergy effects, which may lend parsley cytoprotective effects. Modern research demonstrates antioxidant effects and protection against heavy metal toxicity, protecting cell membrane lipids from oxidation.[145] Apigenin is also found in chamomile and other medicinal plants and may be able to help protect the lungs from fibrosis,[146] suggesting that parsley may also be useful in formulas for COPD.

Phyllanthus emblica • Amla

The leaves and fruits (which are also called Indian gooseberry) of this plant are traditionally used in India to prepare a rejuvenation tonic known as amalaka, credited with cooling and drying effects. Amla is believed to balance the three bodily energies—the doshas in Ayurvedic medicine—and is used to treat asthma and cough. The plant is also combined with several species of *Terminalia* in a traditional Ayurvedic digestive and energy tonic known as triphala. The slightly sweet and astringent fruits are referred to as yuganzi in China and used for clearing heat from the throat and moistening lungs for arresting cough in TCM. Amla leaves contain polyphenols credited with numerous antioxidant and anti-inflammatory effects able to reduce alveolar damage and macrophage infiltration and may help repair pulmonary damage and attenuate fibrotic processes. Modern research suggests amla may have the ability to optimize myocardial contractile dynamics[147] and may thereby be a supportive herb in formulas for dyspnea-related cardiopulmonary disease. This plant is also known as *Emblica officinalis*, and other common names are amala and amalaki.

Phytolacca americana • Pokeroot

Phytolacca root preparations are often used as an adjuvant herb in formulas for sore throats, tonsillitis, dyspnea due to fluid stagnation, mucous congestion, and swollen tissues and lymph nodes. *Phytolacca americana* is considered to be synonymous with *P. decandra*.

Picrorhiza kurroa • Kutki

This Himalayan native plant is most well known as a hepatoprotective agent and is shown to be active against hepatitis viruses, but it is also active against respiratory viruses. In addition to its use as a bitter tonic and as an antiperiodic, kutki has also historically been used to treat asthma, allergic airway diseases, and atopic conditions. Picrosides are iridoid glycosides in *Picrorhiza* that have immunomodulating activities contributing to the ability to reduce airway inflammation. The phenolic glycoside androsin is bronchodilating and noted to be a mast cell stabilizer. Due to overharvest, *Picrorhiza kurroa* has become endangered. *Neopicrorhiza scrophulariiflora* is a possible substitute and cell cultures are being developed to produce picrosides.

Pimpinella anisum • Anise

Anise seeds are used in teas and tinctures to open the airways. The seeds are high in essential oils, including carvone and anethol, which have bronchodilating, expectorating, antimicrobial, antispasmodic, and anti-inflammatory properties. Thus the oils are useful to include in inhalants and steams or to use topically over the chest to relax bronchospasms, quiet coughs, and relax tight painful muscles. Anise has a pleasant taste, and anise tinctures, oils, and syrups can be used to improve the flavor of cough formulas, especially for children. Anise is also helpful to treat digestive upset that may accompany viral infections.

Pinellia ternata • Crow Dipper

The dried tubers of *Pinellia* are used in TCM to help remove phlegm from the respiratory tract and dampness from the body in general, especially when associated with obesity, tissue congestion, and stagnation. In Korea, *Pinellia* is combined with an equal part of dried orange peel to treat asthma. Modern research suggests that *Pinellia* may promote thermogenesis and fatty oxidation, explaining the traditional ideas for its use in cold and deficient subjects.[148] The tubers are often processed in alum, licorice, ginger, or a combination thereof to reduce any possible toxicity, because *Pinellia* tubers contain calcium oxalate crystals. The plant also contains alkaloids, including ephedrine, choline, cavidine, and immunomodulating lectins and polysaccharides.

Pinus pinaster • Maritime Pine

Maritime pine bark is the source of Pycnogenol, a trade name for the combined flavonoids, catechins, proanthocyanidins, and phenolic acids including ferulic and caffeic acid found in the bark. Pyconogenol may have antioxidant and antiallergy activity when consumed on a regular basis due to its stabilizing effects on mast cells and other antioxidant and anti-inflammatory mechanisms. Several clinical trials have shown Pycnogenol to be effective for asthma and allergic airway disorders including allergic rhinitis and eosinophilic bronchitis. Pycnogenol has also demonstrated antiviral and antibacterial properties, including activity against common respiratory pathogens. Pycnogenol is safe and gentle, suitable for children or to complement pharmaceuticals and other medicines used in the treatment of hyperreactivity of the airways. Other species of pine, including *P. brutia*, *P. nigra*, *P. sylvestris*, and *P. pinea*, also contain high amounts of polyphenolic compounds and can be used as an alternate raw material for manufacturing Pycnogenol.

Piper longum • Long Pepper

Also called pipli, pipali, and pippali, this tropical vining pepper is used as a seasoning spice like other peppers. It is commonly used in Ayurvedic medicine as a digestive carminative, and at one time it was popular for intermittent fevers as an adjuvant to quinine, especially as an alkali extract of the piperin resin. Like common black pepper (*Piper nigrum*), the ripe seeds contain the alkaloid piperine, studied to enhance the intestinal assimilation of curcumin and many other nutrients and medicinal compounds. Another alkaloid, piperlongumine is credited with anticancer effects. It seems likely that the inclusion of *Piper longum* in formulas increases the absorption of other molecules in the formula. Piperine and piperlongumine also have broad anti-inflammatory and antioxidative properties, and some of the immunomodulating effects may be due to piperine's ability to increase the permeabilty of mitochondrial transition pores.[149] *Piper cubeba* (cubeb pepper) is also used medicinally.

Piper methysticum • Kava Kava

The large resin-rich roots of kava kava can be included in formulas for muscle tension in the chest and pain in the ribs due to vigorous coughing. Kava kava is a nervine and musculoskeletal antispasmodic but is included in some formulas in this chapter for its anodyne and anti-inflammatory effects. Kava kava may allay nerve pain such as intercostal neuralgia resulting from severe and prolonged coughing. Kava kava may also lend some anti-inflammatory and immunomodulating effects to respiratory formulas and has shown anticancer and chemopreventive effects in some tissues. Nuclear factor-κB (NF-κB) is a transcription factor that plays an essential role in cancer development, including lung adenoma, and is suppressed by kava lactones such as methysticin.

Plectranthus forskohlii • Coleus

Plectranthus forskohlii, also known as *Coleus forskohlii*, is a tropical Lamiaceae family perennial and the source of forskolin, a diterpene compound that has been well established as a tool for researching signal transduction at a variety of cell types, due to its ability to promote cyclic adenosine monophosphate (cAMP) and thereby alter cell membrane reactivity. Coleus roots are the exclusive source of forskolin and have been used traditionally since ancient times in Ayurvedic medicine to treat asthma and hypertension. Coleus has a mast cell–stabilizing effect, useful in treating asthma and dyspnea, and reducing hyperreactivity and respiratory epithelial reactivity.

Rosmarinic is also found in the plant and credited with antiallergy and immunomodulating effects.

Polygala spp. • Milkwort

Polygala species, also called snakeroot, belong to the Polygalaceae family, and are flowering plants whose roots often have wintergreen-like aromas due to the presence of salicylates. The common name milkwort is due to a galatogogue activity in cattle, and the name snakeroot, because many species are used to treat snakebites. One North American species, *Polygala senega*, or Seneca snakeroot, was specifically indicated for relaxation of the respiratory mucus, with a cough, rales, or a deep and hoarse voice, and laryngeal congestion. *Polygala* species are indicated for excessive respiratory secretions and paroxsysmal coughing in asthma and croup. Small doses make a useful expectorant in the early stages of acute infections such as croup or in chronic catarrh associated with pneumonia, asthma, and chronic bronchitis, or to help drain chronically congested tonsils and sore throats. The roots are diaphoretic expectorants, and large doses can become irritating and emetic, as well as cathartic and diuretic when continued. Many related species are used in a similar manner. An Asian species, *P. karensium* is used in TCM and found to contain anti-inflammatory, immunomodulating xanthones.

Polygonum • See *Persicaria* and *Reynoutria*

Prunus serotina • Wild Cherry

The aged bark of wild cherry has an antispasmodic effect, useful for coughs and spastic or irritated respiratory passages. As with almonds and other rose family fruit pits, cherry bark contains amygdalin and may be too toxic when fresh, so the bark is aged to become physiologically appropriate. *Prunus* is traditionally prepared into syrups, but is also available as a dried herb for tea and as a tincture. *Prunus* may relax bronchospasms and quiet irritated mucous membranes where the cough reflex is repeatedly stimulated. *Prunus* may be used in formulas for colds, pleurisy, pneumonia, and bronchitis and as a sedative antispasmodic in formulas for throat constriction or a tickle in the throat.

Pueraria montana var. lobata • Kudzu

Pueraria roots have many cardiovascular effects and may be included in formulas to help protect the lungs and pulmonary blood vessels from oxidative stress, fibrosis, and loss of elastic recoil and function.

Punica granatum • Pomegranate

The fruit rind and the tree bark of pomegranate are boiled and used to prepare medicines traditionally emphasized to treat diarrhea and intestinal parasites (especially tapeworms) due to strong astringent effects, while the flavonoid-rich fruit may help to treat vascular inflammation. Rind decoctions can also be used as gargles for colds and respiratory infections to break up mucus and astringe swollen pharyngeal tissues. Such mouth rinses and gargles may also be effective for oral candidiasis[150] and periodontal disease. Various forms of vascular inflammation may be indications for *Punica* fruit and juice preparations, including heart disease, diabetes, and erectile dysfunction. Current research has shown *Punica* to have antihyperglycemic, hepatoprotective, and antihyperlipidemic effects.[151] Ellagic acid, found in pomegranate and other fruits, is shown to be a mast cell stabilizer, contributing to the anti-inflammatory effects.

Raphanus sativus var. niger • Spanish Black Radish

Spanish black radishes can be included in the diet and are available in tinctures and encapsulations, which may be included in formulas for asthma, bronchitis, and other chest complaints. Sulfur compounds in radishes may help thin and expectorate mucus. Spanish black radish has been used in the folk medicine of China, Iran, Turkey, and other regions in the treatment of asthma, bronchitis, and other chest complaints. Modern research has suggested that Spanish black radish may help reduce pulmonary fibrosis via effects on transforming growth factor (TGF-β1).

Reynoutria japonica • Japanese Knotweed

Also known as *Polygonum cuspidatum* and *Fallopia japonica*, the plant contains emodin, which is shown to reduce collagen deposition in the airways following silica exposure. *Reynoutria* is also a source of resveratrol, and the related polydatin, that are credited with powerful anti-inflammatory and lung protective effects.

Rhamnus purshiana • Cascara

Cascara is not classically thought of as a respiratory herb, but rather an irritant laxative. However, it is noted that laxative therapies and enemas may help some lung conditions where pulmonary hypertension is improved by reducing intestinal pressure. Furthermore, emodin, which is found in this herb, is being explored to reduce collagen deposition in the airways following silica exposure. Therefore, *Rhamnus* may be an adjuvant herb in inflammatory conditions of the lungs, but due to its laxative effects can be consumed only in small doses.

Rheum palmatum • Chinese Rhubarb

Writings about the medicinal virtues of this relative of common rhubarb (*Rheum rhabarbarum*) can be dated to as far back as 2700 BC, including mention as a remedy against fever. The fresh leaves were used topically to treat fever, and root preparations were taken internally, while various teas and medicines are credited with antibacterial and anti-inflammatory effects. *Rheum* species contain emodin, noted to attenuate collagen deposition in the airways in various inflammatory conditions, but it can have a laxative effect if taken at too large of a dose. Current research on emodin suggests it may have antifibrotic effects on organs, even when applied topically over an inflamed organ. Emodin contained in the roots is shown to reduce collagen deposition in the airways following silica exposure. This species is also called turkey rhubarb, and related species include *R. emodi* and *R. officinale*.

Ricinus communis • Castor

While castor beans contain ricin, one of the most potent toxins known, the poison is not fat-soluble and hence castor oil made from the seed oil is safe to apply to the skin. Castor oil is even sometimes consumed orally in small amounts and for short duration to treat constipation. Castor oil is high in lectins credited with immunomodulating activity, and because lectins can bind to glycoprotein receptors on cell surfaces, they may also affect cell adhesion to allow the oil to penetrate deeply into the skin or possibly enter interstitial spaces and lymphatic circulation. Castor oil is used in some of the topical formulas in this chapter to act as a vehicle to assist other herbal substances to be drawn and absorbed more deeply into the body. While the research on castor oil itself is not robust, the research on the immunological effects of lectins, both endogenous and exogenous (from foods and medicinal herbs), *is* robust. Castor oil has been used traditionally topically over swellings and tumors and to stimulate lymphatic circulation and drive other substances more deeply into the tissues.

Rosa canina • Dog Rose

The hips (fruits) of the rose are excellent sources of vitamin C and bioflavonoids, and they make enjoyable additions to teas due to their slightly sweet and tart flavor and rosy color. The seeds are processed into quality oils used in skin products, but their oral consumption may contribute to the plant's anti-inflammatory and connective tissue supportive effects. *Rosa canina* can be considered a medicinal food and included as a tonic ingredient in a wide variety of herbal formulas.

Rosmarinus officinalis • Rosemary

Rosemary has antiallergy, antimicrobial, and anti-inflammatory effects and the leaves can be included in teas and tinctures for colds, sinus infections, bronchitis, asthma, and coughs. The plant has drying and warming effects, so it is especially useful to remedy abundant mucous secretions and to help astringe productive coughs. Rosemary essential oil may also be used topically in chest rubs and liniments to help treat allergic airway diseases. The plant contains rosmarinic acid, which is shown to have numerous antiallergy effects.

Rumex crispus • Yellow Dock

Rumex species contain emodin, which is shown to reduce collagen deposition in the airways following silica exposure. *Rumex crispus* is a traditional alterative herb also used for sore throats with congested lymphatic glands and to astringe and soothe laryngeal irritation causing tickling coughs. The emerging leaves of the young spring growth are highly mucilaginous and can be used to prepare soothing teas.

Salvia officinalis • Sage

Salvia leaves have drying, astringent effects on inflamed throats and tonsils and may be included in cough formulas when there is abundant expectoration to help tone swollen, boggy mucous membranes. *Salvia* leaf tea makes an excellent gargle for tonsillitis and pharyngitis when there is a large amount of exudate, for a tickle in the throat due to mucosal congestion, or to astringe postnasal drip.

Sambucus canadensis • Elderberry

Elderberries, as well as the leaves and flowers, are traditional cold and flu remedies, and modern research has demonstrated activity against respiratory viruses. The berries are often prepared into syrups used around the world for influenza and respiratory infections, and a small glass of warmed elderberry wine is a traditional preventive remedy used when colds and other respiratory viruses are prevalent in a community. *Sambucus* is specifically indicated for childhood illnesses that are accompanied by skin rashes, particularly for weeping lesions, where the skin is swollen and the epidermis separates and crusts over. *Sambucus* medications are pleasant tasting and gentle enough to use for infants with colds and congestion. The leaves and flowers have a diaphoretic effect when consumed as a hot tea and can be consumed while soaking in a hot bath to promote sweating as a method of speeding resolution of a respiratory infection.

Sanguinaria canadensis • Bloodroot

Sanguinaria is a potentially irritating herb and can be caustic at a high dose, so it is usually used in small proportions in formulas compared to other herbs. *Sanguinaria* has stimulating and expectorant effects and is specifically indicated for damp, swollen oral, nasal, and pharyngeal mucosa, especially when chronic, and associated with poor tone and underlying circulatory insufficiency, such as in long-term smokers or the elderly. Yet *Sanguinaria* is also reported to reduce excessive secretions. Due to its irritating effects, it is not usually used for the initial stages of an acute respiratory infection, but rather for low-grade, chronic, and subacute situations, or to promote expectoration after the acute phase has passed. *Sanguinaria* may help decongest the eustachian tubes and tonsils when associated with swollen and congested mucous membranes and is specifically indicated for a tickling or burning sensation in the throat that triggers coughing. The most effective dose is just a drop or two; typically, only a few milliliters are used when making a 60-milliliter quantity of a formula, in order to avoid nausea and irritation.

Saussurea costus • Saw Wort

Saussurea roots contain sesquiterpene lactones,[152] as do the Asteraceae relatives *Tanacetum* and *Petasites* better known in Western herbalism. *Saussurea* has been traditionally used for its antiulcer, anticonvulsant, anticancer, hepatoprotective, antiarthritic, and antiviral activities. *Saussurea* is a thistle and has been used traditionally in India, China, and Japan and reportedly traded in ancient Rome. Modern research shows immunomodulating effects that may reduce hypersensitivity and atopy[153] and activity against tuberculosis.[154] Hispidulin, a constituent also found in the respiratory herbs *Salvia* and *Grindelia*, is credited with antifungal, anti-inflammatory, antimutagenic, and antineoplastic properties.[155] These plants may be included in formulas for lung cancer, asthma, and fibrotic pathologies. This species is also known as *S. lappa*, and another common name is saw thistle. The plant is so widely used that it is becoming endangered, and cell cultures are being explored to produce the active compounds.

Schisandra chinensis • Magnolia Vine

Schisandra is native to East Asia, where it goes by the name wu wei zi (which translates as "five flavor fruit") because the ripe berries are purported to taste sweet, sour, sour, spicy, and bitter all at once. Magnolia vine has been traditionally used in China for gastrointestinal tract and liver disorders, respiratory failure, and cardiovascular diseases and is reported to improve fatigue and weakness, excessive sweating, and insomnia. *Schisandra* fruits are decocted and included in formulas for immune support and to treat cough and upper respiratory infections. Including the fruits in winter teas may help stave off common colds and infections.

Scutellaria baicalensis • Scute

This Asian species of *Scutellaria* is considered to be one of the 50 fundamental medicines. *S. baicalensis* leaves, stems, and especially the roots, the latter being referred to as huángqín or huang qin, are all credited with broad anti-inflammatory effects. *Scutellaria* may help reduce asthma and allergic airway disease when taken on a regular basis, for at least several months. *Scutellaria* may also be active against respiratory viruses. *Scutellaria baicalensis* has antiviral activity as baicalein promotes interferon-driven virucidal effects, as well as impedes viral attachment to host cells. *S. baicalensis* may be used in China as an adjuvant to chemotherapy in the treatment of primary bronchial pulmonary squamous cell carcinoma and non-small-cell lung cancers and is reported to improve outcome and prolong life.[156] The flavones baicalin, wogonoside, and their aglycones baicalein and wogonin are major active compounds in *Scutellaria* roots and are credited with the anti-inflammatory and anticancer effects of huang qin. These compounds have also been extensively studied for antiallergy effects and have mast cell–stabilizing and other beneficial effects on the airways, helping to prevent vascular and airway damage in acute inflammatory distress in the lungs, including asthma, allergic bronchitis, and COPD. The plant should not be confused with *Scutellaria lateriflora* (skullcap) used in Western herbalism as a nervine.

Selaginella uncinata • Spikemoss

Selaginella species are tropical fernlike club mosses. This and a variety of different species of *Selaginella* genus plants are used medicinally for antinociceptive, anti-inflammatory, antimutagenic, antispasmodic, cytotoxic, immune, and antiretroviral properties. *Selaginella uncinata* has been included in traditional cough and lung infection formulas, and modern research has shown activity against respiratory syncytial virus.

Senna spp. • Cassia

Cassia is one possible source of emodin, which is noted to attenuate collagen deposition in the airways in various inflammatory conditions. This is not a traditional use of this plant in Western herbalism, but with research mounting on emodin's utility in fibrotic processes, it may be a possible tool in pulmonary fibrosis and related pathologies. The seed pods have traditionally been used as a constipation remedy and to stimulate intestinal motility. Because emodin is also an irritant laxative, the dose must always be on the low side. Emodin has been shown to deter fibrotic changes in abdominal organs when topically applied, so topical application over the lungs may also be of use and warrants further research. *Senna alata* contains anthraquinones and flavonoids credited with mast cell–stabilizing effects. Some plants in the *Senna* genus are also known as *Cassia*.

Silybum marianum • Milk Thistle

Most well known as a liver tonic, milk thistle seeds are increasingly being shown to reduce inflammatory processes in tissues other than the liver. The flavanols, collectively referred to as silymarin, have mast cell–stabilizing effects, explaining milk thistle's traditional use to treat respiratory allergy and inflammation. One of the flavanols in silymarin, silybinin, is shown to reduce NF-κB activation, a key molecule involved with inflammatory and infectious disease of the lungs, as well as cancer initiation. This and other mechanisms can help protect the lungs from permanent damage in cases of acute respiratory distress. Milk thistle is also being shown to exert anticancer effects on various organs including the lungs.

Stemona tuberosa, S. japonica, S. sessilifolia • Bai Bu

Stemona species are traditional Chinese medicines used to moisten the lungs in cases of dry cough. The dried tuberous roots are also used for spastic cough and wheezing, pleurisy, influenza, and asthma. *Stemona* alkaloids bind muscarinic receptors and block other stimulants, leading to relaxation of bronchial smooth muscles. *Stemona* is used both topically and internally and is noted to improve coughs by relaxing the respiratory spasm and lowering blood pressure. Taken as a decoction, *Stemona* also has antimicrobial effects, including acivity against tuberculosis and common respiratory pathogens. Modern research has shown *Stemona* to inhibit a variety of inflammatory mediators involved with asthma, pneumonia, whooping cough, and inflammatory diseases of the airways. Alkaloids in the plant are being explored for their contribution to the antitussive and other medicinal effects and include stemonine, stenine, neostenine, tuberstemonine, and neotuberostemonine. Animal models of COPD show stemonine to significantly alleviate lung injury by decreasing the levels of enzymes and cytokines associated with inflammation and oxidative stress.[157] *Stemona* can cause digestive upset in those with irritable bowel and is best used in small to moderate amounts in formulas, paired with soothing or demulcent herbs.

Sticta pulmonaria • Lungwort

This epiphytic lichen of deciduous trees has also gone by the genus name *Lobaria* and is a traditional remedy for lung pain, as well as musculoskeletal pain in the back and shoulders. *Sticta* is a traditional medicine for cough, tuberculosis, and blood in the sputum, and used by Native Americans throughout northwest United States and Canada to treat asthma, hemorrhage, and skin rashes and injuries. *Sticta* is useful for cough by reducing irritation in the pneumogastric nerve, helping to allay chronic coughing, hacking, and wheezing. *Sticta* can also be included in formulas to treat upper respiratory infection and hay-fever symptoms including nasal congestions, conjunctivitis, and sneezing and may be included in formulas for acute influenza, tuberculosis, bronchitis, laryngitis, whooping cough, croup, and asthma, especially humid asthma with catarrh, expectoration, and chest pain. *Sticta* is specifically indicated for soreness in the chest that is aggravated by coughing or for sharp chest pains with coughing. *Sticta* is also specific for lung infections accompanied by musculoskeletal soreness extending from the upper back to the occiput and shoulders. Respiratory infections accompanied by fever and chills, night sweats, and headaches are additional traditional indications for using *Sticta*. As with other lichens, *Sticta* is sensitive to pollution and world populations are declining. *Sticta pulmonaria* is synonymous with *Lobaria pulmonaria*.

Stillingia sylvatica • Queen's Root

Stillingia roots are a traditional remedy for cough, croup, and lung disease, taken internally and prepared into topical chest liniments. *Stillingia* is a stimulating expectorant best for dry cough, laryngitis, and chronic bronchitis. Some early American herbal authors report the herb to be alterative and supportive to the lymphatic system, helpful for children with chronic otitis media when pallor, anemia, poor appetite, and debility accompany. *Stillingia* was traditionally mixed with *Lobelia* and essential oil, such as eucalyptus oil, and prepared into liniments to use over the throat or chest in cases of croup, as well as taken internally as a syrup. Specific indications include lymphatic congestion with scant, but chronic mucus in the throat and airways. While an emetic at high doses, *Stillingia* can also become emetic when small dosages are repeated frequently.

Symphytum officinale • Comfrey

Comfrey leaves and especially the roots are soothing mucilaginous substances, and the plant may be used in small amounts for short periods of time to soothe throat, digestive, and respiratory mucous membranes. Its pyrrollizidine alkaloids (PAs), however, limit its use and safety. *Symphytum* is contraindicated in pregnancy and lacation, for infants, and for anyone who has liver disease or is taking hepatotoxic medication. PA-free products are becoming available.

Symplocarpus foetidus • Skunk Cabbage

The roots are strong expectorants, but potentially irritating and very caustic and can only be used in diluted or skillful preparations. Skunk cabbage was included in nineteenth century pharmacopoeias under the name "dracontium" and was recommended for respiratory complaints.

Syzygium aromaticum • Cloves

Also known as *Eugenia caryophyllata* and *Caryophyllum aromaticum*, *Syzygium aromaticum* is the currently accepted taxonomy for this tropical tree. The emerging flower buds are highly aromatic and dried to use as a culinary spice, mouthwash ingredient, and herbal medicine. One of the dominant volatile oils in cloves, eugenol, has an anesthetic quality in the mouth and may also help to ease sore throat and painful coughing. Due to the warming, stimulating qualities of clove teas, tinctures, and chest rubs are specific for those with cold and damp constitutions. Clove is contraindicated for those with acute heat symptoms. Use clove essential oil in throat sprays for tonsillar pain and in warming expectorant formulas, in oil-based chest rubs, or in steams and inhalants for lung pain and congestion, taking care to properly dilute the fiery oil. Modern research has shown eugenol to reduce proinflammatory cytokines that contribute to airway inflammation. Use clove powder and whole dried buds in medicinal teas for COPD and chronic lung congestion and inflammation, or use clove teas or tincture as ingredients in formulas for acute respiratory infections.

Other taxonomical classifications for cloves have included *Eugenia aromatica*, *Jambosa caryophyllus*, and *Syzygium jambolanum*.

Tanacetum parthenium • Feverfew

Most well studied for headaches, *Tanacetum* leaves and flowering tops are also useful for treating allergies and reactive airway diseases. Feverfew can be included in formulas for dyspnea, including wheezing and cough, and is especially indicated when respiratory symptoms are due to underlying atopic reactivity. Due to antihistamine and anti-inflammatory effects, *Tanacetum parthenium* can be used for hay fever and allergy symptoms.

Taraxacum officinale • Dandelion

Dandelion roots are most well known as gentle alterative agents, but *Taraxacum* may be included in lung formulas for antiviral effects. *Taraxacum* inhibits influenza due to inhibition of viral nucleoprotein RNA levels and polymerase activity. Dandelion leaf is a nutritive diuretic and may be used to treat edema that occurs with heart failure and advanced pulmonary disease.

Tephrosia purpurea • Wild Indigo

While some species of *Tephrosia* are also known as *Polygonum* and *Senna* or *Cassia* species, all plants are known to contain emodin. Some species of *Tephrosia* are also dye plants and may also be known as indigo, matching the common name of wild indigo. The roots are used to treat chronic inflammatory diseases including asthma. The roots contain tephrosin, a flavonoid with bronchodilating effects. Tephrosin is toxic to fish, but harmless to mammals.

Terminalia chebula • Myrobalan

The fruits, sometimes referred to as myrobalans or haritaki, from this Southeast Asian tree are used as traditional anti-inflammatory medicines and are noted to contain ellagic acid, tannins, and chebulagic acid. The fruits are used in Ayurvedic medicine as an antitussive and are included in the well-known daily tonic called triphala formula. Mast cell–stabilizing effects have been reported, and several compounds in *Terminalia* have been shown to bind to glycosaminoglycan receptors of respiratory mucosal cells, disrupting the ability of viruses to bind at the same location. These mechanisms may explain the traditional usage for respiratory allergies and infections.

Terminalia may deter influenza and respiratory syncytial viruses in this manner. Chebulinic acid in the plant is shown to prevent the excessive migration of vascular smooth muscle cells—a known contributor to pulmonary obstructive disorders and blood vessel damage in the lungs—explaining its traditional use for cardiac coughs. *Terminalia* is therefore useful in formulas for treating dyspnea resulting from atherosclerotic disease in the lungs and other cardiopulmonary disorders.[158]

Thymus vulgaris • Thyme

Thyme makes a wonderful respiratory herb because it is drying to excessive mucus, antispasmodic to the bronchial passages, and a broad-acting antimicrobial. Include thyme as an ingredient in formulas for cough, bronchitis, asthma, pneumonia, and hay fever with profuse mucus and in convalescent teas to support recovery from pneumonia, chest colds, and allergic airway disorders. Thyme essential oil is extremely irritating on the skin, making eucalyptus, camphor, mint, and other essential oils safer and better choices in topical chest liniments. However, a single drop of thyme oil in steam inhalants is useful. While not a common commercial tea, thyme infusions are reasonably palatable and a great addition to respiratory tea blends. Thyme is also useful as a tincture, and it is specifically indicated for abundant mucus and spastic coughing or wheezing. *Thymus* species have activity against *Pseudomonas aeruginosa*.

Tinospora cordifolia • Guduchi

Tinospora goes by the common names guduchi or amrita in Ayurvedic medicine where it is used in many chronic inflammatory diseases and credited with immunomodulating activity. In TCM, related *Tinospora* species may go by the common names kuan jin teng (*T. sinensis*) and jin guo lan (*T. sagittata*, also known as *T. capillipes*), where they are used in formulas to clear toxins from the body and to treat common colds, headache, pharyngitis, and fever, among other inflammatory conditions. Tinosporin in the plant is credited with mast cell–stabilizing effects, explaining its traditional use in treating hay fever and respiratory symptoms. Several studies have suggested that *Tinospora* may treat allergic disorders of the airways as effectively as prescription antihistamines, but without the drowsiness side effect.

Trichosanthes kirilowii • Chinese Cucumber

This curcubit is on the list of the 50 fundamental medicines in TCM. It is used to drain heat and generate fluids, including in the lungs where it is noted to transform old and stagnant phlegm and help to better moisten dryness, heat, and inflammation and thereby "unbind the chest." The fruits, leaves, and occasionally other plant parts are used medicinally and noted to interfere with the protein coat of some virons, and antiviral activity is documented, supporting the folkloric use for respiratory and other infections. The plant has been traditionally used to resolve abcesses and dissipate nodules and is being explored for use against various cancers. The flavonoid trichosanthin is a ribosomal inactivating protein with antiviral effects, and trichosanthin inhibits the pro-inflammatory cytokine IL-17, often upregulated in inflammatory airway diseases.

Tussilago farfara • Coltsfoot

Tussilago means "cough dispeller" in Latin, and the roots and young emerging stalks and leaves are mucilaginous and antispasmodic, useful in both teas and tinctures to soothe irritated respiratory tissues and coughs. The fresh roots can be coarsely chopped and covered with honey, which will thin down as the honey draws out the mucilage and other constituents in the roots. Once strained, the honey makes a useful base to combine with other respiratory herbs for cough syrups. However, the pyrrolizidine alkaloids in the plant are hepatotoxic and must be avoided in infants and pregnant and nursing women, as well as in patients with liver disease or taking other liver-toxin medicines. It should never be used as a long-term treatment. Pyrrollizidine alkaloid-free varieties have been developed.

Tylophora asthmatica • Indian Ipecac

Sometimes called Indian ipecac, *Tylophora* is a potentially toxic plant used historically to poison vermin; however, the leaf has been used traditionally in Southeast Asia for asthma. Medicinal preparations are a strong expectorant at low doses and an emetic with higher doses. The plant contains tylophorine with mast cell–stabilizing and other anti-inflammatory effects on the airways. *Tylophora* may also improve adrenal steroid output and reduce cellular immune hypersensitivity reactions. The related species *Tylophora indica* also contains tylophorine and is used medicinally.

Ulmus fulva • Slippery Elm

Slippery elm is "slippery" due to the high-quality mucilage found in the bark; the plant has been used for hundreds of years to soothe irritated mucous membranes. Slippery elm tincture and teas can be used for painful cough and dry hacking or burning and tickling sensations in the throat and esophagus. Slippery elm lozenges are commercially available and may be one prong of an overall protocol for treating throat and chest pain due to respiratory infections.

Uncaria tomentosa • Cat's Claw

Cat's claw is a traditional South American vine and traditional medicine of the Amazon where both the leaves and the bark are used to treat cancer, infection, and inflammation. The plant contains the alkaloid mitraphylline, which is noted to inhibit pro-inflammatory cytokines, often elevated in inflammatory conditions of the lungs, and to attenuate the ability of bacterial endotoxins to promote excessive white blood cell activation. Mitraphylline is also active in the central nervous system, and neurotransmitter effects may contribute to the plant's anodyne properties. My teachers among the Amazonian peoples have told me that the plant is "very hot" so must be blended with cooling herbs such as demulcents. *Uncaria* also has immunomodulating activity credited to tetracyclic and pentacyclic oxindole alkaloids and to phenolic acids. *U. rhynchophylla* is also used medicinally in a similar manner.

Usnea barbata • Old Man's Beard

Also known as tree moss, *Usnea* is said to be antimicrobial, and it is traditionally used in teas and cough syrups for URIs. Astringent effects were emphasized by some early American authors, recommending that the green coating of oven-dried *Usnea* be ground into a powder in a mortar and pestle to treat hemoptysis, or "the spitting of blood." Other species are used medicinally, including *U. hirta*. Some folkloric texts suggest efficacy for throbbing head pain and heavy menstruation.

Vaccinium myrtillus • Blueberry

This genus also includes cranberries and other species of edible berries. Berries are not specific to respiratory complaints; however, berries and grapes are among the best dietary sources of flavonoids credited with anti-inflammatory, antioxidant, tissue-stabilizing, and antifibrotic effects. *Vaccinium* is available in the form of tinctures and syrups, and dried berries and berry powders can be included in teas to treat chronic respiratory complaints. *Vaccinium* would not work quickly for allergies or respiratory infections or provide bronchodilating effects, but rather may help reduce allergic reactivity over time. When consumed on a regular basis for many months or even years, berries may help protect the vasculature and lungs from fibrosis. Include blueberries and other berries in oatmeal, smoothies, salads, and dressings, and include *Vaccinium* in herbal formulas intended for long-term use to manage asthma, cystic fibrosis, emphysema, pulmonary hypertension, and other chronic inflammatory diseases of the lungs.

Verbascum thapsus • Mullein

Mullein leaves are soothing and demulcent agents to include in teas and tinctures for lung complaints, including cough, hoarseness, and lingering congestion in the airways following simple infections and colds. A mullein leaf decoction with a bit of fresh lemon juice is a traditional antitussive agent. As a traditional remedy for ear infections, fresh mullein flowers are macerated in olive or other quality oil directly in the sun and placed in the ears, when colds and upper respiratory catarrh are impeding eustachian tube drainage.

Vitex negundo • Chinese Chaste Tree

This Asian species has been used widely in Southeast Asia for treating coughs, and bronchodilating compounds have been identified, including casticin, isoorientin, chrysophenol D, and luteolin. Chaste tree berry (*Vitex agnus-castus*) may be better known to Western herbalists.

Vitis vinifera • Common Grape Vine

Grapes are an excellent source of resveratrol, well studied for numerous antioxidant and anti-inflammatory effects, useful to protect the lungs, heart, blood vessels, and other tissues from fibrotic, inflammatory, and allergic processes. Among its many mechanisms of action, resveratrol is a mast cell stabilizer, useful for asthma and other inflammatory conditions in the lungs.

Withania somnifera • Ashwagandha

Withania is an adaptogenic herb whose roots are widely used in Ayurvedic medicine for calming effects on mental-emotional stress and symptoms, and research has shown the plants' constituents can bind to GABA receptors. *Withania*'s adaptogenic effects make it useful in a wide variety of formulas where stress, exhaustion, and anxiety contribute to respiratory illness. *Withania* can improve endocrine balance, optimize the hypothalamic-pituitary-adrenal axis, and support general vitality. *Withania* may improve sleep and may improve immune function where chronic colds and respiratory infections occur.

Xanthium cavanillesii • Cocklebur

The leaves and roots have been traditionally used for cough, fever, and tuberculosis, as well as for hay fever and sinusitis. The plant contains xanthatin, a sesquiterpene lactone shown to be a mast cell stabilizer useful in formulas for allergies, asthma, and hypersensitivity of the airways.

Zanthoxylum clava-herculis • Southern Prickly Ash

Zanthoxylum is a warming, stimulating expectorant, specifically indicated for atonic respiratory conditions, bringing heat, stimulating circulation and salivation, and helping to move old thick secretions out of the lungs. *Zanthoxylum* bark preparations produce a warming, prickly sensation in the mouth, hence the name prickly ash. (Other common names are pepperwood, Hercules' herb, Hercules' club, and toothache tree.) It is also indicated when there is a tendency to dryness in the airways and a tendency to mucous plugs that dry in hard crusts. The plant is also credited with alterative effects, enhancing digestion and circulation in the stomach, and was traditionally combined with *Hydrastis* and *Capsicum* for sluggish digestion.

Zingiber officinale • Ginger

Ginger root is warming and decongesting to balance conditions of cold and dampness. Ginger has numerous anti-inflammatory effects and is appropriate for allergic conditions and extensive inflammation; it also has antimicrobial effects. Ginger syrups are readily available and can be used as a base for childrens' formulas, and ginger can be included in tea and tincture formulas for a wide variety of acute and chronic respiratory conditions. Ginger capsules are also readily available and can be included in protocols to complement teas and tinctures in the management of chronic or severe lung diseases.

— ACKNOWLEDGMENTS —

I would like to thank my dear friend and colleague Dr. Mary Bove for reading these chapters and sharing her insights on the formulas and commentaries, and especially for her vast clinical knowledge and her grounded, firmly moral, and wise-woman ways with me over the years. And, I would like to thank Ryan Bradley, ND, for his expertise in cardiology and suggestions that improved chapter 2. I would also like to thank Mikailah Grover, ND, class of 2019, and Patryk Madrid, ND, class of 2020, for their admirable skills with word processing and editing support.

Thank you to my editor at Chelsea Green, Fern Marshall Bradley, for her professional prowess and for making this much more polished than it would have been without her guidance. I wish to also thank Margo Baldwin, the publisher at Chelsea Green, for making this text possible at all, as well as Pati Stone and the entire production team for helping to perfect the many small details that producing a reference text of this nature entails.

And I extend warm, fond appreciation to my kind and supportive sweetie, Warren Martin, for his patience, political updates, and willingness to pitch in with the day-to-day chores of running a household and a business.

I would like to thank and honor all my teachers and the herb community for being my beloved tribe, filled with powerful, skillful women and gentle, reverent men, and for all the times that the healing plants have brought us together in beauty, to share our gifts and wisdom with one another. If I have never told you to your face (and to many of you I have), I love you!

And last, but not least, to Pachamama herself for her mystery, abundance, and amazing gifts. May we all Walk in Beauty and return her gifts tenfold. Blessed Be!

— SCIENTIFIC NAMES —
TO COMMON NAMES

The following lists include all of the herbs, medicinal fungi, and homeopathic preparations mentioned in the text of this book.

Achillea millefolium	yarrow
Aconitum carmichaelii	fu zhi
Aconitum napellus	aconite, wolfsbane
Actaea cimicifuga (also known as *Cimicifuga foetida*), *A. dahurica*, *A. heracleifolia*, *Actaea* spp.	shengma
Actaea racemosa (also known as *Cimicifuga racemosa*)	black cohosh, macrotys
Adhatoda vasica (also known as *Justicia adhatoda*)	Malabar nut
Adonis vernalis	pheasant's eye
Aesculus hippocastanum	horse chestnut
Agathosma betulina (also known as *Barosma betulina*)	buchu
Albizia julibrissin	silk tree, mimosa
Albizia lebbeck	siris, lebbeck tree
Alisma plantago-aquatica (also known as *A. orientale*)	Asian water plantain
Allium cepa	onion
Allium sativum	garlic
Alpinia galanga	
Aloe vera, A. barbadensis	aloe
Alstonia scholaris	dita wood, blackboard tree, devil tree
Althaea officinalis	marshmallow
Ammi visnaga, A. majus	khella, bishop's weed
Ananas comosus	pineapple
Andrographis paniculata	king of bitters, andrographis

Anemopsis californica	yerba mansa
Anethum graveolens	dill
Angelica archangelica	garden angelica
Angelica dahurica	bai zhi
Angelica sinensis	dong quai
Annona muricata	custard apple
Anthoxanthum odoratum	sweet vernal grass
Apium graveolens	celery
Apocynum cannabinum	dogbane
Arachis hypogaea	peanut
Arctium lappa	burdock
Ardisia coriacea	beef-wood
Ardisia esculenta	dogberry
Ardisia japonica	marlberry, coralberry
Ardisia tinifolia	ardisia
Armoracia rusticana	horseradish
Arnica montana	leopard's bane
Artemisia annua	sweet Annie
Artemisia caerulescens	artemisia
Artemisia keiskeana	an lu
Artemisia vulgaris	mugwort
Artocarpus altilis	breadfruit
Asclepias tuberosa	pleurisy root, butterfly weed, milkweed
Aspidosperma quebracho	quebracho
Astragalus membranaceus	milk vetch
Atractylodes lancea, A. macrocephala	atractylodes
Atractylodes spp.	bai zhu, atractylodes
Atropa belladonna	belladonna
Avena sativa	oats
Azadirachta indica (also known as *Melia azadirachta*)	neem, bean tree, China tree

Bacopa monnieri	brahmi
Berberis aquifolium (also known as *Mahonia aquifolium*)	Oregon grape, mahonia
Biancaea sappan (also known as *Caesalpinia sappan*)	brezel wood, su mu
Borago officinalis	borage
Brassica nigra (also known as *Sinapis nigra*)	black mustard
Bupleurum chinense	chai hu
Bupleurum falcatum	Chinese thoroughwax, saiko
Bryonia alba	white bryony
Bryonia dioica	bryony
Cactus grandiflorus (also known as *Selenicereus grandiflorus*)	night-blooming cactus
Caesalpinia sappan (also known as *Biancaea sappan*)	brezel wood, su mu
Calendula officinalis	pot marigold
Camellia sinensis	green tea
Capsella bursa-pastoris	shepherd's purse
Capsicum annuum (also known as *Capsicum frutescens*)	cayenne
Carum carvi	caraway
Carum copticum (also known as *Trachyspermum ammi*)	ajwain
Cassia occidentalis (also known as *Senna occidentalis*)	coffee weed
Cassia senna (also known as *Senna alexandrina*)	senna
Castanospermum australe	blackbean tree
Carthamus tinctorius	safflower
Caulophyllum thalictroides	blue cohosh
Ceanothus americanus	red root
Centella asiatica	gotu kola, pennywort
Ceratonia siliqua	carob
Cereus grandiflorus (also known as *Selenicereus grandiflorus*)	night-blooming cactus, night-blooming cereus
Chimaphila umbellata	pipsissewa
Chionanthus virginicus	fringe tree
Cimicifuga foetida (also known as *Actaea cimicifuga*)	shengma
Cimicifuga racemosa (also known as *Actaea racemosa*)	black cohosh
Cinnamomum camphora	camphor tree
Cinnamomum cassia	cassia cinnamon, Chinese cinnamon
Cinnamomum polyadelphum (also known as *C. saigonicum*)	Saigon cinnamon
Cinnamomum verum (also known as *C. zeylanicum*)	Ceylon cinnamon, cinnamon
Cirsium japonicum	Japanese thistle
Cissampelos sympodialis	cissampelos
Cissampelos pareira	abuta, laghu patha
Citrus spp.	citrus
Citrus paradisi	grapefruit
Clerodendrum serratum	glorybower, bagflower, bharangi
Cnicus benedictus	blessed thistle
Cnidium monnieri	snow parsley, osthole
Cocos nucifera	coconut
Coleus forskohlii (also known as *Plectranthus forskohlii*)	coleus
Collinsonia canadensis	stoneroot
Commiphora mukul	guggul
Commiphora myrrha, C. molmol	myrrh
Convallaria majalis	lily of the valley
Coptis chinensis	goldthread, duan e huang lian
Cordyceps sinensis	cordyceps, caterpillar fungus
Coriandrum sativum	coriander
Cornus officinalis	Japanese cornel, shan zhu yu
Corydalis cava	corydalis, turkey corn
Crataegus monogyna, Crataegus oxyacantha, Crataegus laevigata	hawthorn
Crinum glaucum	swamp lily

Croton lechleri	sangre de drago, dragon's blood
Curcuma longa	turmeric
Cuscuta reflexa	giant dodder
Cynara scolymus	artichoke
Cytisus scoparius	Scotch broom, Scot's broom
Datura stramonium	jimson weed
Daucus carota	wild carrot
Desmodium styracifolium	desmodium
Digitalis purpurea	foxglove, digitalis
Dioscorea villosa	wild yam
Dracocephalum rupestre	dragon's head
Drimia maritima (also known as *Urginea maritima*)	sea squill, squill
Drosera rotundifolia	sundew
Echinacea angustifolia	coneflower
Echinacea purpurea	coneflower, purple coneflower
Eclipta prostrata (also known as *E. alba*)	eclipta, bhringraj, yerba de tago
Elaeocarpus serratus (also known as *E. ganitrus*)	bhadrasey, Rudraksha
Elettaria cardamomum	cardamom
Eleutherococcus senticosus	Siberian ginseng
Emblica officinalis (also known as *Phyllanthus emblica*)	amla, amala, Indian gooseberry
Ephedra nevadensis	Mormon tea
Ephedra sinica	ma huang, ephedra
Epimedium grandiflorum	horny goatweed
Epimedium brevicornu (also known as *E. rotundatum*)	horny goatweed, yin yang huo, bishop's hat
Equisetum arvense, E. hyemale	horsetail
Eriodictyon californicum	yerba santa, mountain balm
Eschscholzia californica	California poppy
Eucalyptus globulus	eucalyptus
Eupatorium perfoliatum, E. chinense	boneset
Euphorbia hirta	spurge, asthma weed
Euphorbia trigona	African milk tree, cathedral cactus
Euphorbia pekinensis	Peking spurge
Euphorbia parviflora (also known as *E. pilulifera*)	asthma weed, pill-bearing spurge
Euphrasia officinalis	eyebright
Fagopyrum esculentum	buckwheat
Foeniculum vulgare	fennel
Fritillaria cirrhosa	yellow Himalayan fritillaria, chuan bei mu
Fritillaria sichuanica	hua xi bei mu
Fritillaria taipaiensis	tai bai bei mu
Fritillaria yuzhongensis	yu min bei mu
Fuchsia magellanica	fuchsia
Fucus vesiculosus	kelp, bladderwrack
Galega officinalis	goat's rue, galega, French lilac
Galium aparine	cleavers
Galphimia glauca	calderona amarilla, flor estrella
Ganoderma lucidum	reishi
Gardenia latifolia	gardenia, Indian boxwood
Gardenia jasminoides	cape jasmine, gardenia
Gastrodia elata	tian ma
Gelsemium sempervirens	false jasmine
Gentiana acaulis, G. lutea	gentian
Geranium maculatum	wild geranium
Ginkgo biloba	ginkgo, maidenhair tree
Glycine max	soybean
Glycyrrhiza glabra	licorice
Glycyrrhiza inflata, G. uralensis	Chinese licorice
Gossypium barbadense	cotton
Grifola frondosa	maitake, mushroom, hen of the woods
Grindelia robusta, G. hirsutula (also known as *G. cuneifolia*), *G. squarrosa*	gumweed

Gymnema sylvestre	gurmar, sugar destroyer
Hamamelis virginiana	witch hazel
Helichrysum italicum	immortelle
Hemidesmus indicus	hemidesmus, East Indian sarsaparilla
Hibiscus rosa-sinensis	Chinese hibiscus
Hibiscus sabdariffa	roselle, flor de Jamaica, hibiscus
Hippophae rhamnoides	sea buckthorn
Huperzia serrata	Chinese club moss, qian ceng ta
Hydrangea macrophylla, H. paniculata	hydrangea
Hydrastis canadensis	goldenseal
Hyoscyamus niger	henbane
Hypericum perforatum	St. Johnswort
Hyssopus officinalis (also known as *H. vulgaris*)	hyssop
Inula helenium	elecampane
Inula racemosa	inula
Iris versicolor, I. tenax	wild iris, blue flag, vegetable mercury
Isatis tinctoria (also known as *I. indigotica*)	woad, dyer's woad, ban lan gen
Juniperus communis	juniper
Juniperus excelsa, J. phoenicea	juniper
Juniperus oxycedrus	juniper, cade juniper
Justicia adhatoda (also known as *Adhatoda vasica*)	Malabar nut, adulsa, adhatoda
Lavandula angustifolia	lavender
Lentinula edodes	shiitake mushroom
Leonurus cardiaca	motherwort
Lepidium apetalum	du xing cai
Lepidium latifolium	rompe piedras, pepperweed, dittander
Lepidium meyenii (also known as *L. peruvianum*)	maca
Lepidium sativum	garden cress
Ligusticum striatum (also known as *L. chuanxiong*)	Szechuan lovage, chuanxiong, ligusticum

Ligusticum porteri	osha
Ligusticum striatum	Szechuan lovage, chuanxiong
Linum usitatissimum	flax
Lobelia inflata	pukeweed
Lomatium dissectum	biscuitroot
Lonicera japonica, L. caprifolium	honeysuckle, ren dong teng
Lonicera xylosteum	woodbine, fly woodbine
Lophatherum gracile	dan zhu ye gen
Lycium barbarum, L. chinense	goji berry, wolfberry
Lycopus virginicus	bugleweed
Magnolia officinalis	magnolia
Magnolia obovata	Japanese bigleaf magnolia
Mahonia aquifolium (also known as *Berberis aquifolium*)	Oregon grape
Malus domestica	apple
Marrubium vulgare	horehound, hoarhound
Matricaria chamomilla, M. recutita	chamomile
Medicago sativa	alfalfa
Melaleuca alternifolia	tea tree
Melia azadirachta (also known as *Azadirachta indica*)	China tree, ku lian pi
Melilotus officinalis	sweet clover, yellow sweet clover
Melilotus suaveolens	sweet clover
Melissa officinalis	lemon balm
Mentha piperita	peppermint
Mentha pulegium	pennyroyal
Mentha spicata	spearmint
Mikania glomerata	huaco, guaco
Morella cerifera (also known as *Myrica cerifera*)	Southern wax myrtle, bayberry
Morinda citrifolia	noni
Mucuna pruriens	velvet bean, cowage, cow-itch
Myrica cerifera (also known as *Morella cerifera*)	bayberry, wax myrtle, candleberry

Neopicrorhiza scrophulariiflora (also known as *Picrorhiza scrophulariiflora*)	picrorhiza
Nepeta cataria	catnip, catnep, catmint
Nigella sativa	black cumin, black seed
Ocimum basilicum	sweet basil
Ocimum tenuiflorum (also known as *O. sanctum*)	holy basil, tulsi
Oenothera biennis	evening primrose
Olea europaea	olive
Ophiopogon japonicus	mai men dong, dwarf lilyturf, mondograss
Oplopanax horridus	devil's club
Opuntia ficus-indica	prickly pear
Origanum vulgare	oregano
Paeonia emodi	Himalayan peony
Paeonia lactiflora	white peony, bai shao yao, red peony
Paeonia × suffruticosa, P. anomala	moutan, peony, tree peony
Panax ginseng	ginseng, ren shen
Passiflora incarnata	passionflower
Pelargonium sidoides	African geranium
Perilla frutescens	shiso, perilla, Korean perilla
Persicaria tinctoria (also known as *Polygonum tinctorium*)	Chinese indigo
Petasites hybridus, P. japonicus	butterbur
Petroselinum crispum (also known as *P. sativum*)	parsley
Phoradendron chrysocladon (also known as *Viscum flavens*)	mistletoe
Phyllanthus amarus	chanca piedra, bahupatra, hurricane weed
Phyllanthus emblica (also known as *Emblica officinalis*)	amla, amalaki, Indian gooseberry
Phytolacca american (also known as *P. decandra*)	pokeweed, pokeroot, poke
Picrorhiza kurroa	kutki
Picrorhiza scrophulariiflora (also known as *Neopicrorhiza scrophulariiflora*)	picrorhiza
Pilocarpus alatus, P. jaborandi	jaborandi
Pimpinella anisum	anise, aniseed
Pinellia ternata	crow dipper
Pinus pinaster	maritime pine
Piper cubeba	cubeb pepper
Piper longum	pippali, long pepper
Piper methysticum	kava kava
Piper nigrum	black pepper
Plantago indica (also known as *P. psyllium* and *P. arenaria*)	plantain, psyllium
Polygala spp.	milkwort, snakeroot
Polygala karensium	snakeroot
Polygala paucifolia	fringed snakeroot
Polygala senega	Seneca snakeroot
Polygonum cuspidatum (also known as *Reynoutria japonica*)	hu jiang, hu zhang, Japanese knotweed
Polygonum multiflorum (also known as *Reynoutria multiflora*)	fo ti, he shou wu, Chinese knotweed
Polygonum tinctorium (also known as *Persicaria tinctoria*)	Chinese indigo
Poria cocos	hoelen, fu ling
Prunus dulcis (also known as *P. amygdalis*)	bitter almond
Prunus serotina	wild cherry, black cherry
Pueraria montana var. *lobata*	kudzu, gegen, Japanese arrowroot
Pulmonaria officinalis	lungwort
Punica granatum	pomegranate
Quercus alba	white oak
Quercus robur	common oak
Quercus rubra	red oak
Raphanus sativus var. *niger*	Spanish black radish
Rauvolfia serpentina, R. vomitoria	Indian snakeroot

Ravensara aromatica	clove nutmeg
Reynoutria japonica (also known as *Polygonum cuspidatum*)	Japanese knotweed, hu jiang, hu zhang
Reynoutria multiflora	fo ti, he shou wu, Chinese knotweed
Rhamnus purshiana	cascara
Rhaptopetalum coriaceum	rhaptopetalum
Rheum officinale, R. palmatum, R. emodi	Chinese rhubarb, turkey rhubarb
Rhodiola kirilowii, R. crenulata	rhodiola
Rhodiola quadrifida (also known as *Sedum quadrifidurn*)	rhodiola
Rhodiola rosea	arctic rose, rhodiola
Ribes spp.	currant
Ribes nigrum	black currant
Ricinus communis	castor, castor bean
Rosa canina	dog rose
Rosmarinus officinalis	rosemary
Rumex spp.	dock
Rumex acetosella	sheep sorrel
Rumex crispus	yellow dock
Ruscus aculeatus, R. oblongifolius	butcher's broom
Ruta graveolens	rue
Salix spp.	willow
Salvia miltiorrhiza	red sage, dan shen
Salvia officinalis	sage
Sambucus canadensis, S. nigra	elderberry
Sanguinaria canadensis	bloodroot
Saposhnikovia divaricata	fang feng
Saussurea costus (also known as *S. lappa*)	saw wort, saw thistle
Schisandra chinensis	magnolia vine, wu wei zi, five flavor fruit
Scutellaria baicalensis	scute, huang qin, baical
Scutellaria lateriflora	skullcap
Selaginella uncinata	spikemoss
Selenicereus grandiflorus (also known as *Cereus grandiflorus*)	night-blooming cactus, night-blooming cereus, queen of the night
Senna alata	candle bush
Senna insularis	senna
Senna occidentalis (also known as *Cassia occidentalis*)	coffee weed
Senna alexandrina (also known as *Cassia senna*)	senna
Silybum marianum	milk thistle
Sinapis alba	white mustard
Sinapis nigra (also known as *Brassica nigra*)	black mustard
Smilax officinalis, S. aristolochiifolia, S. ornata, S. regelii	sarsasparilla
Stellaria media	chickweed
Stemona tuberosa, S. japonica, S. sessilifolia	bai bu
Stevia rebaudiana	sweet leaf, stevia
Sticta pulmonaria (also known as *Lobaria pulmonaria*)	lungwort
Stillingia sylvatica	queen's root
Strophanthus hispidus	strophanthus, kombé
Symphytum officinale	comfrey
Symplocarpus foetidus	skunk cabbage
Syzygium aromaticum (also known as *Eugenia caryophyllata*)	cloves
Syzygium cumini (also known as *S. jambolanum*), *S. jambos*	roseapple, jambul, jambolanum
Tanacetum parthenium	feverfew
Taraxacum officinale	dandelion
Tephrosia purpurea	wild indigo
Terminalia arjuna	arjuna
Terminalia chebula	myrobalan, haritaki
Theobroma cacao	cacao, cocoa tree, cocoa
Thuja occidentalis	Northern white cedar
Thuja plicata	Western red cedar
Thymus vulgaris	thyme
Tilia × europaea	linden
Tinospora cordifolia	guduchi, amrita
Tinospora sinensis	kuan jin teng

Tinospora sagittata (also known as *T. capillipes*)	jin guo lan
Trachyspermum ammi (also known as *Carum copticum*)	ajwain
Trichosanthes kirilowii	Chinese cucumber
Trifolium pratense	red clover
Trigonella foenum-graecum	fenugreek
Tussilago farfara	coltsfoot
Tylophora asthmatica, T. indica	Indian ipecac
Ulmus fulva	slippery elm
Uncaria rhynchophylla, U. tomentosa	cat's claw, uno de gato
Urginea maritima (also known as *Drimia maritima*)	squill, sea onion, maritime squill
Urtica dioica, U. urens	nettle, stinging nettle
Usnea barbata, U. hirta	old man's beard, tree moss
Vaccinium myrtillus	bilberry, blueberry
Valeriana officinalis	valerian
Valeriana sitchensis	Sitka valerian

Veratrum album	white hellebore
Veratrum viride	false hellebore, corn lily
Verbascum thapsus	mullein
Viburnum opulus	crampbark, guelder rose, snowball bush
Viburnum prunifolium	blackhaw
Vinca minor	lesser periwinkle
Viscum album	mistletoe
Viscum articulatum	leafless mistletoe
Viscum flavens (also known as *Phoradendron chrysocladon*)	mistletoe
Vitex negundo	Chinese chaste tree
Vitis vinifera	common grape vine
Withania somnifera	ashwagandha
Xanthium cavanillesii	cocklebur
Zanthoxylum clava-herculis	Southern prickly ash, Hercules' club, toothache tree
Zingiber officinale	ginger
Ziziphus jujuba	Chinese date, jujube
Ziziphus spinosa	suan zao ren

— COMMON NAMES —
TO SCIENTIFIC NAMES

The following lists include all of the herbs, medicinal fungi, and homeopathic preparations mentioned in the text of this book.

abuta	*Cissampelos pareira*
aconite	*Aconitum napellus*
adhatoda	*Justicia adhatoda* (also known as *Adhatoda vasica*)
adulsa	*Justicia adhatoda* (also known as *Adhatoda vasica*)
African geranium	*Pelargonium sidoides*
African milk tree	*Euphorbia trigona*
ajwain	*Trachyspermum ammi* (also known as *Carum copticum*)
alfalfa	*Medicago sativa*
aloe	*Aloe vera, A. barbadensis*
amala	*Emblica officinalis* (also known as *Phyllanthus emblica*)
amalaki	*Phyllanthus emblica* (also known as *Emblica officinalis*)
amla	*Phyllanthus emblica* (also known as *Emblica officinalis*)
amrita	*Tinospora cordifolia*
an lu	*Artemisia keiskeana*
andrographis	*Andrographis paniculata*
anise	*Pimpinella anisum*
aniseed	*Pimpinella anisum*
apple	*Malus domestica*
arctic rose	*Rhodiola rosea*
ardisia	*Ardisia tinifolia*
arjuna	*Terminalia arjuna*
artemisia	*Artemisia caerulescens*
artichoke	*Cynara scolymus*
ashwagandha	*Withania somnifera*

Asian water plantain	*Alisma plantago-aquatica* (also known as *A. orientale*)
asthma weed	*Euphorbia hirta, Euphorbia parviflora* (also known as *E. pilulifera*)
atractylodes	*Atractylodes* spp., *A. lancea, A. macrocephala*
bagflower	*Clerodendrum serratum*
bahupatra	*Phyllanthus amarus*
bai bu	*Stemona tuberosa, S. japonica, S. sessilifolia*
bai shao yao	*Paeonia lactiflora*
bai zhi	*Angelica dahurica*
bai zhu	*Atractylodes* spp.
baical	*Scutellaria baicalensis*
ban lan gen	*Isatis tinctoria* (also known as *I. indigotica*)
bayberry	*Morella cerifera* (also known as *Myrica cerifera*)
bean tree	*Azadirachta indica* (also known as *Melia azadirachta*)
beef-wood	*Ardisia coriacea*
belladonna	*Atropa belladonna*
bhadrasey	*Elaeocarpus serratus* (also known as *E. ganitrus*)
bharangi	*Clerodendrum serratum*
bhringraj	*Eclipta prostrata* (also known as *E. alba*)
bilberry	*Vaccinium myrtillus*
biscuitroot	*Lomatium dissectum*
bishop's hat	*Epimedium brevicornu* (also known as *E. rotundatum*)
bitter almond	*Prunus dulcis* (also known as *P. amygdalis*)

black cherry	*Prunus serotina*
black cohosh	*Actaea racemosa* (also known as *Cimicifuga racemosa*)
black cumin	*Nigella sativa*
black currant	*Ribes nigrum*
black mustard	*Brassica nigra* (also known as *Sinapis nigra*)
black pepper	*Piper nigrum*
black seed	*Nigella sativa*
blackbean tree	*Castanospermum australe*
blackboard tree	*Alstonia scholaris*
blackhaw	*Viburnum prunifolium*
bladderwrack	*Fucus vesiculosus*
blessed thistle	*Cnicus benedictus*
bloodroot	*Sanguinaria canadensis*
blue cohosh	*Caulophyllum thalictroides*
blue flag	*Iris versicolor, I. tenax*
blueberry	*Vaccinium myrtillus*
boneset	*Eupatorium perfoliatum, E. chinense*
borage	*Borago officinalis*
brahmi	*Bacopa monnieri*
breadfruit	*Artocarpus altilis*
brezel wood	*Caesalpinia sappan* (also known as *Biancaea sappan*)
bryony	*Bryonia dioica*
buchu	*Agathosma betulina*
buckwheat	*Fagopyrum esculentum*
bugleweed	*Lycopus virginicus*
burdock	*Arctium lappa*
butcher's broom	*Ruscus aculeatus*
butterbur	*Petasites hybridus, P. japonicus*
butterfly weed	*Asclepias tuberosa*
cacao	*Theobroma cacao*
cade juniper	*Juniperus oxycedrus*
calderona amarilla	*Galphimia glauca*
California poppy	*Eschscholzia californica*
camphor tree	*Cinnamomum camphora*
candle bush	*Senna alata*

candleberry	*Myrica cerifera* (also known as *Morella cerifera*)
cape jasmine	*Gardenia jasminoides*
caraway	*Carum carvi*
cardamom	*Elettaria cardamomum*
carob	*Ceratonia siliqua*
cascara	*Rhamnus purshiana*
cassia cinnamon	*Cinnamomum cassia*
castor	*Ricinus communis*
castor bean	*Ricinus communis*
cat's claw	*Uncaria rhynchophylla, U. tomentosa*
caterpillar fungus	*Cordyceps sinensis*
cathedral cactus	*Euphorbia trigona*
catmint	*Nepeta cataria*
catnep	*Nepeta cataria*
catnip	*Nepeta cataria*
cayenne	*Capsicum annuum* (also known as *Capsicum frutescens*)
celery	*Apium graveolens*
Ceylon cinnamon	*Cinnamomum verum* (also known as *C. zeylanicum*)
chai hu	*Bupleurum chinense*
chamomile	*Matricaria chamomilla, M. recutita*
chanca piedra	*Phyllanthus amarus*
chickweed	*Stellaria media*
China tree	*Azadirachta indica* (also known as *Melia azadirachta*)
Chinese chaste tree	*Vitex negundo*
Chinese cinnamon	*Cinnamomum cassia*
Chinese club moss	*Huperzia serrata*
Chinese cucumber	*Trichosanthes kirilowii*
Chinese date	*Ziziphus jujuba*
Chinese hibiscus	*Hibiscus rosa-sinensis*
Chinese indigo	*Persicaria tinctoria* (also known as *Polygonum tinctorium*)
Chinese knotweed	*Polygonum multiflorum* (also known as *Reynoutria multiflora*)
Chinese licorice	*Glycyrrhiza inflata, G. uralensis*

Chinese rhubarb	*Rheum officinale, R. palmatum, R. emodi*
Chinese thoroughwax	*Bupleurum falcatum*
chuan bei mu	*Fritillaria cirrhosa*
chuanxiong	*Ligusticum striatum* (also known as *L. Chuanxiong*)
cinnamon	*Cinnamomum verum* (also known as *C. zeylanicum*)
cissampelos	*Cissampelos sympodialis*
citrus	*Citrus* spp.
cleavers	*Galium aparine*
clove nutmeg	*Ravensara aromatica*
cloves	*Syzygium aromaticum* (also known as *Eugenia caryophyllata*)
cocoa	*Theobroma cacao*
cocklebur	*Xanthium cavanillesii*
cocoa tree	*Theobroma cacao*
coconut	*Cocos nucifera*
coffee weed	*Senna occidentalis* (also known as *Cassia occidentalis*)
coleus	*Coleus forskohlii* (also known as *Plectranthus forskohlii*)
coltsfoot	*Tussilago farfara*
comfrey	*Symphytum officinale*
common grape vine	*Vitis vinifera*
common oak	*Quercus robur*
coneflower	*Echinacea angustifolia, E. purpurea*
coralberry	*Ardisia japonica*
cordyceps	*Cordyceps sinensis*
coriander	*Coriandrum sativum*
corn lily	*Veratrum viride*
corydalis	*Corydalis cava*
cotton	*Gossypium barbadense*
cow-itch	*Mucuna pruriens*
cowage	*Mucuna pruriens*
crampbark	*Viburnum opulus*
crow dipper	*Pinellia ternata*
cubeb pepper	*Piper cubeba*
currant	*Ribes* spp.
custard apple	*Annona muricata*
dan shen	*Salvia miltiorrhiza*
dan zhu ye gen	*Lophatherum gracile*
dandelion	*Taraxacum officinale*
desmodium	*Desmodium styracifolium*
devil tree	*Alstonia scholaris*
devil's club	*Oplopanax horridus*
digitalis	*Digitalis purpurea*
dill	*Anethum graveolens*
dita wood	*Alstonia scholaris*
dittander	*Lepidium latifolium*
dock	*Rumex* spp.
dog rose	*Rosa canina*
dogbane	*Apocynum cannabinum*
dogberry	*Ardisia esculenta*
dong quai	*Angelica sinensis*
dragon's blood	*Croton lechleri*
dragon's head	*Dracocephalum rupestre*
du xing cai	*Lepidium apetalum*
duan e huang lian	*Coptis chinensis*
dwarf lilyturf	*Ophiopogon japonicus*
dyer's woad	*Isatis tinctoria* (also known as *I. indigotica*)
East Indian sarsaparilla	*Hemidesmus indicus*
eclipta	*Eclipta prostrata* (also known as *E. alba*)
elderberry	*Sambucus canadensis, S. nigra*
elecampane	*Inula helenium*
ephedra	*Ephedra sinica*
eucalyptus	*Eucalyptus globulus*
evening primrose	*Oenothera biennis*
eyebright	*Euphrasia officinalis*
false hellebore	*Veratrum viride*
false jasmine	*Gelsemium sempervirens*
fang feng	*Saposhnikovia divaricata*
fennel	*Foeniculum vulgare*
fenugreek	*Trigonella foenum-graecum*

feverfew	*Tanacetum parthenium*
five flavor fruit	*Schisandra chinensis*
flax	*Linum usitatissimum*
flor de Jamaica	*Hibiscus sabdariffa*
flor estrella	*Galphimia glauca*
fly woodbine	*Lonicera xylosteum*
fo ti	*Polygonum multiflorum* (also known as *Reynoutria multiflora*)
foxglove	*Digitalis purpurea*
French lilac	*Galega officinalis*
fringe tree	*Chionanthus virginicus*
fringed snakeroot	*Polygala paucifolia*
fu ling	*Poria cocos*
fu zhi	*Aconitum carmichaelii*
fuchsia	*Fuchsia magellanica*
galega	*Galega officinalis*
garden cress	*Lepidium sativum*
garden angelica	*Angelica archangelica*
gardenia	*Gardenia jasminoides, Gardenia latifolia*
garlic	*Allium sativum*
gegen	*Pueraria montana* var. *lobata*
gentian	*Gentiana acaulis, G. lutea*
giant dodder	*Cuscuta reflexa*
ginger	*Zingiber officinale*
ginkgo	*Ginkgo biloba*
ginseng	*Panax ginseng*
glorybower	*Clerodendrum serratum*
goat's rue	*Galega officinalis*
goji berry	*Lycium barbarum, L. chinense*
goldenseal	*Hydrastis canadensis*
goldthread	*Coptis chinensis*
gotu kola	*Centella asiatica*
grapefruit	*Citrus paradisi*
green tea	*Camellia sinensis*
guaco	*Mikania glomerata*
guduchi	*Tinospora cordifolia*
guelder rose	*Viburnum opulus*
guggul	*Commiphora mukul*
gumweed	*Grindelia robusta, G. hirsutula* (also known as *G. cuneifolia*), *G. squarrosa*
gurmar	*Gymnema sylvestre*
haritaki	*Terminalia chebula*
hawthorn	*Crataegus monogyna, Crataegus oxyacantha, Crataegus laevigata*
he shou wu	*Polygonum multiflorum* (also known as *Reynoutria multiflora*)
hemidesmus	*Hemidesmus indicus*
hen of the woods	*Ginfola frondosa*
henbane	*Hyoscyamus niger*
Hercules' club	*Zanthoxylum clava-herculis*
hibiscus	*Hibiscus sabdariffa*
Himalayan peony	*Paeonia emodi*
hoarhound	*Marrubium vulgare*
hoelen	*Poria cocos*
holy basil	*Ocimum tenuiflorum* (also known as *O. sanctum*)
honeysuckle	*Lonicera japonica, L. caprifolium*
horehound	*Marrubium vulgare*
horny goatweed	*Epimedium brevicornu* (also known as *E. rotundatum*), *Epimedium grandiflorum*
horse chestnut	*Aesculus hippocastanum*
horseradish	*Armoracia rusticana*
horsetail	*Equisetum arvense, E. hyemale*
hu jiang	*Reynoutria japonica* (also known as *Polygonum cuspidatum*)
hu zhang	*Reynoutria japonica* (also known as *Polygonum cuspidatum*)
hua xi bei mu	*Fritillaria sichuanica*
huaco	*Mikania glomerata*
huang qin	*Scutellaria baicalensis*
hurricane weed	*Phyllanthus amarus*
hydrangea	*Hydrangea macrophylla, H. paniculata*
hyssop	*Hyssopus officinalis* (also known as *H. vulgaris*)

immortelle	*Helichrysum italicum*
Indian boxwood	*Gardenia latifolia*
Indian gooseberry	*Phyllanthus emblica* (also known as *Emblica officinalis*)
Indian ipecac	*Tylophora asthmatica, T. indica*
Indian snakeroot	*Rauvolfia serpentina, R. vomitoria*
inula	*Inula racemosa*
jaborandi	*Pilocarpus alatus, P. jaborandi*
jambolanum	*Syzygium cumini* (also known as *S. jambolanum*), *S. jambos*
jambul	*Syzygium cumini* (also known as *S. jambolanum*), *S. jambos*
Japanese arrowroot	*Pueraria hiontana* var. *lobata*
Japanese bigleaf magnolia	*Magnolia obovata*
Japanese cornel	*Cornus officinalis*
Japanese knotweed	*Reynoutria japonica* (also known as *Polygonum cuspidatum*)
Japanese thistle	*Cirsium japonicum*
jimson weed	*Datura stramonium*
jin guo lan	*Tinospora sagittata* (also known as *T. capillipes*)
jujube	*Ziziphus jujuba*
juniper	*Juniperus communis, J. excelsa, J. phoenicea, J. oxycedrus*
kava kava	*Piper methysticum*
kelp	*Fucus vesiculosus*
khella, bishop's weed	*Ammi visnaga, A. majus*
king of bitters	*Andrographis paniculata*
kombé	*Strophanthus hispidus*
Korean perilla	*Perilla frutescens*
ku lian pi	*Melia azadirachta* (also known as *Azadirachta indica*)
kuan jin teng	*Tinospora sinensis*
kudzu	*Pueraria montana* var. *lobata*
kutki	*Picrorhiza kurroa*
laghu patha	*Cissampelos pareira*
lavender	*Lavandula angustifolia*
leafless mistletoe	*Viscum articulatum*
lebbeck tree	*Albizia lebbeck*
lemon balm	*Melissa officinalis*
leopard's bane	*Arnica montana*
lesser periwinkle	*Vinca minor*
licorice	*Glycyrrhiza glabra*
ligusticum	*Ligusticum striatum* (also known as *L. chuanxiong*)
lily of the valley	*Convallaria majalis*
linden	*Tilia × europaea*
long pepper	*Piper longum*
lungwort	*Sticta pulmonaria* (also known as *Lobaria pulmonaria*), *Pulmonaria officinalis*
ma huang	*Ephedra sinica*
maca	*Lepidium meyenii* (also known as *L. peruvianum*)
macrotys	*Actaea racemosa* (also known as *Cimicifuga racemosa*)
magnolia	*Magnolia officinalis*
magnolia vine	*Schisandra chinensis*
mahonia	*Berberis aquifolium* (also known as *Mahonia aquifolium*)
mai men dong	*Ophiopogon japonicus*
maidenhair tree	*Ginkgo biloba*
maitake mushroom	*Grifola frondosa*
Malabar nut	*Justicia adhatoda* (also known as *Adhatoda vasica*)
maritime pine	*Pinus pinaster*
maritime squill	*Urginea maritima* (also known as *Drimia maritima*)
marlberry	*Ardisia japonica*
marshmallow	*Althaea officinalis*
milk thistle	*Silybum marianum*
milk vetch	*Astragalus membranaceus*
milkweed	*Asclepias tuberosa*
milkwort	*Polygala* spp.
mimosa	*Albiza julibrissin*
mistletoe	*Phoradendron chrysocladon* (also known as *Viscum flavens*), *Viscum album*

mondograss	*Ophiopogon japonicus*
Mormon tea	*Ephedra nevadensis*
motherwort	*Leonurus cardiaca*
mountain balm	*Eriodictyon californicum*
moutan	*Paeonia × suffruticosa, P. anomala*
mugwort	*Artemisia vulgaris*
mullein	*Verbascum thapsus*
myrobalan	*Terminalia chebula*
myrrh	*Commiphora myrrha, C. molmol*
neem	*Azadirachta indica* (also known as *Melia azadirachta*)
nettle	*Urtica dioica, U. urens*
night-blooming cactus	*Selenicereus grandiflorus* (also known as *Cereus grandiflorus* and *Cactus grandiflorus*)
night-blooming cereus	*Selenicereus grandiflorus* (also known as *Cereus grandiflorus*)
noni	*Morinda citrifolia*
Northern white cedar	*Thuja occidentalis*
oats	*Avena sativa*
old man's beard	*Usnea barbata, U. hirta*
olive	*Olea europaea*
onion	*Allium cepa*
oregano	*Origanum vulgare*
Oregon grape	*Berberis aquifolium* (also known as *Mahonia aquifolium*)
osha	*Ligusticum porteri*
osthole	*Cnidium monnieri*
parsley	*Petroselinum crispum* (also known as *P. sativum*)
passionflower	*Passiflora incarnata*
peanut	*Arachis hypogaea*
Peking spurge	*Euphorbia pekinensis*
pennyroyal	*Mentha pulegium*
pennywort	*Centella asiatica*
peony	*Paeonia × suffruticosa, P. anomala*
peppermint	*Mentha piperita*
pepperweed	*Lepidium latifolium*

perilla	*Perilla frutescens*
pheasant's eye	*Adonis vernalis*
picrorhiza	*Neopicrorhiza scrophulariiflora* (also known as *Picrorhiza scrophulariiflora*)
pill-bearing spurge	*Euphorbia parviflora* (also known as *E. pilulifera*)
pineapple	*Ananas comosus*
pippali	*Piper longum*
pipsissewa	*Chimaphila umbellata*
plantain	*Plantago indica* (also known as *P. psyllium* and *P. arenaria*)
pleurisy root	*Asclepias tuberosa*
poke	*Phytolacca american* (also known as *P. decandra*)
pokeroot	*Phytolacca american* (also known as *P. decandra*)
pokeweed	*Phytolacca american* (also known as *P. decandra*)
pomegranate	*Punica granatum*
pot marigold	*Calendula officinalis*
prickly pear	*Opuntia ficus-indica*
psyllium	*Plantago indica* (also known as *P. psyllium* and *P. arenaria*)
pukeweed	*Lobelia inflata*
purple coneflower	*Echinacea purpurea*
qian ceng ta	*Huperzia serrata*
quebracho	*Aspidosperma quebracho*
queen of the night	*Selenicereus grandiflorus* (also known as *Cereus grandiflorus*)
queen's root	*Stillingia sylvatica*
red clover	*Trifolium pratense*
red oak	*Quercus rubra*
red peony	*Paeonia lactiflora*
red root	*Ceanothus americanus*
red sage	*Salvia miltiorrhiza*
reishi	*Ganoderma lucidum*
ren dong teng	*Lonicera japonica, L. caprifolium*
ren shen	*Panax ginseng*
rhaptopetalum	*Rhaptopetalum coriaceum*

rhodioa	*Rhodiola rosea*
rhodiola	*Rhodiola kirilowii, R. crenulata, R. quadrifida* (also known as *Sedum quadrifidum*)
rompe piedras	*Lepidium latifolium*
roseapple	*Syzygium cumini* (also known as *S. jambolanum*), *S. jambos*
roselle	*Hibiscus sabdariffa*
rosemary	*Rosmarinus officinalis*
rudraksha	*Elaeocarpus serratus* (also known as *E. ganitrus*)
rue	*Ruta graveolens*
safflower	*Carthamus tinctorius*
sage	*Salvia officinalis*
Saigon cinnamon	*Cinnamomum polyadelphum* (also known as *C. saigonicum*)
saiko	*Bupleurum falcatum*
sangre de drago	*Croton lechleri*
sarsasparilla	*Smilax officinalis, S. aristolochiifolia, S. ornata, S. regelii*
saw thistle	*Saussurea costus* (also known as *S. lappa*)
saw wort	*Saussurea costus* (also known as *S. lappa*)
Scot's broom	*Cytisus scoparius*
Scotch broom	*Cytisus scoparius*
scute	*Scutellaria baicalensis*
sea buckthorn	*Hippophae rhamnoides*
sea onion	*Urginea maritima* (also known as *Drimia maritima*)
sea squill	*Drimia maritima* (also known as *Urginea maritima*)
Seneca snakeroot	*Polygala senega*
senna	*Cassia senna* (also known as *Senna alexandrina*), *Senna insularis*
shan zhu yu	*Cornus officinalis*
sheep sorrel	*Rumex acetosella*
shengma	*Actaea cimicifuga* (also known as *Cimicifuga foetida*), *A. dahurica, A. heracleifolia, Actaea* spp.
shepherd's purse	*Capsella bursa-pastoris*

shiitake mushroom	*Lentinula edodes*
shiso	*Perilla frutescens*
Siberian ginseng	*Eleutherococcus senticosus*
silk tree	*Albizia julibrissin*
siris	*Albizia lebbeck*
Sitka valerian	*Valeriana sitchensis*
skullcap	*Scutellaria lateriflora*
skunk cabbage	*Symplocarpus foetidus*
slippery elm	*Ulmus fulva*
snakeroot	*Polygala karensium, Polygala* spp.
snow parsley	*Cnidium monnieri*
snowball bush	*Viburnum opulus*
Southern prickly ash	*Zanthoxylum clava-herculis*
Southern wax myrtle	*Morella cerifera* (also known as *Myrica cerifera*)
soybean	*Glycine max*
Spanish black radish	*Raphanus sativus* var. *niger*
spearmint	*Mentha spicata*
spikemoss	*Selaginella uncinata*
spurge	*Euphorbia hirta*
squill	*Drimia maritima* (also known as *Urginea maritima*)
St. Johnswort	*Hypericum perforatum*
stevia	*Stevia rebaudiana*
stinging nettle	*Urtica dioica, U. urens*
stoneroot	*Collinsonia canadensis*
strophanthus	*Strophanthus hispidus*
su mu	*Caesalpinia sappan* (also known as *Biancaea sappan*)
suan zao ren	*Ziziphus spinosa*
sugar destroyer	*Gymnema sylvestre*
sundew	*Drosera rotundifolia*
swamp lily	*Crinum glaucum*
sweet Annie	*Artemisia annua*
sweet basil	*Ocimum basilicum*
sweet clover	*Melilotus officinalis, Melilotus suaveolens*

sweet leaf	*Stevia rebaudiana*
sweet vernal grass	*Anthoxanthum odoratum*
Szechuan lovage	*Ligusticum striatum* (also known as *L. chuanxiong*)
tai bai bei mu	*Fritillaria taipaiensis*
tea tree	*Melaleuca alternifolia*
thyme	*Thymus vulgaris*
tian ma	*Gastrodia elata*
toothache tree	*Zanthoxylum clava-herculis*
tree moss	*Usnea barbata, U. hirta*
tree peony	*Paeonia × suffruticosa, P. anomala*
tulsi	*Ocimum tenuiflorum* (also known as *O. sanctum*)
turkey corn	*Corydalis cava*
turkey rhubarb	*Rheum officinale, R. palmatum, R. emodi*
turmeric	*Curcuma longa*
uno de gato	*Uncaria rhynchophylla, U. tomentosa*
valerian	*Valeriana officinalis*
vegetable mercury	*Iris versicolor, I. tenax*
velvet bean	*Mucuna pruriens*
wax myrtle	*Myrica cerifera* (also known as *Morella cerifera*)
Western red cedar	*Thuja plicata*
white bryony	*Bryonia alba*
white hellebore	*Veratrum album*
white mustard	*Sinapis alba*
white oak	*Quercus alba*
white peony	*Paeonia lactiflora*
wild carrot	*Daucus carota*
wild cherry	*Prunus serotina*
wild geranium	*Geranium maculatum*
wild indigo	*Tephrosia purpurea*
wild iris	*Iris versicolor, I. tenax*
wild yam	*Dioscorea villosa*
willow	*Salix* spp.
witch hazel	*Hamamelis virginiana*
woad	*Isatis tinctoria* (also known as *I. indigotica*)
wolfberry	*Lycium barbarum, L. chinense*
wolfsbane	*Aconitum napellus*
woodbine	*Lonicera xylosteum*
wu wei zi	*Schisandra chinensis*
yarrow	*Achillea millefolium*
yellow dock	*Rumex crispus*
yellow Himalayan fritillaria	*Fritillaria cirrhosa*
yellow sweet clover	*Melilotus officinalis*
yerba de tago	*Eclipta prostrata* (also known as *E. alba*)
yerba mansa	*Anemopsis californica*
yerba santa	*Eriodictyon californicum*
yin yang huo	*Epimedium brevicornu* (also known as *E. rotundatum*)
yu min bei mu	*Fritillaria yuzhongensis*

— GLOSSARY —
OF THERAPEUTIC TERMS

Abortifacient. An agent capable of promoting the expulsion of a developing fetus.

Absorbent. A drug that promotes the absorption of medicinal compounds.

Acidifier. An agent imparting acidity to body fluids, especially blood and urine.

Acute. A condition that has a new onset, comes on suddenly, and is relatively short-lasting in its entire duration.

Aerial parts. The parts of a plant that grow above ground.

Alkalinizer. An agent that increases the alkalinity of bodily fluids, especially the blood and urine.

Allopathic. A term applying to conventional, modern Western medicine. *Allo* refers to "opposite," and in this case, means to oppose pathology. For example: In cases of fever, an antipyretic is used; to treat inflammation, an anti-inflammatory is used; and to treat an infection, antimicrobials are used.

Alterative. An agent that favorably "alters" an individual's health. Alteratives stimulate digestive and absorptive functions while enhancing elimination of wastes. Alteratives are also traditionally said to "purify" the blood and optimize metabolic functions.

Analgesic. An agent that is pain-relieving.

Anaphrodisiac. An agent that diminishes sexual drive or function.

Anesthetic. An agent that diminishes pain and tactile sensations temporarily.

Anhydrotic. An agent that diminishes excessive sweating.

Anodyne. An agent that is pain-relieving.

Antacid. An agent that diminishes stomach acid.

Antagonist. An agent that opposes the action of some other medicine, usually a poison or toxic alkaloid.

Anthelmintic. An agent used to combat intestinal worms.

Antidote. A remedy to counteract the action of poisons or other strong actions.

Antiemetic. An agent that allays nausea and vomiting.

Antigalactogogue. An agent that diminishes lactation.

Antihemorrhagic. An agent that helps control excessive bleeding.

Anti-inflammatory. An agent that reduces inflammatory processes by a variety of mechanisms, reducing oxidative stress and protecting tissues from stress and damage.

Antilithic. An agent used to reduce the formation of stones and calculi in the body.

Antioxidant. An agent capable of accepting electrons or highly reactive molecules that could damage body membranes if left free in circulation.

Antiperiodic. An agent used to combat the periodic fevers of malaria.

Antiphlogistic. An agent used to reduce fever and inflammation.

Antipyretic. An agent used to reduce fever.

Antiscorbutic. An agent used to provide vitamin C and prevent or treat scurvy.

Antiseptic. An agent having antimicrobial capacity for the prevention of sepsis.

Antisialagogue. An agent capable of reducing salivation.

Antispasmodic. An agent capable of reducing painful spasms in muscles and hollow organs.

Antitussive. An agent used to diminish coughing.

Aperient. A gentle nonirritating laxative.

Aphrodisiac. An agent used to stimulate the libido.

Aromatic. An agent with a strong fragrance to be inhaled or absorbed through the skin.

Astringent. An agent that dries, condenses, and shrinks inflamed or suppurative tissues.

Bitter. An agent that has a bitter flavor and is used to stimulate gastrointestinal tone and secretions. Bitters prepare mucosa for food, stimulate appetite, and enhance digestion.

Blood mover. A term from Traditional Chinese Medicine (TCM) used to refer to agents capable of improving circulation and relieving blood stagnation and tissue congestion.

Cardiotonic. An agent that improves heart function.

Carminative. An agent that reduces gas, bloating, flatulence, and associated pain.

Cathartic. A strong, potentially harsh laxative.

Caustic. An agent having a corrosive action on tissues.

Chi tonic. A term from TCM in which chi refers to the body's vital energy. Chi tonics are herbs purported to increase and support vitality, longevity, stamina, fertility, and other aspects of the vital force. Chi deficiency manifests as low energy and stamina, a weak voice, and coldness, as well as exercise intolerance, general fatigue, shortness of breath, and dizziness.

Cholagogue. An agent that increases gallbladder tone and the flow of bile from the gallbladder.

Choleretic. An agent that increases the production of bile.

Chronic. A condition that develops slowly over time and becomes persistent and sometimes permanent.

Corrigent. An agent that balances a harsh or strong action of another agent, a corrective.

Counterirritant. An agent that irritates local tissues to enhance blood flow to the area. Counterirritants are used to induce temporary hyperemia in chronic conditions in an attempt to relieve pain, promote healing, and reduce inflammation.

Dacryagogue. An agent that promotes the flow of tears (lacrimation).

Dampness. A term used in TCM and other energetic descriptions of physiologic tissue states that refers to fluid stagnation when evidenced by a coated tongue, chronic phlegm in the mucous membranes, fluid stagnation, and an increased tendency to opportunistic infections.

Deficient. Referring to low energy, low vitality, and poor functioning tissues (herbal medicine). In Traditional Chinese Medicine, the term chi deficiency is used when the entire body is in a weakened state. The term may also be used to indicate a poorly functioning organ or biochemical state, such as digestive deficiency, circulatory deficiency, or metabolic deficiency.

Demulcent. A cooling, soothing, mucilagenous substance used internally or topically to emolliate abraded, inflamed, or irritated mucosal tissues.

Depurant. Any agent aimed at purifying, such as a liver depurant, a renal depurant, a blood depurant, and so on. Depurants have a purifying effect by promoting the elimination of wastes from the body.

Diaphoretic. An agent capable of inducing perspiration and often a temporary fever.

Diuretic. An agent that stimulates the production and flow of urine.

Ecbolic. An agent that stimulates childbirth (parturition).

Emetic. An agent that causes vomiting (emesis).

Emmenagogue. An agent that promotes menstrual flow.

Emollient. An agent that soothes and softens the skin and mucosal tissues.

Errhine. An agent that irritates the nasal mucosa and promotes sneezing and secretions.

Escharotic. Any caustic substances applied topically to diseased tissues to kill the cells and promote sloughing away. The word *eschar* means to cast off.

Essential oil. See Volatile oil.

Excess. Indicates a condition beyond normal range, such as too much heat, overstimulated bowel or muscle tone, or other situations of excess in various physiologic functions.

Excitant. An agent that causes excitation of nervous, circulatory, or motor functions; however, in Latin America, the term is more often used to refer to an aphrodisiac, or sexual excitant.

Exhilarant. An agent that causes excitation of psychic functions and promotes euphoria.

Expectorant. An agent that promotes the flow of secretions from the respiratory tract.

Febrifuge. An agent used to bring down the temperature in cases of fever.

Fibrinolytic. An agent capable of breaking down fibrin, which may be deposited in vein and artery walls, as well as numerous tissues, in response to inflammatory processes.

Galactogogue. An agent that stimulates lactation.

Hematic, hematinic. An agent that improves the quality of the blood, especially in cases of anemia, but may be used in other situations.

Hemostatic. An agent that reduces blood flow and promotes clotting in cases of hemorrhage, trauma, and internal bleeding.

Hepatic. An agent that improves the function of the liver.

Hydragogue. An agent that promotes watery secretions.

Inotropic. An agent that supports ion flow in electrically active cardiac muscle and improves the contractile force of the heart, slowing and strengthening the heartbeat and improving circulation.

Irritant. An agent applied locally for the purposes of intentionally causing local hyperemia. See also Counterirritant.

Laxative. An agent that promotes a mild and painless evacuation of the bowels.

Lithotriptic. An agent aimed at dissolving calculi within the body.

Lipotropic. Literally translates as "fat mover," and used to refer to various alterative and cholagogue herbs, as well as substances such as choline that promote bile flow and biliary function, and thereby improve liver function, and hepatic clearance of lipids, carbohydrates, hormones, toxins, and chemicals.

Material dose. A term used by herbalists and alternative medicine practitioners to distinguish between a highly diluted or homeopathic preparation of a substance and the substance given in a more substantial or "material dose."

Miotic, myotic. An agent that causes the pupil to contract (miosis).

Mydriatic. An agent that promotes dilation of the pupil (mydriasis).

Narcotic. A drug that promotes stupor or sleep and is used to relieve pain or diminish consciousness.

Nervine. An agent having a tonifying effect on the nervous system, usually only used in the context of herbal medicine, with some herbs being referred to as nervine herbs.

Nutriant, nutrient, nutritive. An agent that enhances assimilation, metabolism, and nutrition.

Oxytocic. An agent that promotes uterine contractions and hastens childbirth.

Parturifacient. An agent that facilitates childbirth when taken during labor.

Partus preparator. An agent taken in the last months of pregnancy to tone the uterus and optimize labor and delivery.

Purgative. A strong laxative that may be irritating and cause cramping.

Refrigerant. An agent capable of imparting a cooling sensation when applied topically.

Revulsive. An agent used to enhance the blood flow to a particular body part (hand, foot) in order to draw it away from a congested, engorged area (head, uterus).

Rubefacient. An agent that promotes reddening or hyperemia of the tissues.

Sedative. An agent that calms in cases of nervousness, insomnia, and mania, and may be stronger and less tonifying than a nervine.

Sialagogue, salivant. An agent that increases the flow of saliva.

Simple. A term used to refer to a single herb, not mixed with other herbs or used in a formula, but rather used as "a simple."

Specific. An agent thought to be of specific value for a collection of symptoms.

Sternutatory. An agent that promotes sneezing when inhaled.

Styptic. A strong astringent agent capable of reducing bleeding when applied topically.

Sudorific. An agent capable of inducing perspiration and regarded as being stronger than a diaphoretic.

Synergist. An agent that duplicates, enhances, or pulls together the action of a group of medicinal substances.

Taenicide. An agent that kills or weakens tapeworms.

Tonic. An agent that has a positive effect on the function of an organ or tissue and suggests an ability to restore normal function be it excess or deficient, atonic or hypertonic, overstimulated or understimulated. Tonic supports the optimal physiologic state.

Toxicity. A deranged, inflammatory, or otherwise corrupted or polluted biochemical state in the body.

Vasoconstrictor. An agent that constricts the blood vessels.

Vasodepressant. An agent that slows the pulse rate and lowers the pressure.

Vasodilator. An agent used to dilate the vasculature, usually used in cases of hypertension.

Vermifuge. An agent that promotes the expulsion of intestinal worms.

Vesicant. An agent that promotes blistering or vesication of the skin.

Volatile oil. An essential oil. Aromatic plants are high in volatile oils, so named due to the fact that they are small, light molecules that readily volatilize into the air, contributing to the aromatic quality. Volatile oils are often distilled out of aromatic plants such as mint, thyme, citrus, and numerous others and sold in small bottles to use in aromatherapy, to make body products, and for other purposes.

— NOTES —

Chapter 2:
Creating Herbal Formulas for Cardiovascular, Peripheral Vascular, and Pulmonary Conditions

1. Sara S. Al Disi et al., "Anti-hypertensive Herbs and Their Mechanisms of Action: Part I," *Frontiers in Pharmacology* 6 (2015): 323, https://doi.org/10.3389/fphar.2015.00323.

2. Asiful Islam et al., "Dietary Phytochemicals: Natural Swords Combating Inflammation and Oxidation-Mediated Degenerative Diseases," *Oxidative Medicine and Cellular Longevity* 2016 (2016), https://doi.org/10.1155/2016/5137431.

3. Ramachandran Vinayagam and Baojun Xu, "Antidiabetic Properties of Dietary Flavonoids: A Cellular Mechanism Review," *Nutrition & Metabolism* 12 (2015): 60, https://doi.org/10.1186/s12986-015-0057-7.

4. Gui-lin Xiao et al., "Clinical Observation on Treatment of *Russula subnigricans* Poisoning Patients by *Ganoderma lucidum* Decoction," *Zhongguo Zhong Xi Yi Jie He Za Zhi Zhongguo Zhongxiyi Jiehe Zazhi (Chinese Journal of Integrated Traditional and Western Medicine)* 23, no. 4 (2003): 278–280, https://www.ncbi.nlm.nih.gov/pubmed/12764911; Sissi Wachtel-Galor et al., "*Ganoderma Lucidum* ('Lingzhi'); Acute and Short-Term Biomarker Response to Supplementation," *International Journal of Food Sciences and Nutrition* 55, no. 1 (2004): 75–83, https://doi.org/10.1080/09637480310001642510.

5. Sissi Wachtel-Galor et al., "*Ganoderma lucidum* ('Lingzhi'), a Chinese Medicinal Mushroom: Biomarker Responses in a Controlled Human Supplementation Study," *The British Journal of Nutrition* 91, no. 2 (2004): 263–269, https://doi.org/10.1079/BJN20041039.

6. N. P. Sudheesh et al., "Therapeutic Potential of *Ganoderma lucidum* (Fr.) P. Karst. against the Declined Antioxidant Status in the Mitochondria of Post-Mitotic Tissues of Aged Mice," *Clinical Nutrition* 29, no. 3 (2010): 406–412, https://doi.org/10.1016/j.clnu.2009.12.003; Yu-hong You and Zhi-bin Lin, "Antioxidant Effect of Ganoderma Polysaccharide Peptide," *Yao Xue Xue Bao (Acta Pharmaceutica Sinica)* 38, no. 2 (2003): 85–88, https://www.ncbi.nlm.nih.gov/pubmed/12778739.

7. N. P. Sudheesh et al., "*Ganoderma lucidum* Ameliorate Mitochondrial Damage in Isoproterenol-Induced Myocardial Infarction in Rats by Enhancing the Activities of TCA Cycle Enzymes and Respiratory Chain Complexes," *International Journal of Cardiology* 165, no. 1 (2013): 117–25, https://doi.org/10.1016/j.ijcard.2011.07.103.

8. N. P. Sudheesh et al., "*Ganoderma lucidum* (Fr.) P. Karst Enhances Activities of Heart Mitochondrial Enzymes and Respiratory Chain Complexes in the Aged Rat," *Biogerontology* 10, no. 5 (2009): 627–36, https://doi.org/10.1007/s10522-008-9208-9.

9. T. V. Lasukova et al., "*Ganoderma lucidum* Extract in Cardiac Diastolic Dysfunction and Irreversible Cardiomyocytic Damage in Ischemia and Reperfusion of the Isolated Heart," *Patologicheskaia Fiziologiia I Eksperimental'naia Terapiia*, no. 1 (2008): 22–25, https://www.ncbi.nlm.nih.gov/pubmed/18411655.

10. Kar-Lok Wong et al., "Antioxidant Activity of *Ganoderma lucidum* in Acute Ethanol-Induced Heart Toxicity," *Phytotherapy Research: PTR* 18, no. 12 (2004): 1024–26, https://doi.org/10.1002/ptr.1557.

11. Hua Xue et al., "Effect of *Ganoderma lucidum* Polysaccharides on Hemodynamic and Antioxidation in T2DM Rats," *Zhongguo Zhong Yao Za Zhi = Zhongguo Zhongyao Zazhi (China Journal of Chinese Materia Medica)* 35, no. 3 (2010): 339–43, https://www.ncbi.nlm.nih.gov/pubmed/20423001.

12. C. R. Pace-Asciak et al., "Wines and Grape Juices as Modulators of Platelet Aggregation in Healthy Human Subjects," *Clinica Chimica Acta (International Journal of Clinical Chemistry)* 246, no. 1–2 (1996): 163–82, https://www.ncbi.nlm.nih.gov/pubmed/8814965.

13. F. Orsini et al., "Isolation, Synthesis, and Antiplatelet Aggregation Activity of Resveratrol 3-O-Beta-D-Glucopyranoside and Related Compounds," *Journal of Natural Products* 60, no. 11 (1997): 1082–87, https://doi.org/10.1021/np970069t.

14. D. Tsi and B. K. Tan, "The Mechanism Underlying the Hypocholesterolaemic Activity of Aqueous Celery Extract, Its Butanol and Aqueous Fractions in Genetically Hypercholesterolaemic RICO Rats," *Life Sciences* 66, no. 8 (2000): 755–67, https://www.ncbi.nlm.nih.gov/pubmed/10680583.

15. D. Tsi et al., "Effects of Aqueous Celery (*Apium graveolens*) Extract on Lipid Parameters of Rats Fed a High Fat Diet," *Planta Medica* 61, no. 1 (1995): 18–21, https://doi.org/10.1055/s-2006-957990.

16. Mira Popović et al., "Effect of Celery (*Apium graveolens*) Extracts on Some Biochemical Parameters of Oxidative Stress in Mice Treated with Carbon Tetrachloride," *Phytotherapy Research: PTR* 20, no. 7 (2006): 531–37, https://doi.org/10.1002/ptr.1871.

17. Chin-Chuan Chen et al., "Viscolin Inhibits In Vitro Smooth Muscle Cell Proliferation and Migration and Neointimal Hyperplasia In Vivo," *PloS ONE* 11, no. 12 (2016): e0168092, https://doi.org/10.1371/journal.pone.0168092.

18. Amy G. W. Gong et al., "Ferulic Acid Orchestrates Anti-Oxidative Properties of Danggui Buxue Tang, an Ancient Herbal Decoction: Elucidation by Chemical Knock-Out Approach," *PloS ONE* 11, no. 11 (2016): e0165486, https://doi.org/10.1371/journal.pone.0165486.

19. Wenwen Zhao et al., "Dihydrotanshinone I Attenuates Atherosclerosis in ApoE-Deficient Mice: Role of NOX4/NF-κB Mediated Lectin-Like Oxidized LDL Receptor-1 (LOX-1) of the Endothelium," *Frontiers in Pharmacology* 7 (2016), https://doi.org/10.3389/fphar.2016.00418.

20. Michele Cavalera et al., "Rose Hip Supplementation Increases Energy Expenditure and Induces Browning of White Adipose Tissue," *Nutrition & Metabolism* 13 (2016), https://doi.org/10.1186/s12986-016-0151-5.

21. Tianhua Xu et al., "Effect of Omega-3 Fatty Acid Supplementation on Serum Lipids and Vascular Inflammation in Patients with End-Stage Renal Disease: A Meta-Analysis," *Scientific Reports* 6 (2016): 39346, https://doi.org/10.1038/srep39346.

22. Srujana Rayalam et al., "Anti-Obesity Effects of Xanthohumol plus Guggulsterone in 3T3-L1 Adipocytes," *Journal of Medicinal Food* 12, no. 4 (2009): 846–53, https://doi.org/10.1089/jmf.2008.0158.

23. H. Sumiyoshi et al., "New Pharmacological Activities of Garlic and Its Constituents," *Nihon Yakurigaku Zasshi (Folia Pharmacologica Japonica)* 110 Suppl 1 (1997): 93P–97P, https://www.ncbi.nlm.nih.gov/pubmed/9503413.

24. B. Ramesh and K. V. Pugalendi, "Antihyperlipidemic and Antidiabetic Effects of Umbelliferone in Streptozotocin Diabetic Rats," *The Yale Journal of Biology and Medicine* 78, no. 4 (2005): 189–96, https://www.ncbi.nlm.nih.gov/pubmed/16720013.

25. Hiroshi Ogawa et al., "Effects of Osthol on Blood Pressure and Lipid Metabolism in Stroke-Prone Spontaneously Hypertensive Rats," *Journal of Ethnopharmacology* 112, no. 1 (2007): 26–31, https://doi.org/10.1016/j.jep.2007.01.028.

26. D. Tsi and B. K. Tan, "The Mechanism Underlying the Hypocholesterolaemic Activity of Aqueous Celery Extract, Its Butanol and Aqueous Fractions in Genetically Hypercholesterolaemic RICO Rats," *Life Sciences* 66, no. 8 (2000): 755–67, https://www.ncbi.nlm.nih.gov/pubmed/10680583.

27. C. Wang et al., "Hypolipidemic Action of Soy Fiber and Its Effects on Platelet Aggregation and Coagulation Time in Rats," *Zhonghua Yu Fang Yi Xue Za Zhi (Chinese Journal of Preventive Medicine)* 30, no. 4 (1996): 205–8, https://www.ncbi.nlm.nih.gov/pubmed/9388894.

28. Dongfang Yang et al., "Hypolipidemic Agent Z-Guggulsterone: Metabolism Interplays with Induction of Carboxylesterase and Bile Salt Export Pump," *Journal of Lipid Research* 53, no. 3 (2012): 529–39, https://doi.org/10.1194/jlr.M014688.

29. C. von Schacky et al., "The Effect of Dietary Omega-3 Fatty Acids on Coronary Atherosclerosis. A Randomized, Double-Blind, Placebo-Controlled Trial," *Annals of Internal Medicine* 130, no. 7 (1999): 554–62, https://www.ncbi.nlm.nih.gov/pubmed/10189324.

30. A. Bordia et al., "Effect of Garlic (*Allium sativum*) on Blood Lipids, Blood Sugar, Fibrinogen and Fibrinolytic Activity in Patients with Coronary Artery Disease," *Prostaglandins, Leukotrienes, and Essential Fatty Acids* 58, no. 4 (1998): 257–63, https://www.ncbi.nlm.nih.gov/pubmed/9654398.

31. D. J. Maslin et al., "Nitric Oxide—a Mediator of the Effects of Garlic?," *Biochemical Society Transactions* 25, no. 3 (1997): 408S, https://www.ncbi.nlm.nih.gov/pubmed/9388638.

32. M. Ali, "Mechanism by Which Garlic (*Allium sativum*) Inhibits Cyclooxygenase Activity. Effect of Raw versus Boiled Garlic Extract on the Synthesis of Prostanoids," *Prostaglandins, Leukotrienes, and Essential Fatty Acids* 53, no. 6 (1995): 397–400, https://www.ncbi.nlm.nih.gov/pubmed/8821119.

33. K. Breithaupt-Grögler et al., "Protective Effect of Chronic Garlic Intake on Elastic Properties of Aorta in the Elderly," *Circulation* 96, no. 8 (1997): 2649–55, https://www.ncbi.nlm.nih.gov/pubmed/9355906.

34. I. Das et al., "Potent Activation of Nitric Oxide Synthase by Garlic: A Basis for Its Therapeutic Applications," *Current Medical Research and Opinion* 13, no. 5 (1995): 257–63, https://doi.org/10.1185/03007999509111550.

35. A. A. Mateen et al., "Pharmacodynamic Interaction Study of *Allium sativum* (Garlic) with Cilostazol in Patients with Type II Diabetes Mellitus," *Indian Journal of Pharmacology* 43, no. 3 (2011): 270–74, https://doi.org/10.4103/0253-7613.81514.

36. Davood Soleimani et al., "Effect of Garlic Powder Consumption on Body Composition in Patients with Nonalcoholic Fatty Liver Disease: A Randomized, Double-Blind, Placebo-Controlled Trial," *Advanced Biomedical Research* 5 (2016): 2, https://doi.org/10.4103/2277-9175.174962.

37. J. P. Terwari et al., "Pharmacologic Studies of *Achillea millefolium* Linn," *Indian J. Med Sci* 28, no. 8 (1974): 31–36, https://www.ncbi.nlm.nih.gov/pubmed/4443016.

38. A. S. Goldberg et al., "Isolation of the Anti-Inflammatory Principles from *Achillea millefolium* (Compositae)," *Journal of Pharmaceutical Sciences* 58, no. 8 (1969): 938–41, https://www.ncbi.nlm.nih.gov/pubmed/4310133.

39. D. Heber et al., "Cholesterol-Lowering Effects of a Proprietary Chinese Red-Yeast-Rice Dietary Supplement," *The American Journal of Clinical Nutrition* 69, no. 2 (1999): 231–36, https://www.ncbi.nlm.nih.gov/pubmed/9989685.

40. Hui-Ting Yang et al., "Acute Administration of Red Yeast Rice (*Monascus purpureus*) Depletes Tissue Coenzyme Q(10) Levels in ICR Mice," *The British Journal of Nutrition* 93, no. 1 (2005): 131–35, https://www.ncbi.nlm.nih.gov/pubmed/15705235.

41. Liliana Vercelli et al., "Chinese Red Rice Depletes Muscle Coenzyme Q10 and Maintains Muscle Damage after Discontinuation of Statin Treatment," *Journal of the American Geriatrics Society* 54, no. 4 (2006): 718–20, https://doi.org/10.1111/j.1532-5415.2006.00668_7.x.

42. Yinhua Li et al., "A Meta-Analysis of Red Yeast Rice: An Effective and Relatively Safe Alternative Approach for Dyslipidemia," *PLoS ONE* 9, no. 6 (2014): e98611, https://doi.org/10.1371/journal.pone.0098611.

43. Yinhua Li et al., "A Meta-Analysis of Red Yeast Rice: An Effective and Relatively Safe Alternative Approach for Dyslipidemia," *PLoS ONE* 9, no. 6 (2014): e98611, https://doi.org/10.1371/journal.pone.0098611.

44. Veronique Verhoeven et al., "Red Yeast Rice Lowers Cholesterol in Physicians—a Double Blind, Placebo Controlled Randomized Trial," *BMC Complementary and Alternative Medicine* 13 (2013): 178, https://doi.org/10.1186/1472-6882-13-178.

45. Patrick M. Moriarty et al., "Effects of Xuezhikang in Patients with Dyslipidemia: A Multicenter, Randomized, Placebo-Controlled Study," *Journal of Clinical Lipidology* 8, no. 6 (2014): 568–75, https://doi.org/10.1016/j.jacl.2014.09.002.

46. Arrigo F. G. Cicero et al., "Red Yeast Rice Improves Lipid Pattern, High-Sensitivity C-Reactive Protein, and Vascular Remodeling Parameters in Moderately Hypercholesterolemic Italian Subjects," *Nutrition Research* 33, no. 8 (2013): 622–28, https://doi.org/10.1016/j.nutres.2013.05.015.

47. Mei Ding et al., "Red Yeast Rice Repairs Kidney Damage and Reduces Inflammatory Transcription Factors in Rat Models of Hyperlipidemia," *Experimental and Therapeutic Medicine* 8, no. 6 (2014): 1737–44, https://doi.org/10.3892/etm.2014.2035.

48. Mei Ding et al., "Red Yeast Rice Repairs Kidney Damage and Reduces Inflammatory Transcription Factors in Rat Models of Hyperlipidemia," *Experimental and Therapeutic Medicine* 8, no. 6 (2014): 1737–44, https://doi.org/10.3892/etm.2014.2035.

49. Shariq Shamim et al., "Red Yeast Rice for Dysipidemia," *Missouri Medicine* 110, no. 4 (2013): 349–54, https://www.ncbi.nlm.nih.gov/pubmed/24003656.

50. Sun Ha Lim et al., "Plant-Based Foods Containing Cell Wall Polysaccharides Rich in Specific Active Monosaccharides Protect against Myocardial Injury in Rat Myocardial Infarction Models," *Scientific Reports* 6 (2016), https://doi.org/10.1038/srep38728.

51. Theresa A. Nicklas et al., "Consumption of Various Forms of Apples Is Associated with a Better Nutrient Intake and Improved Nutrient Adequacy in Diets of Children: National Health and Nutrition Examination Survey 2003–2010," *Food & Nutrition Research* 59 (2015): 25948, https://www.ncbi.nlm.nih.gov/pubmed/26445211.

52. Athanasios Koutsos et al., "Apples and Cardiovascular Health—Is the Gut Microbiota a Core Consideration?," *Nutrients* 7, no. 6 (2015): 3959–98, https://doi.org/10.3390/nu7063959.

53. Sun Ha Lim et al., "Apple Pectin, a Dietary Fiber, Ameliorates Myocardial Injury by Inhibiting Apoptosis in a Rat Model of Ischemia/Reperfusion," *Nutrition Research and Practice* 8, no. 4 (2014): 391–97, https://doi.org/10.4162/nrp.2014.8.4.391.

54. Jingting Jiang et al., "Apple-Derived Pectin Modulates Gut Microbiota, Improves Gut Barrier Function, and Attenuates Metabolic Endotoxemia in Rats with Diet-Induced Obesity," *Nutrients* 8, no. 3 (2016): 126, https://doi.org/10.3390/nu8030126.

55. Jeong-Hyun Yoo et al., "Hawthorn Fruit Extract Elevates Expression of Nrf2/HO-1 and Improves Lipid Profiles in Ovariectomized Rats," *Nutrients* 8, no. 5 (2016), https://doi.org/10.3390/nu8050283.

56. Wei Zhou et al., "Systems Pharmacology Exploration of Botanic Drug Pairs Reveals the Mechanism for Treating Different Diseases," *Scientific Reports* 6 (2016): 36985, https://doi.org/10.1038/srep36985.

57. Liping Huang et al., "Pharmacological Profile of Essential Oils Derived from *Lavandula angustifolia* and *Melissa officinalis* with Anti-Agitation Properties: Focus on Ligand-Gated Channels," *The Journal of Pharmacy and Pharmacology* 60, no. 11 (2008): 1515–22, https://doi.org/10.1211/jpp/60.11.0013.

58. Siyavash Joukar et al., "Efficacy of *Melissa officinalis* in Suppressing Ventricular Arrhythmias Following Ischemia-Reperfusion of the Heart: A Comparison with Amiodarone," *Medical Principles and Practice: International Journal of the Kuwait University, Health Science Centre* 23, no. 4 (2014): 340–45, https://doi.org/10.1159/000363452.

59. Fatemeh Alijaniha et al., "Heart Palpitation Relief with *Melissa officinalis* Leaf Extract: Double Blind, Randomized, Placebo Controlled Trial of Efficacy and Safety," *Journal of Ethnopharmacology* 164 (2015): 378–84, https://doi.org/10.1016/j.jep.2015.02.007.

60. Young-In I. Kwon et al., "Evaluation of Clonal Herbs of Lamiaceae Species for Management of Diabetes and Hypertension," *Asia Pacific Journal of Clinical Nutrition* 15, no. 1 (2006): 107–18, https://www.ncbi.nlm.nih.gov/pubmed/16500886.

61. Dae Young Yoo et al., "Effects of *Melissa officinalis l.* (Lemon Balm) Extract on Neurogenesis Associated with Serum Corticosterone and GABA in the Mouse Dentate Gyrus," *Neurochemical Research* 36, no. 2 (2011): 250–57, https://doi.org/10.1007/s11064-010-0312-2.

62. Alvin Ibarra et al., "Effects of Chronic Administration of *Melissa officinalis l.* Extract on Anxiety-like Reactivity and on Circadian and Exploratory Activities in Mice," *Phytomedicine: International Journal of Phytotherapy and Phytopharmacology* 17, no. 6 (2010): 397–403, https://doi.org/10.1016/j.phymed.2010.01.012; Rosalie Awad et al., "Bioassay-Guided Fractionation of Lemon Balm (*Melissa officinalis l.*) Using an in Vitro Measure of GABA Transaminase Activity," *Phytotherapy Research: PTR* 23, no. 8 (2009): 1075–81, https://doi.org/10.1002/ptr.2712.

63. Mee-Ra Rhyu et al., "*Radix angelica* Elicits Both Nitric Oxide-Dependent and Calcium Influx-Mediated Relaxation in Rat Aorta," *Journal of Cardiovascular Pharmacology* 46, no. 1 (2005): 99–104, https://www.ncbi.nlm.nih.gov/pubmed/15965361.

64. T. Mencherini et al., "An Extract of *Apium graveolens var. dulce* Leaves: Structure of the Major Constituent, Apiin, and Its Anti-Inflammatory Properties," *The Journal of Pharmacy and Pharmacology* 59, no. 6 (2007): 891–97, https://doi.org/10.1211/jpp.59.6.0016.

65. J. Chen et al., "Effect of the Plant-Extract Osthole on the Relaxation of Rabbit Corpus Cavernosum Tissue in Vitro," *The Journal of Urology* 163, no. 6 (2000): 1975–80, https://www.ncbi.nlm.nih.gov/pubmed/10799242.

66. Tommaso Simoncini et al., "Activation of Nitric Oxide Synthesis in Human Endothelial Cells by Red Clover Extracts," *Menopause* 12, no. 1 (2005): 69–77, https://www.ncbi.nlm.nih.gov/pubmed/15668603.

67. Khalijah Awang et al., "Cardiovascular Activity of Labdane Diterpenes from *Andrographis paniculata* in Isolated Rat Hearts," *BioMed Research International* (2012), https://doi.org/10.1155/2012/876458.

68. Min Tao et al., "Relationship of Polyunsaturated Fatty Acid Intake to Peripheral Neuropathy among Adults with Diabetes in the National Health and Nutrition Examination Survey (NHANES) 1999–2004," *Diabetes Care* 31, no. 1 (2008): 93–95, https://doi.org/10.2337/dc07-0931.

69. Guan-Yu Ren et al., "Effect of Flaxseed Intervention on Inflammatory Marker C-Reactive Protein: A Systematic Review and Meta-Analysis of Randomized Controlled Trials," *Nutrients* 8, no. 3 (2016): 136, https://doi.org/10.3390/nu8030136.

70. Jiarong Lan et al., "Meta-Analysis of the Effect and Safety of Berberine in the Treatment of Type 2 Diabetes Mellitus, Hyperlipemia and Hypertension," *Journal of Ethnopharmacology* 161 (2015): 69–81, https://doi.org/10.1016/j.jep.2014.09.049.

71. Atsuko Itoh et al., "Indole Alkaloids and Other Constituents of *Rauwolfia serpentine*," *Journal of Natural Products* 68, no. 6 (2005): 848–52, https://doi.org/10.1021/np058007n.

72. Satyanarayana Sreemantula et al., "Reserpine Methonitrate, a Novel Quaternary Analogue of Reserpine Augments Urinary Excretion of VMA and 5-HIAA without Affecting HVA in Rats," *BMC Pharmacology* 4 (2004): 30, https://doi.org/10.1186/1471-2210-4-30.

73. Sandy D. Shamon and Marco I. Perez, "Blood Pressure Lowering Efficacy of Reserpine for Primary Hypertension," *The Cochrane Database of Systematic Reviews*, no. 4 (2009): CD007655. https://doi.org/10.1002/14651858.CD007655.pub2.

74. F. López-Muñoz et al., "Historical Approach to Reserpine Discovery and Its Introduction in Psychiatry," *Actas Espanolas De Psiquiatria* 32, no. 6 (2004): 387–95, https://www.ncbi.nlm.nih.gov/pubmed/15529229.

75. Shaima Al-Bloushi et al., "Green Tea Modulates Reserpine Toxicity in Animal Models," *The Journal of Toxicological Sciences* 34, no. 1 (2009): 77–87, https://www.ncbi.nlm.nih.gov/pubmed/19182437.

76. A. Bordia et al., "Effect of Garlic (*Allium sativum*) on Blood Lipids, Blood Sugar, Fibrinogen and Fibrinolytic Activity in Patients with Coronary Artery Disease," *Prostaglandins, Leukotrienes, and Essential Fatty Acids* 58, no. 4 (1998): 257–63, https://www.ncbi.nlm.nih.gov/pubmed/9654398.

77. M. Tognolini et al., "Comparative Screening of Plant Essential Oils: Phenylpropanoid Moiety as Basic Core for Antiplatelet Activity," *Life Sciences* 78, no. 13 (2006): 1419–32, https://doi.org/10.1016/j.lfs.2005.07.020.

78. Feng-Nien Ko et al., "Vasorelaxation of Rat Thoracic Aorta Caused by Osthole Isolated from *Angelica pubescens*," *European Journal of Pharmacology* 219, no. 1 (1992): 29–34, https://doi.org/10.1016/0014-2999(92)90576-P.

79. J. Durate et al., "Effects of Visnadine on Rat Isolated Vascular Smooth Muscles," *Planta Medica* 63, no. 3 (1997): 233–36, https://www.ncbi.nlm.nih.gov/pubmed/9225605.

80. Feng-Nien Ko et al., "Vasorelaxation of Rat Thoracic Aorta Caused by Osthole Isolated from *Angelica pubescens*," *European Journal of Pharmacology* 219, no. 1 (1992): 29–34, https://doi.org/10.1016/0014-2999(92)90576-P.

81. S. M. Yu et al., "Cinnamophilin, a Novel Thromboxane A2 Receptor Antagonist, Isolated from *Cinnamomum philippinense*," *European Journal of Pharmacology* 256, no. 1 (1994): 85–91, https://www.ncbi.nlm.nih.gov/pubmed/8026563.

82. Sheng Nan Wu et al., "Inhibitory Effect of the Plant-Extract Osthole on L-Type Calcium Current in NG108-15 Neuronal Cells," *Biochemical Pharmacology* 63, no. 2 (2002): 199–206, https://www.ncbi.nlm.nih.gov/pubmed/11841794.

83. A. H. Gilani et al., "Hypotensive Action of Coumarin Glycosides from *Daucus carota*," *Phytomedicine: International Journal of Phytotherapy and Phytopharmacology* 7, no. 5 (2000): 423–26, https://www.ncbi.nlm.nih.gov/pubmed/11081994.

84. Muhammad Nabeel Ghayur and Anwarul Hassan Gilani, "Ginger Lowers Blood Pressure through Blockade of Voltage-Dependent Calcium Channels," *Journal of Cardiovascular Pharmacology* 45, no. 1 (2005): 74–80, https://www.ncbi.nlm.nih.gov/pubmed/15613983.

85. Cheng-Zhi Chai et al., "Protective Effects of Sheng-Mai-San on Right Ventricular Dysfunction during Chronic Intermittent Hypoxia in Mice," *Evidence-Based Complementary and Alternative Medicine* (2016), https://doi.org/10.1155/2016/4682786.

86. Andrew P. Ambrosy et al., "The Use of Digoxin in Patients with Worsening Chronic Heart Failure: Reconsidering an Old Drug to Reduce Hospital Admissions," *Journal of the American College of Cardiology* 63, no. 18 (2014): 1823–32, https://doi.org/10.1016/j.jacc.2014.01.051.

87. Syed Jalal Khundmiri, "Advances in Understanding the Role of Cardiac Glycosides in Control of Sodium Transport in

Renal Tubules," *The Journal of Endocrinology* 222, no. 1 (2014): R11–24, https://doi.org/10.1530/JOE-13-0613.

88. Deok Ho Choi et al., "The Positive Inotropic Effect of the Aqueous Extract of *Convallaria keiskei* in Beating Rabbit Atria," *Life Sciences* 79, no. 12 (2006): 1178–85, https://doi.org/10.1016/j.lfs.2006.03.019.

89. H. Wagner and J. Grevel, "New Cardioactive Drugs II, Detection and Isolation of Cardiotonic Amines with Ionpair-HPLC," *Planta Medica* 44, no. 1 (1982): 36–40, https://doi.org/10.1055/s-2007-971397.

90. Kerry J. Welsh et al., "Rapid Detection of the Active Cardiac Glycoside Convallatoxin of Lily of the Valley Using LOCI Digoxin Assay," *American Journal of Clinical Pathology* 142, no. 3 (2014): 307–12, https://doi.org/10.1309/AJCPCOXF0O5XXTKD.

91. Taro Higano et al., "Convallasaponin A, a New 5beta-Spirostanol Triglycoside from the Rhizomes of *Convallaria majalis*," *Chemical & Pharmaceutical Bulletin* 55, no. 2 (2007): 337–39, https://www.ncbi.nlm.nih.gov/pubmed/17268112.

92. B. Kopp and W. Kubelka, "New Cardenolides from *Convallaria majalis*," *Planta Medica* 45, no. 4 (1982): 195–202, https://doi.org/10.1055/s-2007-971372.

93. Luis Alcocer, "Challenges and Treatment for Stroke Prophylaxis in Patients with Atrial Fibrillation in Mexico: A Review," *American Journal of Cardiovascular Drugs: Drugs, Devices, and Other Interventions* 16, no. 3 (2016): 171–82, https://doi.org/10.1007/s40256-016-0163-6.

94. A. Greiner, "Hypotensive Effect of *Viscum album*," *Orvosi Hetilap* 94, no. 3 (1953): 80–81, https://www.ncbi.nlm.nih.gov/pubmed/13037384.

95. J. X. Wu et al., "Effect of *Viscum coloratum* Flavonoids on Fast Response Action Potentials of Hearts," *Zhongguo Yao Li Xue Bao (Acta Pharmacologica Sinica)* 15 (1994): 169–72, https://www.ncbi.nlm.nih.gov/pubmed/?term=8010115.

96. J. X. Wu et al., "Experimental Study on Cellular Electrophysiology of *Viscum coloratum* Flavonoid in Treating Tachyarrhythmias," *Zhongguo Zhong Xi Yi Jie He Za Zhi Zhongguo Zhongxiyi Jiehe Zazhi (Chinese Journal of Integrated Traditional and Western Medicine)* 14, no. 7 (1994): 421–23, https://www.ncbi.nlm.nih.gov/pubmed/7950230.

97. D. Deliorman et al., "Studies on the Vascular Effects of the Fractions and Phenolic Compounds Isolated from *Viscum album* ssp. *album*," *Journal of Ethnopharmacology* 72, no. 1–2 (2000): 323–29, https://www.ncbi.nlm.nih.gov/pubmed/10967490.

98. F. B. O. Mojiminiyi et al., "The Vasorelaxant Effect of *Viscum album* Leaf Extract Is Mediated by Calcium-Dependent Mechanism," *Nigerian Journal of Physiological Sciences: Official Publication of the Physiological Society of Nigeria* 23, no. 1–2 (2008): 115–20, https://www.ncbi.nlm.nih.gov/pubmed/19434226.

99. Tenorio López et al., "*Viscum album* Aqueous Extract Induces Inducible and Endothelial Nitric Oxide Synthases Expression in Isolated and Perfused Guinea Pig Heart. Evidence of the Coronary Vasodilation Mechanism," *Archivos De Cardiologia De Mexico* 76, no. 2 (2006): 130–39, https://www.ncbi.nlm.nih.gov/pubmed/16859209.

100. Mirjana Radenkovic et al., "Effects of Mistletoe (*Viscum album l.*, Loranthaceae) Extracts on Arterial Blood Pressure in Rats Treated with Atropine Sulfate and Hexocycline," *Clinical and Experimental Hypertension* 31, no. 1 (2009): 11–19, https://doi.org/10.1080/10641960802409820.

101. Tenorio López et al., "*Viscum album* Aqueous Extract Induces Inducible and Endothelial Nitric Oxide Synthases Expression in Isolated and Perfused Guinea Pig Heart. Evidence of the Coronary Vasodilation Mechanism," *Archivos De Cardiologia De Mexico* 76, no. 2 (2006): 130–39, https://www.ncbi.nlm.nih.gov/pubmed/16859209.

102. Vedat Sekeroğlu et al., "*Viscum album l.* Extract and Quercetin Reduce Cyclophosphamide-Induced Cardiotoxicity, Urotoxicity and Genotoxicity in Mice," *Asian Pacific Journal of Cancer Prevention: APJCP* 12, no. 11 (2011): 2925–31, https://www.ncbi.nlm.nih.gov/pubmed/22393965.

103. R. Klopp et al., "Influence of Complementary *Viscum album* (Iscador) Administration on Microcirculation and Immune System of Ear, Nose and Throat Carcinoma Patients Treated with Radiation and Chemotherapy," *Anticancer Research* 25, no. 1B (2005): 601–10, https://www.ncbi.nlm.nih.gov/pubmed/15816634.

104. A. Panossian et al., "Pharmacological Activity of Phenylpropanoids of the Mistletoe, *Viscum album l.*, Host: *Pyrus caucasica Fed*," *Phytomedicine: International Journal of Phytotherapy and Phytopharmacology* 5, no. 1 (1998): 11–17, https://doi.org/10.1016/S0944-7113(98)80053-6.

105. A. V. Timoshenko and H. J. Gabius, "Efficient Induction of Superoxide Release from Human Neutrophils by the Galactoside-Specific Lectin from *Viscum album*," *Biological Chemistry Hoppe-Seyler* 374, no. 4 (1993): 237–43, https://www.ncbi.nlm.nih.gov/pubmed/8392351.

106. A. Zschäbitz et al., "Characterization of Glycoconjugate Expression during Development of Meckel's Cartilage in the Rat," *Anatomy and Embryology* 191, no. 1 (1995): 47–49, https://www.ncbi.nlm.nih.gov/pubmed/7717533.

107. A. M. Gray and P. R. Flatt, "Insulin-Secreting Activity of the Traditional Antidiabetic Plant *Viscum album* (Mistletoe)," *The Journal of Endocrinology* 160, no. 3 (1999): 409–14, https://www.ncbi.nlm.nih.gov/pubmed/10076186.

108. Min Jung Kim et al., "The Supplementation of Korean Mistletoe Water Extracts Reduces Hot Flushes, Dyslipidemia, Hepatic Steatosis, and Muscle Loss in Ovariectomized Rats," *Experimental Biology and Medicine* 240, no. 4 (2015): 477–87, https://doi.org/10.1177/1535370214551693.

109. Ki-Wook Kim et al., "Protein Fractions from Korean Mistletoe (*Viscum album Coloratum*) Extract Induce Insulin Secretion from Pancreatic Beta Cells," *Evidence-Based Complementary and Alternative Medicine* 2014 (2014), https://doi.org/10.1155/2014/703624.

110. Kurian John Poruthukaren et al., "Clinical Evaluation of *Viscum album* Mother Tincture as an Antihypertensive: A Pilot Study," *Journal of Evidence-Based Complementary & Alternative Medicine* 19, no. 1 (2014): 31–35, https://doi.org/10.1177/2156587213507726.

111. Feng-Nien Ko et al., "Vasorelaxation of Rat Thoracic Aorta Caused by Osthole Isolated from *Angelica pubescens*," *European Journal of Pharmacology* 219, no. 1 (1992): 29–34, https://doi.org/10.1016/0014-2999(92)90576-P.

112. J. R. Casley-Smith, "The Actions of the Benzopyrones on the Blood-Tissue-Lymph System," *Folia angiol. (Pisa)* 24, no. 7 (1976), https://scholar.google.com/scholar_lookup?title=The%20actions%20of%20the%20benzopyrones%20on%20the%20blood-tissue%20%E2%80%94%20Lymph%20System&author=JR.%20Casley-Smith&journal=Folia%20Angiol&volume=24&pages=7-22&publication_year=1976.

113. Mee-Ra Rhyu et al., "*Radix angelica* Elicits Both Nitric Oxide-Dependent and Calcium Influx-Mediated Relaxation in Rat Aorta," *Journal of Cardiovascular Pharmacology* 46, no. 1 (2005): 99–104, https://www.ncbi.nlm.nih.gov/pubmed/15965361.

114. S. J. Sjeu, et al., "Analysis and Processing of Chinese Herbal Drugs; VI The Study of Angelica Radix," *Planta Medica* (1987): 377–8, https://doi.org/10.1055/s-2006-962742; C. P. Sung et al., "Effect of Extracts of *Angelica polymorpha* on Reaginic Antibody Production," *Journal of Natural Products* 45, no. 4 (1982): 398–406, https://www.ncbi.nlm.nih.gov/pubmed/7130985.

115. D. P. Zhu, "Dong Quai," *The American Journal of Chinese Medicine* 15, no. 3–4 (1987): 117–25, https://doi.org/10.1142/S0192415X87000151.

116. J. H. Guh et al., "Antiproliferative Effect in Rat Vascular Smooth Muscle Cells by Osthole, Isolated from *Angelica pubescens*," *European Journal of Pharmacology* 298, no. 2 (1996): 191–97, https://www.ncbi.nlm.nih.gov/pubmed/8867108.

117. Yong Yook Lee et al., "Platelet Anti-Aggregatory Effects of Coumarins from the Roots of *Angelica genuflexa* and *A. gigas*," *Archives of Pharmacal Research* 26, no. 9 (2003): 723–26, https://www.ncbi.nlm.nih.gov/pubmed/14560920.

118. B. Ramesh and K. V. Pugalendi, "Antihyperlipidemic and Antidiabetic Effects of Umbelliferone in Streptozotocin Diabetic Rats," *The Yale Journal of Biology and Medicine* 78, no. 4 (2005): 189–96, https://www.ncbi.nlm.nih.gov/pubmed/16720013.

119. Bin Yu et al., "Synergic Effect of Borneol and Ligustrazine on the Neuroprotection in Global Cerebral Ischemia/Reperfusion Injury: A Region-Specificity Study," *Evidence-Based Complementary and Alternative Medicine: ECAM* 2016 (2016), https://doi.org/10.1155/2016/4072809.

120. Bo Liu et al., "Gastrodin Ameliorates Subacute Phase Cerebral Ischemia-Reperfusion Injury by Inhibiting Inflammation and Apoptosis in Rats," *Molecular Medicine Reports* 14, no. 5 (2016): 4144–52, https://doi.org/10.3892/mmr.2016.5785.

121. Zhu Zhang et al., "Effect of *Herba centellae* on the Expression of HGF and MCP-1," *Experimental and Therapeutic Medicine* 6, no. 2 (2013): 427–34, https://doi.org/10.3892/etm.2013.1146.

122. Amala Soumyanath et al., "*Centella asiatica* Accelerates Nerve Regeneration upon Oral Administration and Contains Multiple Active Fractions Increasing Neurite Elongation In-Vitro," *The Journal of Pharmacy and Pharmacology* 57, no. 9 (2005): 1221–29, https://doi.org/10.1211/jpp.57.9.0018.

123. Uwe Wollina et al., "A Review of the Microcirculation in Skin in Patients with Chronic Venous Insufficiency: The Problem and the Evidence Available for Therapeutic Options," *The International Journal of Lower Extremity Wounds* 5, no. 3 (2006): 169–80, https://doi.org/10.1177/1534734606291870.

124. M. R. Cesarone et al., "Evaluation of Treatment of Diabetic Microangiopathy with Total Triterpenic Fraction of *Centella asiatica*: A Clinical Prospective Randomized Trial with a Microcirculatory Model," *Angiology* 52 Suppl 2 (2001): S49-54, https://www.ncbi.nlm.nih.gov/pubmed/11666124.

125. Hong-Feng Zhang et al., "An Overview of Systematic Reviews of *Ginkgo biloba* Extracts for Mild Cognitive Impairment and Dementia," *Frontiers in Aging Neuroscience* 8 (2016), https://doi.org/10.3389/fnagi.2016.00276.

126. Xiaoling Chen and Kewei Wang, "The Fate of Medications Evaluated for Ischemic Stroke Pharmacotherapy over the Period 1995–2015," *Acta Pharmaceutica Sinica B* 6, no. 6 (2016): 522–30, https://doi.org/10.1016/j.apsb.2016.06.013.

127. Florensia Nailufar et al., "Thrombus Degradation by Fibrinolytic Enzyme of *Stenotrophomonas Sp.* Originated from Indonesian Soybean-Based Fermented Food on Wistar Rats," *Advances in Pharmacological Sciences* 2016 (2016), https://doi.org/10.1155/2016/4206908.

128. K. K. Pulicherla and Mahendra Kumar Verma, "Targeting Therapeutics across the Blood Brain Barrier (BBB), Prerequisite towards Thrombolytic Therapy for Cerebrovascular Disorders-an Overview and Advancements," *AAPS PharmSciTech* 16, no. 2 (2015): 223–33, https://doi.org/10.1208/s12249-015-0287-z.

129. Joanna M. Wardlaw et al., "Thrombolysis for Acute Ischaemic Stroke," *The Cochrane Database of Systematic Reviews*, no. 7 (2014): CD000213, https://doi.org/10.1002/14651858.CD000213.pub3.

130. Cheng-Chieh Chang et al., "Oxidative Stress and *Salvia miltiorrhiza* in Aging-Associated Cardiovascular Diseases," *Oxidative Medicine and Cellular Longevity* 2016 (2016), https://doi.org/10.1155/2016/4797102.

131. Sai-Wang Seto et al., "Angiogenesis in Ischemic Stroke and Angiogenic Effects of Chinese Herbal Medicine," *Journal of Clinical Medicine* 5, no. 6 (2016), https://doi.org/10.3390/jcm5060056.

132. Yuwei Pan et al., "Synergistic Effect of Ferulic Acid and Z-Ligustilide, Major Components of A. Sinensis, on

Regulating Cold-Sensing Protein TRPM8 and TPRA1 In Vitro," *Evidence-Based Complementary and Alternative Medicine* 2016 (2016), https://doi.org/10.1155/2016/3160247.

133. Andrew M. Prentice et al., "Dietary Strategies for Improving Iron Status: Balancing Safety and Efficacy," *Nutrition Reviews* 75, no. 1 (2017): 49–60, https://doi.org/10.1093/nutrit/nuw055.

134. P. F. Smith et al., "The Neuroprotective Properties of the *Ginkgo biloba* Leaf: A Review of the Possible Relationship to Platelet-Activating Factor (PAF)," *Journal of Ethnopharmacology* 50, no. 3 (1996): 131–39, https://www.ncbi.nlm.nih.gov/pubmed/8691847.

135. Larry E. Becker and George B. Skipworth, "Ginkgo-Tree Dermatitis, Stomatitis, and Proctitis," *JAMA* 231, no. 11 (1975): 1162–63, https://doi.org/10.1001/jama.1975.03240230036017.

136. J. Kleijnen and P. Knipschild, "*Ginkgo biloba* for Cerebral Insufficiency," *British Journal of Clinical Pharmacology* 34, no. 4 (1992): 352–58, https://www.ncbi.nlm.nih.gov/pmc/articles/PMC1381419/.

137. G. Mussgnug and J. Alemany, "Studies on Peripheral Arterial Blood Circulation Disorders. XV. On the Problems of Conservative Therapy of Obliterating Peripheral Blood Circulation Disorders Demonstrated on Tincture and Extract from *Ginkgo biloba l.*," *Arzneimittel-Forschung* 18, no. 5 (1968): 543–50, https://www.ncbi.nlm.nih.gov/pubmed/5755881.

138. K. Schaffler and P. W. Reeh, "Double Blind Study of the Hypoxia Protective Effect of a Standardized *Ginkgo biloba* Preparation after Repeated Administration in Healthy Subjects," *Arzneimittel-Forschung* 35, no. 8 (1985): 1283–86, https://www.ncbi.nlm.nih.gov/pubmed/3907639.

139. S. Reuse-Blom and K. Drieu, "Effect of *Ginkgo bilobo* Extract on Arteriorlar Spasm in Rabbits," *Presse Med* 15, no. 31 (1986): 1520–1523, https://www.ncbi.nlm.nih.gov/pubmed/2947092.

140. Gregory J. Stoddard et al., "Ginkgo and Warfarin Interaction in a Large Veterans Administration Population," *AMIA Annual Symposium Proceedings* 2015 (2015): 1174–83, https://www.ncbi.nlm.nih.gov/pmc/articles/PMC4765589/.

141. Ying Zhang et al., "Continuing Treatment with *Salvia miltiorrhiza* Injection Attenuates Myocardial Fibrosis in Chronic Iron-Overloaded Mice," *PloS ONE* 10, no. 4 (2015): e0124061, https://doi.org/10.1371/journal.pone.0124061.

142. Shengjiang Guan et al., "Danshen (*Salvia miltiorrhiza*) Injection Suppresses Kidney Injury Induced by Iron Overload in Mice," *PLoS ONE* 8, no. 9 (2013): e74318, https://doi.org/10.1371/journal.pone.0074318.

143. Heather C. Hatcher et al., "Synthetic and Natural Iron Chelators: Therapeutic Potential and Clinical Use," *Future Medicinal Chemistry* 1, no. 9 (2009): 1643–70, https://doi.org/10.4155/fmc.09.121.

144. Heather C. Hatcher et al., "Synthetic and Natural Iron Chelators: Therapeutic Potential and Clinical Use," *Future Medicinal Chemistry* 1, no. 9 (2009): 1643–70, https://doi.org/10.4155/fmc.09.121.

145. Heather C. Hatcher et al., "Synthetic and Natural Iron Chelators: Therapeutic Potential and Clinical Use," *Future Medicinal Chemistry* 1, no. 9 (2009): 1643–70, https://doi.org/10.4155/fmc.09.121.

146. Hadi Darvishi Khezri et al., "Potential Effects of Silymarin and Its Flavonolignan Components in Patients with β-Thalassemia Major: A Comprehensive Review in 2015," *Advances in Pharmacological Sciences* 2016 (2016), https://doi.org/10.1155/2016/3046373.

147. Heather C. Hatcher et al., "Synthetic and Natural Iron Chelators: Therapeutic Potential and Clinical Use," *Future Medicinal Chemistry* 1, no. 9 (2009): 1643–70, https://doi.org/10.4155/fmc.09.121.

148. Heather C. Hatcher et al., "Synthetic and Natural Iron Chelators: Therapeutic Potential and Clinical Use," *Future Medicinal Chemistry* 1, no. 9 (2009): 1643–70, https://doi.org/10.4155/fmc.09.121.

149. Chak-Lam Cho et al., "Novel Insights into the Pathophysiology of Varicocele and Its Association with Reactive Oxygen Species and Sperm DNA Fragmentation," *Asian Journal of Andrology* 18, no. 2 (2016): 186–93, https://doi.org/10.4103/1008-682X.170441.

150. S. Visudhiphan et al., "The Relationship between High Fibrinolytic Activity and Daily Capsicum Ingestion in Thais," *The American Journal of Clinical Nutrition* 35, no. 6 (1982): 1452–58, https://www.ncbi.nlm.nih.gov/pubmed/7081126.

151. A. K. Bordia et al., "Effect of Essential Oil of Garlic on Serum Fibrinolytic Activity in Patients with Coronary Artery Disease," *Atherosclerosis* 28, no. 2 (1977): 155–59, https://www.ncbi.nlm.nih.gov/pubmed/911374.

152. K. I. Baghurst et al., "Onions and Platelet Aggregation," *Lancet* 1, no. 8002 (1977): 101, https://www.ncbi.nlm.nih.gov/pubmed/63702.

153. K. C. Srivas, "Effects of Aqueous Extracts of Onion, Garlic and Ginger on Platelet Aggregation and Metabolism of Arachidonic Acid in the Blood Vascular System: In Vitro Study," *Prostaglandins, Leukotrienes, and Medicine* 13, no. 2 (1984): 227–35, https://www.ncbi.nlm.nih.gov/pubmed/6425866.

154. H. Ako et al., "Isolation of a Fibrinolysis Enzyme Activator from Commercial Bromelain," *Archives Internationales De Pharmacodynamie Et De Therapie* 254, no. 1 (1981): 157–67, https://www.ncbi.nlm.nih.gov/pubmed/7199897.

155. I. Jantan et al., "Inhibitory Effect of Compounds from Zingiberaceae Species on Human Platelet Aggregation," *Phytomedicine: International Journal of Phytotherapy and Phytopharmacology* 15, no. 4 (2008): 306–9, https://doi.org/10.1016/j.phymed.2007.08.002.

156. Gianni Belcaro et al., "French Oak Wood (*Quercus robur*) Extract (Robuvit) in Primary Lymphedema: A Supplement, Pilot, Registry Evaluation," *The International Journal of Angiology: Official Publication of the International College of Angiology, Inc* 24, no. 1 (2015): 47–54, https://doi.org/10.1055/s-0034-1395982.

157. Kapil Aggrawal et al., "Efficacy of a Standardized Herbal Preparation (Roidosanal®) in the Treatment of Hemorrhoids: A Randomized, Controlled, Open-Label Multicentre Study," *Journal of Ayurveda and Integrative Medicine* 5, no. 2 (2014): 117–24, https://doi.org/10.4103/0975-9476.131732.

158. Angela V. Ghatnekar et al., "Novel Wound Healing Powder Formulation for the Treatment of Venous Leg Ulcers," *The Journal of the American College of Certified Wound Specialists* 3, no. 2 (2011): 33–41, https://doi.org/10.1016/j.jcws.2011.09.004.

159. K. Greeske and B. K. Pohlmann, "Horse Chestnut Seed Extract—An Effective Therapy Principle in General Practice. Drug Therapy of Chronic Venous Insufficiency," *Fortschritte Der Medizin* 114, no. 15 (1996): 196–200, https://www.ncbi.nlm.nih.gov/pubmed/8767939.

160. Yong Yook Lee et al., "Platelet Anti-Aggregatory Effects of Coumarins from the Roots of *Angelica genuflexa* and *A. gigas*," *Archives of Pharmacal Research* 26, no. 9 (2003): 723–26, https://www.ncbi.nlm.nih.gov/pubmed/14560920.

161. Lai-Hao Wang and Si-Yun Jiang, "Simultaneous Determination of Urinary Metabolites of Methoxypsoralens in Human and Umbelliferae Medicines by High-Performance Liquid Chromatography," *Journal of Chromatographic Science* 44, no. 8 (2006): 473–78, https://www.ncbi.nlm.nih.gov/pubmed/16959122.

162. D. J. Runciman et al., "Dicoumarol Toxicity in Cattle Associated with Ingestion of Silage Containing Sweet Vernal Grass (*Anthoxanthum odoratum*)," *Australian Veterinary Journal* 80, no. 1–2 (2002): 28–32, https://www.ncbi.nlm.nih.gov/pubmed/12180874.

163. Mahmoud F. Elsebai et al., "Cynaropicrin: A Comprehensive Research Review and Therapeutic Potential as an Anti-Hepatitis C Virus Agent," *Frontiers in Pharmacology* 7 (2016), https://doi.org/10.3389/fphar.2016.00472.

164. Yan Zhang et al., "Therapeutic Effect of Osthole on Hyperlipidemic Fatty Liver in Rats," *Acta Pharmacologica Sinica* 28, no. 3 (2007): 398–403, https://doi.org/10.1111/j.1745-7254.2007.00533.x.

165. Fang Song et al., "Experimental Study of Osthole on Treatment of Hyperlipidemic and Alcoholic Fatty Liver in Animals," *World Journal of Gastroenterology* 12, no. 27 (2006): 4359–63, https://www.ncbi.nlm.nih.gov/pubmed/16865778.

166. R. L. Huang et al., "Osthole Increases Glycosylation of Hepatitis B Surface Antigen and Suppresses the Secretion of Hepatitis B Virus in Vitro," *Hepatology* 24, no. 3 (1996): 508–15, https://doi.org/10.1002/hep.510240307.

167. Subathra Devi Chandrasekaran et al., "Exploring the In Vitro Thrombolytic Activity of Nattokinase From a New Strain *Pseudomonas aeruginosa* CMSS," *Jundishapur Journal of Microbiology* 8, no. 10 (2015), https://doi.org/10.5812/jjm.23567.

168. Fuming Zhang et al., "Interactions between Nattokinase and Heparin/GAGs," *Glycoconjugate Journal* 32, no. 9 (2015): 695–702, https://doi.org/10.1007/s10719-015-9620-8.

169. Ja-Young Jang et al., "Nattokinase Improves Blood Flow by Inhibiting Platelet Aggregation and Thrombus Formation," *Laboratory Animal Research* 29, no. 4 (2013): 221–25, https://doi.org/10.5625/lar.2013.29.4.221.

170. Yuko Kurosawa et al., "A Single-Dose of Oral Nattokinase Potentiates Thrombolysis and Anti-Coagulation Profiles," *Scientific Reports* 5 (2015), https://doi.org/10.1038/srep11601.

171. Maqsood M. Elahi et al., "Consequence of Patient Substitution of Nattokinase for Warfarin after Aortic Valve Replacement with a Mechanical Prosthesis," *Proceedings (Baylor University. Medical Center)* 28, no. 1 (2015): 81–82, https://www.ncbi.nlm.nih.gov/pubmed/25552810.

172. Masafumi Funamoto et al., "Highly Absorptive Curcumin Reduces Serum Atherosclerotic Low-Density Lipoprotein Levels in Patients with Mild COPD," *International Journal of Chronic Obstructive Pulmonary Disease* 11 (2016): 2029–34, https://doi.org/10.2147/COPD.S104490.

173. Qingmei Liu et al., "Salvianolic Acid B Attenuates Experimental Pulmonary Fibrosis through Inhibition of the TGF-β Signaling Pathway," *Scientific Reports* 6 (2016), https://doi.org/10.1038/srep27610.

174. Marzyeh Amini et al., "Eosinophil Count Is a Common Factor for Complex Metabolic and Pulmonary Traits and Diseases: The LifeLines Cohort Study," *PLoS ONE* 11, no. 12 (2016): e0168480, https://doi.org/10.1371/journal.pone.0168480.

175. W. F. Chiou et al., "Vasorelaxing Effect of Coumarins from *Cnidium monnieri* on Rabbit Corpus Cavernosum," *Planta Medica* 67, no. 3 (2001): 282–84, https://doi.org/10.1055/s-2001-12013.

176. J. Chen et al., "Effect of the Plant-Extract Osthole on the Relaxation of Rabbit Corpus Cavernosum Tissue in Vitro," *The Journal of Urology* 163, no. 6 (2000): 1975–80, https://www.ncbi.nlm.nih.gov/pubmed/10799242.

177. Ze-wei Zhou and Pei-xun Liu, "Progress in Study of Chemical Constituents and Anti-tumor Activities of *Cnidium monnieri*," *Zhongguo Zhong Yao Za Zhi = Zhongguo Zhongyao Zazhi (China Journal of Chinese Materia Medica)* 30, no. 17 (2005): 1309–13, https://www.ncbi.nlm.nih.gov/pubmed/16323535.

178. J. H. Guh et al., "Antiproliferative Effect in Rat Vascular Smooth Muscle Cells by Osthole, Isolated from *Angelica pubescens*," *European Journal of Pharmacology* 298, no. 2 (1996): 191–97, https://www.ncbi.nlm.nih.gov/pubmed/8867108.

179. W. F. Chiou et al., "Vasorelaxing Effect of Coumarins from *Cnidium monnieri* on Rabbit Corpus Cavernosum," *Planta Medica* 67, no. 3 (2001): 282–84, https://doi.org/10.1055/s-2001-12013.

180. Juanli Yuan et al., "Effects of Osthol on Androgen Level and Nitric Oxide Synthase Activity in Castrate Rats," *Zhong Yao Cai = Zhongyaocai (Journal of Chinese Medicinal Materials)* 27, no. 7 (2004): 504–6, https://www.ncbi.nlm.nih.gov/pubmed/15551966.

181. Qin Chen et al., "Treatment and Prevention of Inflammatory Responses and Oxidative Stress in Patients with Obstructive Sleep Apnea Hypopnea Syndrome Using Chinese Herbal Medicines," *Experimental and Therapeutic Medicine* 12, no. 3 (2016): 1572–78, https://doi.org/10.3892/etm.2016.3484.

182. M. R. Cesarone et al., "Increase in Echogenicity of Echolucent Carotid Plaques after Treatment with Total Triterpenic Fraction of *Centella asiastica*: a Prospective, Placebo-Controlled, Randomized Trial," *Angiology* 52, supplement 2 (2001): S19–25.

183. Francesco Di Pierro et al., "Preliminary Study about the Possible Glycemic Clinical Advantage in Using a Fixed Combination of *Berberis aristata* and *Silybum marianum* Standardized Extracts versus Only *Berberis aristata* in Patients with Type 2 Diabetes," *Clinical Pharmacology: Advances and Applications* 5 (2013): 167–74, https://doi.org/10.2147/CPAA.S54308.

184. Hui Li Tan et al., "*Rhizoma coptidis*: A Potential Cardiovascular Protective Agent," *Frontiers in Pharmacology* 7 (2016): 362, https://doi.org/10.3389/fphar.2016.00362.

185. Vajiheh Rangboo et al., "The Effect of Artichoke Leaf Extract on Alanine Aminotransferase and Aspartate Aminotransferase in the Patients with Nonalcoholic Steatohepatitis," *International Journal of Hepatology* 2016 (2016), https://doi.org/10.1155/2016/4030476.

186. Cen Chen et al., "Natural Products for Antithrombosis," *Evidence-Based Complementary and Alternative Medicine* 2015 (2015), https://doi: 10.1155/2015/876426.

187. Bo Liu et al., "Gastrodin Ameliorates Subacute Phase Cerebral Ischemia-Reperfusion Injury by Inhibiting Inflammation and Apoptosis in Rats," *Molecular Medicine Reports* 14, no. 5 (2016): 4144–52, https://doi.org/10.3892/mmr.2016.5785.

188. Marta Guasch-Ferré et al., "Olive Oil Intake and Risk of Cardiovascular Disease and Mortality in the PREDIMED Study," *BMC Medicine* 12 (2014): 78, https: doi: 10.1186/1741-7015-12-78.

189. Gianni Belcaro et al., "French Oak Wood (*Quercus robur*) Extract (Robuvit) in Primary Lymphedema: A Supplement, Pilot, Registry Evaluation," *The International Journal of Angiology: Official Publication of the International College of Angiology, Inc* 24, no. 1 (2015): 47–54, https://doi.org/10.1055/s-0034-1395982.

190. Chun-Mei Wang et al., "Schisandra Polysaccharide Inhibits Hepatic Lipid Accumulation by Downregulating Expression of SREBPs in NAFLD Mice," *Lipids in Health and Disease* 15 (2016), https://doi.org/10.1186/s12944-016-0358-5.

191. K. Aftab et al., "Traditional Medicine *Cassia absus l.* (Chaksu)-Pharmacological Evaluation," *Phytomedicine: International Journal of Phytotherapy and Phytopharmacology* 2, no. 3 (1996): 213–19, https://doi.org/10.1016/S0944-7113(96)80045-6.

192. Ales Vidlar et al., "The Safety and Efficacy of a Silymarin and Selenium Combination in Men after Radical Prostatectomy—a Six Month Placebo-Controlled Double-Blind Clinical Trial," *Biomedical Papers of the Medical Faculty of the University Palacky, Olomouc, Czechoslovakia* 154, no. 3 (2010): 239–44, https://www.ncbi.nlm.nih.gov/pubmed/21048810.

193. I. S. Chen et al., "Hepatoprotection of Silymarin against Thioacetamide-Induced Chronic Liver Fibrosis," *Journal of the Science of Food and Agriculture* 92, no. 7 (2012): 1441–47, https://doi.org/10.1002/jsfa.4723.

194. Mahmood Bahmani et al., "*Silybum marianum*: Beyond Hepatoprotection," *Journal of Evidence-Based Complementary & Alternative Medicine* 20, no. 4 (2015): 292–301, https://doi.org/10.1177/2156587215571116.

195. Natasa Milosević et al., "phytotherapy and NAFLD—from Goals and Challenges to Clinical Practice," *Reviews on Recent Clinical Trials* 9, no. 3 (2014): 195–203, https://www.ncbi.nlm.nih.gov/pubmed/25514914.

196. Hadi Darvishi Khezri et al., "Potential Effects of Silymarin and Its Flavonolignan Components in Patients with β-Thalassemia Major: A Comprehensive Review in 2015," *Advances in Pharmacological Sciences* (2016), https://doi.org/10.1155/2016/3046373.

197. S. Dwivedi and R. Jauhari, "Beneficial Effects of *Terminalia arjuna* in Coronary Artery Disease," *Indian Heart Journal* 49, no. 5 (1997): 507–10, https://www.ncbi.nlm.nih.gov/pubmed/9505018.

198. G. J. Amabeoku et al., "Antimicrobial and Anticonvulsant Activities of *Viscum capense*," *Journal of Ethnopharmacology* 61, no. 3 (1998): 237–41, https://www.ncbi.nlm.nih.gov/pubmed/9705015.

Chapter 3: Creating Herbal Formulas for Respiratory Conditions

1. S. Franova et al., "Effects of Flavin7 on Allergen Induced Hyperreactivity of Airways," *European Journal of Medical Research* 14, no. 4 (2009): 78–81, doi:10.1186/2047-783X-14-S4-78.

2. Yan Xiao-Dan et al., "Polydatin Protects the Respiratory System from PM2.5 Exposure," *Scientific Reports* 7, no. 40030 (2017), doi:10.1038/srep40030.

3. Zhaoyu Wang et al., "Enzymatic Synthesis of Sorboyl-Polydatin Prodrug in Biomass-Derived 2-Methyltetrahydrofuran and Antiradical Activity of the Unsaturated Acylated Derivatives," *Biomedical Research International* 2016 (2016), doi:10.1155/2016/4357052.

4. Jinpei Lin et al., "Paeoniflorin Attenuated Oxidative Stress in Rat COPD Model Induced by Cigarette Smoke," *Evidence Based Complementary and Alternative Medicine* 2016 (2016), doi:10.1155/2016/1698379.

5. Kalpana S. Paudel et al., "Challenges and Opportunities in Dermal/Transdermal Delivery," *Therapeutic Delivery* 1, no. 1 (2010): 109–131, www.ncbi.nlm.nih.gov/pmc/articles/PMC2995530.

6. Tian Yang et al., "Emodin Suppresses Silica-induced Lung Fibrosis by Promoting Sirt1 Signaling via Direct Contact," *Molecular Medicine Reports* 14, no. 5 (2016): 4643–4649, doi:10.3892/mmr.2016.5838.

7. Meir Mei-Zahav et al., "Topical Propranolol Improves Epistaxis in Patients With Hereditary Hemorrhagic Telangiectasia—A Preliminary Report," *Journal of Otolaryngology—Head and Neck Surgery* 46 (2017): 58, doi:10.1186/s40463-017-0235-x.

8. Khaled Sebei et al., "Chemical Composition and Antibacterial Activities of Seven Eucalyptus Species Essential Oils Leaves," *Biological Research* 48, no. 1 (2015): 7, doi:10.1186/0717-6287-48-7.

9. Maria Paparoupa and Adrian Gillissen, "Is Myrtol® Standardized a New Alternative toward Antibiotics?," *Pharmacognosy Review* 10, no. 20 (2016): 143–146, doi:10.4103/0973-7847.194045.

10. Holger Sudhoff et al., "1,8-Cineol Reduces Mucus-Production in a Novel Human Ex Vivo Model of Late Rhinosinusitis," *PLoS ONE* 10, no. 7 (2015): e0133040, doi:10.1371/journal.pone.0133040.

11. Saranya Sugumar et al., "Eucalyptus Oil Nanoemulsion-Impregnated Chitosan Film: Antibacterial Effects Against a Clinical Pathogen, *Staphylococcus aureus*, in Vitro," *International Journal of Nanomedicine* 10, no. 1 (2015): 67–75, doi:10.2147/IJN.S79982.

12. Alyn Morice and Peter Kardos, "Comprehensive Evidence-Based Review on European Antitussives," *BMJ Open Respiratory Research* 3 (2016): e000137, doi:10.1136/bmjresp-2016-000137.

13. Geoffrey Burnstock, "Purinergic Signaling: Therapeutic Developments," *Frontiers in Pharmacology* 8, (2017): 661, doi:10.3389/fphar.2017.00661.

14. Alexander Yashin et al., "Antioxidant Activity of Spices and Their Impact on Human Health: A Review," *Antioxidants* 6, no. 3 (2017): 70, doi:10.3390/antiox6030070; Xin Jin et al., "Synergistic Apoptotic Effects of Apigenin TPGS Liposomes and Tyroservatide: Implications for Effective Treatment of Lung Cancer," *International Journal of Nanomedicine* 12, (2017): 5109–5118, doi:10.2147/IJN.S140096.

15. P. L. Simmons et al., "Clinical Management of Pain in Advanced Lung Cancer," *Clinical Medicine Insights Oncology* 6, (2012): 331–346, doi:10.4137/CMO.S8360.

16. Yukiko Muroi and Bradley J. Undem, "Targeting Peripheral Afferent Nerve Terminals for Cough and Dyspnea," *Current Opinion in Pharmacology* 11, no. 3 (2011): 254–264, doi:10.1016/j.coph.2011.05.006.

17. Yu-Chin Lin et al., "Antitussive, Anti-Pyretic and Toxicological Evaluation of Ma-Xing-Gan-Shi-Tang in Rodents," *BMC Complementary and Alternative Medicine* 16, (2016): 456, doi:10.1186/s12906-016-1440-2.

18. Gregg Duncan et al., "Microstructural Alterations of Sputum in Cystic Fibrosis Lung Disease," *JCI Insight* 1, no. 18 (2016): e88198, doi:10.1172/jci.insight.88198.

19. Julie R. Kesting et al., "Piperidine and Tetrahydropyridine Alkaloids from *Lobelia siphilitica* and *Hippobroma longiflora*," *Journal of Natural Products* 72, no. 2 (2009): 312–215, doi:10.1021/np800743w.

20. L. P. Dwoskin and P. A. Crooks, "A Novel Mechanism of Action and Potential Use for Lobeline as a Treatment for Psychostimulant Abuse," *Biochemical Pharmacology* 63, no. 2 (2002): 89–98.

21. Ernô Sántha et al., "Multiple Cellular Mechanisms Mediate the Effect of Lobeline on the Release of Norepinephrine," *The Journal of Pharmacology and Experimental Therapeutics* 294, no. 1 (2000): 302–307.

22. A. Subarnas et al., "Pharmacological Properties of Beta-Amyrin Palmitate, a Novel Centrally Acting Compound, Isolated from *Lobelia inflata* Leaves," *Journal of Pharmacy and Pharmacology* 45, no. 6 (1993): 545–50.

23. D. Y. Lim et al., "Influence of Lobeline on Catecholamine Release from the Isolated Perfused Rat Adrenal Gland," *Autonomic Neuroscience* 110, no. 1 (2004): 27–35.

24. S. C. Gandevia et al., "Balancing Acts: Respiratory Sensations, Motor Control and Human Posture," *Clinical and Experimental Pharmacology and Physiology* 29, no. 1–2 (2002): 118–121.

25. S. B. Kästner et al., "Comparison of the Performance of Linear Resistance and Ultrasonic Pneumotachometers at Rest and During Lobeline-Induced Hyperpnoea," *Research in Veterinary Science* 68, no. 2 (2000): 153–159.

26. D. S. Jaju et al., "Comparison of Respiratory Sensations Induced by J Receptor Stimulation with Lobeline in Left Handers & Right Handers," *Indian Journal of Medical Research* 108, (1998): 291–295.

27. D. R. McCrimmon and P. M. Lalley, "Inhibition of Respiratory Neural Discharges by Clonidine and 5-hydroxytryptophan," *The Journal of Pharmacology and Experimental Therapeutics* 222, no. 3 (1982): 771–777; E. van Lunteren et al., "Effects of Dopamine, Isoproterenol, and Lobeline on Cranial and Phrenic Motoneurons," *Journal of Applied Physiology* 56, no. 3 (1984): 737–745; S. Yanaura et al., "A Quantitative Analysis of the Phrenic Nerve Activities during the Cough Reflex," *Nippon Yakurigaku Zasshi* 79, no. 6 (1982): 543–550.

28. V. Deep et al., "Role of Vagal Afferents in the Reflex Effects of Capsaicin and Lobeline in Monkeys," *Respiratory Physiology* 125, no. 3 (2001): 155–168.

29. L. Silva Carvalho et al., "Effect of Propranolol and Pindolol on Carotid Reflexes Induced by Lobeline," *Comptes Rendus des Seances de la Societe de Biologie et de ses Filliales* 175, no. 3 (1981): 416–419.

30. G. A. Dehghani et al., "Presence of Lobeline-like Sensations in Exercising Patients with Left Ventricular Dysfunction," *Respiratory Physiology and Neurobiology* 143, no. 1 (2004): 9–20, doi.org/10.1016/j.resp.2004.07.003; Hans Raj et al., "How Does Lobeline Injected Intravenously Produce a Cough?," *Respiratory Physiology & Neurobiology* 145, no. 1 (2005): 79–90. doi:10.1016/j.resp.2004.09.001.

31. G. Sant'Ambrogio, "Afferent Pathways for the Cough Reflex," *Bulletin European de Physiopathologie Respiratorie* 23, S10 (1987): 19s–23s; Hans Raj et al., "Sensory Origin of Lobeline-induced Sensations: A Correlative Study in Man and Cat," *The Journal of Physiology* 482, no. 1 (1995): 235–246, doi:10.1113/jphysiol.1995.sp020513.

32. Rajesh Kumar Ganjhu et al., "Herbal Plants and Plant Preparations as Remedial Approach for Viral Diseases," *Virusdisease* 26, no. 4 (2015): 225–236, doi:10.1007/s13337 -015-0276-6.

33. Liqin Wang et al., "Effects of Active Components of Fuzi and Gancao Compatibility on Bax, Bcl-2, and Caspase-3 in Chronic Heart Failure Rats," *Evidence Based Complementary and Alternative Medicine* 2016, no. 7686045 (2016), doi:10.1155/2016/7686045.

34. Beomsu Shin et al., "Outcomes of Bronchial Artery Embolization for Life-Threatening Hemoptysis in Patients with Chronic Pulmonary Aspergillosis," *PLoS ONE* 11, no. 12 (2016), doi:10.1371/journal.pone.0168373.

35. Purvi Vyas et al., "Clinical Evaluation of Rasayana Compound as an Adjuvant in the Management of Tuberculosis with Anti-Koch's Treatment," *An International Quarterly Journal of Research in Ayurveda* 33, no. 1 (2012): 38–43, doi:10.4103/0974-8520.100307.

36. Eibhlín McCarthy and Jim M. O'Mahony, "What's in a Name? Can Mullein Weed Beat TB Where Modern Drugs Are Failing?," *Evidence Based Complementary and Alternative Medicine* no. 239237 (2011), http://dx.doi.org/10.1155 /2011/239237.

37. Meir Mei-Zahav et al., "Topical Propranolol Improves Epistaxis in Patients with Hereditary Hemorrhagic Telangiectasia—A Preliminary Report," *Journal of Otolaryngology—Head and Neck Surgery* 46 (2017): 58, doi:10.1186/s40463-017-0235-x.

38. Ceren Günel et al., "Inhibitory Effect of Pycnogenol® on Airway Inflammation in Ovalbumin-Induced Allergic Rhinitis," *Balkan Medical Journal* 33, no. 6 (2016): 620–626, doi:10.5152/balkanmedj.2016.150057.

39. Andreas Schapowal, "Randomised Controlled Trial of Butterbur and Cetirizine for Treating Seasonal Allergic Rhinitis," *British Medical Journal* 324, no. 7330 (2002): 144–146, http://www.ncbi.nlm.nih.gov/pubmed /11799030.

40. G. R. Schinella et al., "Anti-inflammatory Effects of South American *Tanacetum vulgare*," *Journal of Pharmacy and Pharmacology* 50, no. 9 (1998):1069–74; Y. Zhao et al., "Sesquiterpene Lactones Inhibit Advanced Oxidation Protein Product-Induced MCP-1 Expression in Podocytes via an IKK/NF-κB-Dependent Mechanism," *Oxidative Medicine and Cellular Longevity* 2015, no. 934058 (2015), doi:10.1155/2015/934058.

41. Jorge Arroyo-Acevedo et al., "Antiallergic Effect of the Atomized Extract of Rhizome of *Curcuma longa*, Flowers of *Cordia lutea* and Leaves of *Annona muricata*," *Therapeutics*

42. Xudong Liu et al., "Dietary Patterns and the Risk of Rhinitis in Primary School Children: A Prospective Cohort Study," *Scientific Reports* 7, (2017): 44610, doi:10.1038/srep44610.

43. Sung-Wook Hong et al., "Beyond Hygiene: Commensal Microbiota and Allergic Diseases," *Immune Network* 17, no. 1 (2017): 48–59, doi:10.4110/in.2017.17.1.48.

44. Michele Miraglia Del Giudice et al., "*Bifidobacterium* Mixture (*B. longum* BB536, *B. infantis* M-63, *B. breve* M-16V) Treatment in Children with Seasonal Allergic Rhinitis and Intermittent Asthma," *Italian Journal of Pediatrics* 43, no. 1 (2017): 25, doi:10.1186/s13052-017-0340-5.

45. Joon-Ho Hwang et al., "Effects of Chung-Pae Inhalation Therapy on a Mouse Model of Chronic Obstructive Pulmonary Disease," *Evidence Based Complementary and Alternative Medicine* 2015, no. 461295 (2015), doi:10.1155/2015/461295.

46. Andreas Schapowal et al., "Treating Intermittent Allergic Rhinitis: A Prospective, Randomized, Placebo and Antihistamine-controlled Study of Butterbur Extract Ze 339," *Phytotherapy Research* 19, no. 6 (2005): 530–537, doi:10.1002/ptr.1705.

47. D. K. Lee et al., "A Placebo-Controlled Evaluation of Butterbur and Fexofenadine on Objective and Subjective Outcomes in Perennial Allergic Rhinitis," *Clinical Experimental Allergy* 34, no. 4 (2004): 646–49, doi:10.1111/j.1365-2222.2004.1903.x.

48. Andreas Schapowa et al., "Butterbur Ze339 for the Treatment of Intermittent Allergic Rhinitis," *Archives of Otolaryngology—Head & Neck Surgery* 130, no. 12 (2004): 1381, doi:10.1001/archotol.130.12.1381.

49. Ulrich Danesch, "*Petasites hybridus* (Butterbur root) Extract in the Treatment of Asthma—An Open Trial," *Alternative Medicine Review* 9, no. 1 (2004): 54–62.

50. Ruoling Guo et al., "Herbal Medicines for the Treatment of Allergic Rhinitis: A Systematic Review," *Annals of Allergy, Asthma & Immunology* 99, no. 6 (2007): 483–95, doi:10.1016/s1081-1206(10)60375-4.

51. Min-Hee Kim et al., "Efficacy and Safety of So-Cheong-Ryong-Tang in Treatment of Perennial Allergic Rhinitis: Study Protocol for a Double-Blind, Randomised, Parallel-Group, Multicentre Trial," *BMJ Open* 7, no. 9 (2017): e016556, doi:10.1136/bmjopen-2017-016556.

52. Rajesh Kumar Ganjhu et al., "Herbal Plants and Plant Preparations as Remedial Approach for Viral Diseases," *Virusdisease* 26, no. 4 (2015): 225–236, doi:10.1007 /s13337-015-0276-6.

53. Rajesh Kumar Ganjhu et al., "Herbal Plants and Plant Preparations," 225–236.

54. Rajesh Kumar Ganjhu et al., "Herbal Plants and Plant Preparations," 225–236.

55. Rajesh Kumar Ganjhu et al., "Herbal Plants and Plant Preparations," 225–236.

56. Rajesh Kumar Ganjhu et al., "Herbal Plants and Plant Preparations," 225–236.

57. Wenjiao Wu et al., "Quercitin as an Antiviral Agent Inhibits Influenza AVirus (IAV) Entry," *Viruses* 8, no. 1 (2016): 6, doi:10.3390/v8010006.

58. Hoi-Hin Kwok et al., "Anti-Inflammatory Effects of Indirubin Derivatives on Influenza A Virus-Infected Human Pulmonary Microvascular Endothelial Cells," *Scientific Reports* 6, no. 18941 (2016), doi:10.1038/srep18941.

59. Rajesh Kumar Ganjhu et al., "Herbal Plants and Plant Preparations as Remedial Approach for Viral Diseases," *Virusdisease* 26, no. 4 (2015): 225–236, doi:10.1007/s13337 -015-0276-6.

60. Rajesh Kumar Ganjhu et al., "Herbal Plants and Plant Preparations," 225–236.

61. Evelin Tiralongo et al., "Elderberry Supplementation Reduces Cold Duration and Symptoms in Air-Travelers: A Randomized, Double-Blind Placebo-Controlled Clinical Trial," *Nutrients* 8, no. 4 (2016): 182, doi:10.3390/nu8040182.

62. Mallappa Kumara Swamy et al., "Antimicrobial Properties of Plant Essential Oils Against Human Pathogens and Their Mode of Action: An Updated Review," *Evidence Based Complementary and Alternative Medicine* 2016, no. 3012462 (2016), http://dx.doi.org/10.1155/2016/3012462.

63. Madona Khoury et al., "Report on the Medicinal Use of Eleven Lamiaceae Species in Lebanon and Rationalization of Their Antimicrobial Potential by Examination of the Chemical Composition and Antimicrobial Activity of Their Essential Oils," *Evidence Based Complementary and Alternative Medicine* 2016, no. 2547169 (2016), http://dx.doi.org/10.1155/2016/2547169.

64. Rajesh Kumar Ganjhu et al., "Herbal Plants and Plant Preparations as Remedial Approach for Viral Diseases," *Virusdisease* 26, no. 4 (2015): 225–236, doi:10.1007/s13337 -015-0276-6.

65. Zeinab Mohsenipour and Mehdi Hassanshahian, "The Effects of *Allium sativum* Extracts on Biofilm Formation and Activities of Six Pathogenic Bacteria," *Jundishapur Journal of Microbiology* 8, no. 8 (2015): e18971, doi:10.5812 /jjm.18971v2.

66. Daynea Wallock-Richards et al., "Garlic Revisited: Antimicrobial Activity of Allicin-Containing Garlic Extracts against *Burkholderia cepacia* Complex," *PLoS ONE* 9, no. 12 (2014): e112726, doi:10.1371/journal.pone.0112726.

67. Eric Secor et al., "Oral Bromelain Attenuates Inflammation in an Ovalbumin-Induced Murine Model of Asthma," *Evidence Based Complementary and Alternative Medicine* 5, no. 1 (2008): 61–69, doi:10.1093/ecam/nel110.

68. Elizabeth Mazzio et al., "Natural Product HTP Screening for Antibacterial (*E. coli* O157:H7) and Anti-Inflammatory Agents in (LPS from *E. coli* O111:B4) Activated Macrophages and Microglial Cells; Focus on Sepsis," *BMC Complementary and Alternative Medicine* 16, no. 1 (2016): 467, https://doi.org/10.1186/s12906-016-1429-x.

69. Miquéias Lopes-Pacheco, "CFTR Modulators: Shedding Light on Precision Medicine for Cystic Fibrosis," *Frontiers in Pharmacology* 7, (2016): 275, doi:10.3389/fphar .2016.00275.

70. Yaofang Zhang et al., "Identification of Resveratrol Oligomers as Inhibitors of Cystic Fibrosis Transmembrane Conductance Regulator by High-Throughput Screening of Natural Products from Chinese Medicinal Plants," *PLoS ONE* 9, no. 4 (2014): e94302, doi:10.1371/journal .pone.0094302.

71. Lei Chen et al., "Bioactivity-Guided Fractionation of an Antidiarrheal Chinese Herb *Rhodiola kirilowii* (Regel) Maxim Reveals (-)–Epicatechin-3-Gallate and (-)– Epigallocatechin-3-Gallate as Inhibitors of Cystic Fibrosis Transmembrane Conductance Regulator," *PLoS ONE* 10, no. 3 (2015): e0119122, doi:10.1371/journal.pone.0119122.

72. L. Tradtrantip et al., "Crofelemer, an Antisecretory Antidiarrheal Proanthocyanidin Oligomer Extracted from *Croton lechleri*, Targets Two Distinct Intestinal Chloride Channels," *Molecular Pharmacology* 77, no. 1 (2010):69–78, doi:10.1124/mol.109.061051.

73. Lei Chen et al., "Bioactivity-Guided Fractionation of an Antidiarrheal Chinese Herb *Rhodiola kirilowii* (Regel) Maxim Reveals (-)–Epicatechin-3-Gallate and (-)– Epigallocatechin-3-Gallate as Inhibitors of Cystic Fibrosis Transmembrane Conductance Regulator," *PLoS ONE* 10, no. 3 (2015): e0119122, doi:10.1371/journal.pone.0119122.

74. Camille de Seynes et al., "Identification of a Novel Alpha1-Antitrypsin Variant," *Respiratory Medicine Case Reports* 20, (2017): 64–67, doi:10.1016/j.rmcr.2016.11.008.

75. Nazanin Amini et al., "Effect of Nebulized Eucalyptus on Contamination of Microbial Plaque of Endotracheal Tube in Ventilated Patients," *Iranian Journal of Nursing and Midwifery Research* 21, no. 2 (2016): 165–170, doi:10.4103/1735-9066.178242.

76. Maria Paparoupa and Adrian Gillissen, "Is Myrtol® Standardized a New Alternative Toward Antibiotics?," *Pharmacognosy Review* 10, no. 20 (2016): 143–146, doi:10.4103/0973-7847.194045.

77. Nidal Amin Jaradat et al., "The Effect of Inhalation of *Citrus sinensis* Flowers and *Mentha spicata* Leave Essential Oils on Lung Function and Exercise Performance: A Quasi-Experimental Uncontrolled Before-and-After Study," *Journal of the International Society of Sports Nutrition* 13, (2016): 36, doi:10.1186/s12970-016-0146-7.

78. Nidal Amin Jaradat et al., "Effect of Inhalation," 36.

79. Alexandre de Paula Rogerio et al., "Bioactive Natural Molecules and Traditional Herbal Medicine in the Treatment of Airways Diseases," *Evidence Based Complementary Alternative Medicine* no. 9872302 (2016): doi.org/10.1155/2016/9872302.

80. Carlos Nunes et al., "Asthma Costs and Social Impact," *Asthma Research and Practice* 3, no. 1 (2017), doi:10.1186 /s40733-016-0029-3.

81. Thomas Gensollen et al., "How Colonization by Microbiota in Early Life Shapes the Immune System," *Science* 352, no. 6285 (2016): 539–544, doi:10.1126/science.aad9378.

82. Gregory Lanza et al., "Anti-angiogenic Nanotherapy Inhibits Airway Remodeling and Hyper-responsiveness of Dust Mite Triggered Asthma in the Brown Norway Rat," *Theranostics* 7, no. 2 (2017): 377–89, doi:10.7150/thno.16627.

83. Riccardo Sibilano et al., "A TNFRSF14-FcεRI-Mast Cell Pathway Contributes to Development of Multiple Features of Asthma Pathology in Mice," *Nature Communications* 7, (2016): 13696, doi:10.1038/ncomms13696.

84. Ravindra G. Mali and Avinash S. Dhake, "A Review on Herbal Antiasthmatics," *Oriental Pharmacy and Experimental Medicine* 11, no. 2 (2011): 77–90, doi:10.1007/s13596-011-0019-1.

85. Pei-Chia Lo et al., "Risk of Asthma Exacerbation Associated with Nonsteroidal Anti-inflammatory Drugs in Childhood Asthma," *Medicine* 95, no. 41 (2016), doi:10.1097/md.0000000000005109.

86. Ervice Pouokam and Mike Althaus, "Epithelial Electrolyte Transport Physiology and the Gasotransmitter Hydrogen Sulfide," *Oxidative Medicine and Cellular Longevity* 2016, (2016): 1–13, doi:10.1155/2016/4723416.

87. Robert H. Brown et al., "Sulforaphane Improves the Bronchoprotective Response in Asthmatics Through Nrf2-mediated Gene Pathways," *Respiratory Research* 16, no. 1 (2015), doi:10.1186/s12931-015-0253-z.

88. D. F. Finn et al., "Twenty-first Century Mast Cell Stabilizers," *British Journal of Pharmacology* 170, no. 1 (2013): 23–37, doi:10.1111/bph.12138; Ravindra G. Mali and Avinash S. Dhake et al., "A Review on Herbal Antiasthmatics," *Oriental Pharmacy and Experimental Medicine* 11, no. 2 (2011): 77–90, doi:10.1007/s13596-011-0019-1.

89. Akshaya Srikanth Bhagavathula et al., "*Ammi visnaga* in Treatment of Urolithiasis and Hypertriglyceridemia," *Pharmacognosy Research* 7, no. 4 (2015): 397, doi:10.4103/0974-8490.167894.

90. Sayeed Ahmad et al., "Development and Validation of High-Performance Liquid Chromatography and High-Performance Thin-Layer Chromatography Methods for the Quantification of Khellin in *Ammi visnaga* Seed," *Journal of Pharmacy and Bioallied Sciences* 7, no. 4 (2015): 308, doi:10.4103/0975-7406.168033.

91. Yuan Ji et al., "Class III Antiarrhythmic Drugs Amiodarone and Dronedarone Impair KIR2.1 Backward Trafficking," *Journal of Cellular and Molecular Medicine* 21, no. 10 (2017): 2514–2523, doi:10.1111/jcmm.13172.

92. John Wong et al., "Lung Inflammation Caused by Inhaled Toxicants: A Review," *International Journal of Chronic Obstructive Pulmonary Disease* 11 (2016): 1391, doi:10.2147/copd.s106009.

93. Stefen Boehma et al., "MAP3K19 Is Overexpressed in COPD and Is a Central Mediator of Cigarette Smoke-Induced Pulmonary Inflammation and Lower Airway Destruction," *PLoS ONE* 11, no. 12 (2016), doi:10.1371/journal.pone.0167169.

94. Yves Lacasse et al., "Multi-center, Randomized, Placebo-Controlled Trial of Nocturnal Oxygen Therapy in Chronic Obstructive Pulmonary Disease: A Study Protocol for the INOX Trial," *BMC Pulmonary Medicine* 17, no. 1 (2017), doi:10.1186/s12890-016-0343-9.

95. Carlos Machahua et al., "Increased AGE-RAGE Ratio in Idiopathic Pulmonary Fibrosis," *Respiratory Research* 17, no. 1 (2016), doi:10.1186/s12931-016-0460-2.

96. Ailing Liu et al., "The Inhibitory Mechanism of *Cordyceps sinensis* on Cigarette Smoke Extract-Induced Senescence in Human Bronchial Epithelial Cells," *International Journal of Chronic Obstructive Pulmonary Disease* 11, (2016): 1721–1731, doi:10.2147/copd.s107396.

97. Mouna Moutia et al., "*Allium sativum* L. Regulates in Vitro IL-17 Gene Expression in Human Peripheral Blood Mononuclear Cells," *BMC Complementary and Alternative Medicine* 16, no. 1 (2016), doi:10.1186/s12906-016-1365-9.

98. S. P. Guan et al., "Andrographolide Protects Against Cigarette Smoke-Induced Oxidative Lung Injury via Augmentation of Nrf2 Activity," *British Journal of Pharmacology* 168, no. 7 (2013): 1707–1718, doi:10.1111/bph.12054.

99. Fernanda Paula R. Santana et al., "Evidences of Herbal Medicine-Derived Natural Products Effects in Inflammatory Lung Diseases," *Mediators of Inflammation* 2016 (2016): 1–14, doi:10.1155/2016/2348968.

100. Irsa Tahir et al., "Evaluation of Phytochemicals, Antioxidant Activity and Amelioration of Pulmonary Fibrosis with *Phyllanthus emblica* Leaves," *BMC Complementary and Alternative Medicine* 16, no. 1 (2016), doi:10.1186/s12906-016-1387-3.

101. Mohammad Hossein Asghari et al., "Hydro-alcoholic Extract of *Raphanus sativus* L. var *niger* Attenuates Bleomycin-Induced Pulmonary Fibrosis via Decreasing Transforming Growth Factor β1 Level," *Research in Pharmaceutical Sciences* 10, no. 5 (2015): 429–435.

102. Jing Luo et al., "Baicalein Attenuates the Quorum Sensing-Controlled Virulence Factors of *Pseudomonas aeruginosa* and Relieves the Inflammatory Response in *P. aeruginosa*-Infected Macrophages by Downregulating the MAPK and NFκB Signal-transduction Pathways," *Drug Design, Development and Therapy* 10, (2016): 183–203, doi:10.2147/dddt.s97221.

103. Jae-Min Shin et al., "Baicalin Down-Regulates IL-1β-Stimulated Extracellular Matrix Production in Nasal Fibroblasts," *PLoS ONE* 11, no. 12 (2016), doi:10.1371/journal.pone.0168195.

104. Jae-Min Shin et al., "Baicalin Down-Regulates IL-1β-Stimulated Extracellular Matrix Production in Nasal Fibroblasts," *PLoS ONE* 11, no. 12 (2016), doi:10.1371/journal.pone.0168195.

105. Zhong-Shan Yang et al., "Anti-inflammatory Effect of Yu-Ping-Feng-San via TGF-β1 Signaling Suppression in Rat Model of COPD," *Iranian Journal of Basic Medical Sciences* 19, no. 9 (2016): 993–1002.

106. Jun Keng Khoo et al., "Bronchiectasis in the Last Five Years: New Developments," *Journal of Clinical Medicine* 5, no. 12 (2016): 115, doi:10.3390/jcm5120115.

107. Nidal Jaradat et al., "Chemical Composition, Anthelmintic, Antibacterial and Antioxidant Effects of *Thymus bovei* Essential Oil," *BMC Complementary and Alternative Medicine* 16, no. 1 (2016), doi:10.1186/s12906-016-1408-2.

108. Bart C. Moulton and Allison D. Fryer, "Muscarinic Receptor Antagonists, From Folklore to Pharmacology; Finding Drugs That Actually Work in Asthma and COPD," *British Journal of Pharmacology* 163, no. 1 (2011): 44–52, doi:10.1111/j.1476-5381.2010.01190.x.

109. Alcibey Alvarado-Gonzaleza and Isabel Arceb, "Tiotropium Bromide in Chronic Obstructive Pulmonary Disease and Bronchial Asthma," *Journal of Clinical Medical Research* 7, no. 11 (2015): 831–839, doi:10.14740/jocmr2305w.

110. Dejan Radovanovic et al., "The Evidence on Tiotropium Bromide in Asthma: From the Rationale to the Bedside," *Multidisciplinary Respiratory Medicine* 2017, no. 12 (2017), doi:10.1186/s40248-017-0094-3.

111. Miao Liu et al., "Xuan Bai Cheng Qi Formula as an Adjuvant Treatment of Acute Exacerbation of Chronic Obstructive Pulmonary Disease of the Syndrome Type Phlegm-Heat Obstructing the Lungs: A Multicenter, Randomized, Double-Blind, Placebo-Controlled Clinical Trial," *BMC Complementary and Alternative Medicine* 14, no. 1 (2014), doi:10.1186/1472-6882-14-239.

112. N. Chen, "Therapy for Clearing Heat and Resolving Phlegm in Treatment of Systemic Inflammatory Response Syndrome in Acute Deterioration Stage of Chronic Obstructive Pulmonary Disease: A Randomized Controlled Trial," *Journal of Chinese Integrative Medicine* 7, no. 2 (2009): 105–109, doi:10.3736/jcim20090202.

113. Min Xiao et al., "Emodin Ameliorates LPS-Induced Acute Lung Injury, Involving the Inactivation of NFκB in Mice," *International Journal of Molecular Sciences* 15, no. 11 (2014): 19355–19368, doi:10.3390/ijms151119355.

114. Zhengrong Mao et al., "Effects of Xuanbai Chengqi Decoction on Lung Compliance for Patients with Exogenous Pulmonary Acute Respiratory Distress Syndrome," *Drug Design, Development and Therapy* 10, (2016): 793–798, doi:10.2147/dddt.s93165.

115. Ying Huang et al., "Picroside II Protects Against Sepsis via Suppressing Inflammation in Mice," *American Journal of Translational Research* 8, no. 12 (2016): 5519–5531.

116. Jin Choi et al., "Picroside II Attenuates Airway Inflammation by Downregulating the Transcription Factor GATA3 and Th2-Related Cytokines in a Mouse Model of HDM-Induced Allergic Asthma," *PLoS ONE* 12, no. 1 (2017), doi:10.1371/journal.pone.0170832.

117. Soohwan Noh et al., "Neutrophilic Lung Inflammation Suppressed by Picroside II Is Associated with TGF-βSignaling," *Evidence-Based Complementary and Alternative Medicine* 2015, (2015): 1–11, doi:10.1155/2015/897272.

118. Ming-Wei Liu et al., "Effect of *Melilotus suaveolens* Extract on Pulmonary Microvascular Permeability by Downregulating Vascular Endothelial Growth Factor Expression in Rats with Sepsis," *Molecular Medicine Reports* 11, no. 5 (2015): 3308–3316, doi:10.3892/mmr.2015.3146; Ming-Wei Liu et al., "Effect of *Melilotus* Extract on Lung Injury via the Upregulation of Tumor Necrosis Factor-α-induced Protein-8-like 2 in Septic Mice," *Molecular Medicine Reports* 11, no. 3 (2014): 1675–1684, doi:10.3892/mmr.2014.2965.

119. Won Choi, "Novel Pharmacological Activity of Artesunate and Artemisinin: Their Potential as Anti-Tubercular Agents," *Journal of Clinical Medicine* 6, no. 3 (2017): 30, doi:10.3390/jcm6030030.

120. A. Y. Gordien et al., "Antimycobacterial Terpenoids from *Juniperus communis* L. (*Cuppressaceae*)," *Journal of Ethnopharmacology* 126, no. 3 (2009): 500–505, doi:10.1016/j.jep.2009.09.007.

121. Janmejaya Samal, "Ayurvedic Management of Pulmonary Tuberculosis (PTB): A Systematic Review," *Journal of Intercultural Ethnopharmacology* 5, no. 1 (2016): 86, doi:10.5455/jice.20151107020621.

122. Grace Gar-Lee Yeu et al., "New Potential Beneficial Effects of Actein, a Triterpene Glycoside Isolated from *Cimicifuga* species, in Breast Cancer Treatment," *Scientific Reports* 6, no. 1 (2016), doi:10.1038/srep35263.

123. Phuong Quoc Thuc Nguyen et al., "Antiviral Cysteine Knot α-Amylase Inhibitors from *Alstonia scholaris*," *Journal of Biological Chemistry* 290, no. 52 (2015): 31138–31150, doi:10.1074/jbc.m115.654855.

124. S. Akbar, "*Andrographis paniculata*: A Review of Pharmacological Activities and Clinical Effects," *Alternative Medicine Review* 16, no. 1 (2011): 66–77.

125. Sara S. Al Disi et al., "Anti-hypertensive Herbs and Their Mechanisms of Action: Part I," *Frontiers in Pharmacology* 6, (2016): doi:10.3389/fphar.2015.00323.

126. Corinna Herz et al., "Evaluation of an Aqueous Extract from Horseradish Root (*Armoracia rusticana* Radix) Against Lipopolysaccharide-Induced Cellular Inflammation Reaction," *Evidence-Based Complementary and Alternative Medicine* 2017, (2017): 1–10, doi:10.1155/2017/1950692.

127. Madhurima Sarkar et al., "Neem Leaf Glycoprotein Prevents Post-surgical Sarcoma Recurrence in Swiss Mice by Differentially Regulating Cytotoxic T and Myeloid-Derived Suppressor Cells," *PLoS ONE* 12, no. 4 (2017): doi:10.1371/journal.pone.0175540.

128. Rajesh Kumar Ganjhu et al., "Herbal Plants and Plant Preparations as Remedial Approach for Viral Diseases," *VirusDisease* 26, no. 4 (2015): 225–236, doi:10.1007/s13337-015-0276-6.

129. Yan Lui et al., "Regulating Autonomic Nervous System Homeostasis Improves Pulmonary Function in Rabbits with Acute Lung Injury," *BMC Pulmonary Medicine* 17, no. 1 (2017), doi:10.1186/s12890-017-0436-0.

130. Veronique Beng et al., "Multidrug Resistance Bacteria Are Sensitive to *Euphorbia prostrata* and Six Others Cameroonian Medicinal Plant Extracts," *BMC Research Notes* 10, (2017): 321, doi:10.1186/s13104-017-2665-y.

131. M. Yoshikawa et al., "Development of Bioactive Functions in *Hydrangeae Dulcis Folium*. V. on the Antiallergic and Antimicrobial Principles of Hydrangeae Dulcis Folium," *Chemical and Pharmaceutical Bulletin* 44, no. 8 (1996): 1440–1447, doi:10.1248/cpb.44.1440.

132. Yufeng Liu et al., "Aromatic Compounds from an Aqueous Extract of Ban lan gen and Their Antiviral Activities." *Acta Pharmaceutica Sinica B* 7, no. 2 (2017): 179–184, doi:10.1016/j.apsb.2016.09.004.

133. D. F. Finn and J. J. Walsh, "Twenty-first Century Mast Cell Stabilizers," *British Journal of Pharmacology* 170, no. 1 (2013): 23–37, doi:10.1111/bph.12138.

134. A. Camporese, "In Vitro Actirity of *Eucalyptus smithii* and *Juniperus communis* Essential Oils against Bacterial Biofilms and Efficacy Perspectives of Complementary Inhalation Therapy in Chronic and Recurrent Upper Respiratory Tract Infections," *Le Infezioni in Medicina* 21, no. 2 (2013): 117–24.

135. M. Khan et al., "Pharmacological Explanation for the Medicinal Use of *Juniperus excelsa* in Hyperactive Gastrointestinal and Respiratory Disorders," *Journal of Natural Medicines* 66, no. 2 (2012): 292–301, doi:10.1007/s11418-011-0605-z.

136. Kwang Il Park et al., "Polyphenolic Compounds from Korean *Lonicera japonica* Thunb. Induces Apoptosis via AKT and Caspase Cascade Activation in A549 Cells," *Oncology Letters* (2017): 2521–2530, doi:10.3892/ol.2017.5771.

137. Y. Wang et al., "Flavone C-glycosides from the Leaves of *Lophatherum gracile* and Their in Vitro Antiviral Activity," *Planta Medica* 78, no. 1 (2012): 46–51, doi:10.1055/s-0031-1280128.

138. A. Ranjbar et al., "Ameliorative Effect of *Matricaria chamomilla* on Paraquat: Induced Oxidative Damage in Lung Rats," *Pharmacognosy Research* 6, no. 3 (2014): 199–203, doi:10.4103/0974-8490.132595.

139. Ming-Wei Liu et al., "Effect of *Melilotus suaveolens* Extract on Pulmonary Microvascular Permeability by Downregulating Vascular Endothelial Growth Factor Expression in Rats with Sepsis," *Molecular Medicine Reports* 11, no. 5 (2015): 3308–3316, doi:10.3892/mmr.2015.3146.

140. F. Fulanetti et al., "Toxic Effects of the Administration of *Mikania glomerata* Sprengel During the Gestational Period of Hypertensive Rats," *Open Veterinary Journal* 6, no. 1 (2016): 23–29, doi:10.4314/ovj.v6i1.4.

141. Masaaki Minami et al., "Shin'iseihaito (Xinyiqingfeitang) Suppresses the Biofilm Formation of *Streptococcus pneumoniae* In Vitro," *BioMed Research International* (2017): 1–8, doi:10.1155/2017/4575709.

142. M. Shokrzadeh et al., "An Ethanol Extract of *Origanum vulgare* Attenuates Cyclophosphamide-Induced Pulmonary Injury and Oxidative Lung Damage in Mice," *Pharmaceutical Biology* 52, no. 10 (2014): 1229–1236, doi:10.3109/13880209.2013.879908.

143. Jinpei Lin et al., "Paeoniflorin Attenuated Oxidative Stress in Rat COPD Model Induced by Cigarette Smoke," *Evidence-Based Complementary and Alternative Medicine* 2016, (2016): 1–9, doi:10.1155/2016/1698379.

144. Toshiaki Makino et al., "Anti-allergic Effect of Perilla *frutescens* and Its Active Constituents," *Phytotherapy Research* 17, no. 3 (2003): 240–243, doi:10.1002/ptr.1115.

145. Ahmed Allam et al., "Protective Effect of Parsley Juice (*Petroselinum crispum*, Apiaceae) Against Cadmium Deleterious Changes in the Developed Albino Mice Newborns (Mus musculus) Brain," *Oxidative Medicine and Cellular Longevity* 2016, (2016): 1–15, doi:10.1155/2016/2646840.

146. Ling Chen et al., "Apigenin Protects Against Bleomycin-Induced Lung Fibrosis in Rats," *Experimental and Therapeutic Medicine* 11, no. 1 (2015): 230–234, doi:10.3892/etm.2015.2885.

147. Vikas Kumar et al., "*Amalaki rasayana*, a Traditional Indian Drug Enhances Cardiac Mitochondrial and Contractile Functions and Improves Cardiac Function in Rats with Hypertrophy," *Scientific Reports* 7, no. 1 (2017): doi:10.1038/s41598-017-09225-x.

148. Y. J. Kim et al., "Anti-obesity Effect of *Pinellia ternata* Extract in Zucker Rats," *Biological & Pharmaceutical Bulletin* 29, no. 6 (2006): 1278–1281, doi:10.1248/bpb.29.1278.

149. Jia Liu et al., "Piperine Induces Autophagy by Enhancing Protein Phosphotase 2A Activity in a Rotenone-Induced Parkinson's Disease Model," *Oncotarget* 7, no. 38 (2016): 60823–60843, doi:10.18632/oncotarget.11661.

150. P. Madugula et al., "Rhetoric to Reality- Efficacy of *Punica granatum* Peel Extract on Oral Candidiasis: An In Vitro Study," *Journal of Clinical Diagnosis Research* 11, no. 1 (2017): ZC114–ZC117, doi:10.7860/JCDR/2017/22810.9304.

151. Sushil Kumar Middha et al., "A Review on Antihyperglycemic and Antihepatoprotective Activity of Eco-Friendly *Punica granatum* Peel Waste," *Evidence-Based Complementary and Alternative Medicine* 2013, (2013): 1–10, doi:10.1155/2013/656172.

152. Ashish Kumar et al., "UPLC/MS/MS Method for Quantification and Cytotoxic Activity of Sesquiterpene Lactones Isolated from *Saussurea lappa*," *Journal of Ethnopharmacology* 155, no. 2 (2014): 1393–1397, doi:10.1016/j.jep.2014.07.037.

153. Hye-Sun Lim et al., "The Genome-Wide Expression Profile of *Saussurea lappa* Extract on House Dust Mite-Induced Atopic Dermatitis in Nc/Nga Mice," *Molecules and Cells* 38, no. 9 (2015): 765–772, doi:10.14348/molcells.2015.0062.

154. Tzu-Chieh Hung et al., "The Inhibition of Folylpolyglutamate Synthetase (folC) in the Prevention of Drug Resistance in Mycobacterium tuberculosis by Traditional Chinese Medicine," *BioMed Research International* 2014, (2014): 1–14, doi:10.1155/2014/635152.

155. Kanika Patel and Dinesh Kumar Patel, "Medicinal Importance, Pharmacological Activities, and Analytical Aspects of Hispidulin: A Concise Report," *Journal of Traditional and Complementary Medicine* 7, no. 3 (2017): 360–366, doi:10.1016/j.jtcme.2016.11.003.

156. Qing Zhao et al., "*Scutellaria baicalensis*, the Golden Herb from the Garden of Chinese Medicinal Plants," *Science Bulletin* 61, no. 18 (2016): 1391–1398, doi:10.1007/s11434-016-1136-5.

157. Jinbo Zhang et al., "Therapuetic Effects of Stemonine on Particulate Matter 2.5-Induced Chronic Obstructive Pulmonary Disease in Mice," *Experimental and Therapeutic Medicine* 14, no. 5 (2017): 4453–4459, doi:10.3892/etm/2017.5092.

158. In-Sung Song et al., "Chebulinic Acid Inhibits Smooth Muscle Cell Migration by Suppressing PDGF-Rβ phosphorylation and Inhibiting Matrix Metalloproteinase-2 Expression," *Scientific Reports* 7, no. 1 (2017): doi:10.1038/s41598-017-12221-w.

— INDEX —

Note: Page numbers in *italics* refer to figures and illustrations. Page numbers followed by *t* refer to tables.

A

abortifacient, definition of, 202
absorbent, definition of, 202
accentuated breath sounds, 118
Achillea millefolium (yarrow)
  for COPD, 148
  formulas containing
    bronchitis, 133
    coughs, 119
    hemoptysis, 126
    peripheral vascular insufficiency, 61
    Raynaud's syndrome, 67
    stasis ulcers, 60
    venous congestion, 76
  for poor circulation, 58, 63
  specific indications, 89, 156
  for vascular support, 35
acidifier, definition of, 202
acne sample case, 18, *18*
aconite. *See Aconitum napellus* (aconite)
Aconitum carmichaelii (fu zhi), 89
Aconitum napellus (aconite)
  in calming formulas, 20
  formulas containing
    angina, 38
    coughs, 123
    dyspnea, 125
  safety concerns, 20
  specific indications, 89, 156
Actaea cimicifuga (shengma)
  specific indications, 157
  for viral infections, 131
Actaea racemosa (black cohosh)
  formulas containing
    coughs, 123
    dyspnea, 125
  for hyperventilation, 140
  for restless insomnia, 16
  specific indications, 157
actions of herbs, 13–16
acute bronchitis, 130
  See also bronchitis
acute conditions
  definition of, 202
  guidelines for, 20–21, 23
acute respiratory distress syndrome (ARDS)
  Borago officinalis for, 91
  Dracocephalum rupestre for, 163
  Matricaria chamomilla for, 171

  overview, 153–54
  Paeonia lactiflora for, 174
  Persicaria tinctoria for, 175
adhatoda (*Justicia adhatoda*), 169
Adonis vernalis (pheasant's eye)
  for angina, 38
  specific indications, 89
aerial parts, definition of, 202
Aesculus hippocastanum (horse chestnut), *79*
  formulas containing
    cardiopulmonary disease, 85
    hyperlipidemia, 37
    telangiectasias, 66
    varicosities, 74, 75
    venous congestion, 76
  for hypertension, 44
  specific indications, 89
  for varicosities, 74
African geranium (*Pelargonium sidoides*), 131, 175
aging, *Polygonum multiflorum* for, 102
airway inflammation
  Ammi visnaga for, 146
  Eucalyptus globulus for, 112
  herbs for, 144*t*, 149*t*
  Inula helenium for, 168
  Neopicrorhiza scrophulariiflora for, 154
  Ocimum sanctum for, 173
  Ophiopogon japonicus for, 174
  Picrorhiza kurroa for, 176
  Rheum palmatum for, 178
  rhonchi from, 118, 120
  Scutellaria baicalensis for, 150
  stimulating expectorants for, 131
  Syzygium aromaticum for, 181
  Tylophora asthmatica for, 183
  See also allergic rhinosinusitis; asthma;
    bronchitis; chronic obstructive pulmonary
    disorder (COPD); emphysema
Albizia julibrissin (silk tree), 157
Albizia lebbeck (siris)
  bronchodilating properties, 108*t*
  as mast cell stabilizer, 144*t*
  specific indications, 157
alfalfa. *See Medicago sativa* (alfalfa)
Alisma orientale. See Alisma plantago-aquatica
  (Asian Water Plantain)
Alisma plantago-aquatica (Asian water plantain)
  specific indications, 90
  for strokes, 65

alkalinizer, definition of, 202
allergic rhinosinusitis
  Achillea millefolium for, 156
  Allium cepa for, 157
  Ammi visnaga for, 157
  Angelica dahurica for, 158
  Armoracia rusticana for, 159
  Cnidium monnieri for, 161
  Crataegus spp. for, 162
  Curcuma longa for, 162
  Eclipta prostrata for, 163
  Ephedra sinica for, 163–64
  Euphrasia officinalis for, 165
  Foeniculum vulgare for, 166
  formulas for, 127–130
  Inula helenium for, 168
  Juniperus communis for, 169
  Ligusticum striatum for, 169
  Linum usitatissimum for, 169
  Magnolia officinalis for, 171
  nebulized essential oils for, 137
  Oenothera biennis for, 173
  Ophiopogon japonicus for, 174
  overview, 127–28
  Pelargonium sidoides for, 175
  Picrorhiza kurroa for, 176
  Pinus pinaster for, 176
  Rosmarinus officinalis for, 179
  Scutellaria baicalensis for, 180
  Tanacetum parthenium for, 182
  Terminalia chebula for, 182
  Thymus vulgaris for, 182
  Tinospora cordifolia for, 182
allergies
  Alstonia scholaris for, 157
  Angelica dahurica for, 158
  Angelica sinensis for, 158
  Azadirachta indica for, 160
  Bacopa monnieri for, 160
  Camellia sinensis for, 160
  Coleus forskohlii for, 177
  Euphorbia spp. for, 165
  guidelines for alleviating, 128
  Hippophae rhamnoides for, 96
  Ligusticum striatum for, 169
  Linum usitatissimum for, 169
  Perilla frutescens for, 175
  Petasites hybridus for, 175
  Petroselinum crispum for, 175

allergies (*continued*)
  Picrorhiza kurroa for, 176
  Pinus pinaster for, 176
  Rosmarinus officinalis for, 179
  Scutellaria baicalensis for, 103, 180
  Tanacetum parthenium for, 182
  Vitis vinifera for, 184
  Zingiber officinale for, 184
Allium cepa (onion)
  for altered breath sounds, 118
  for atherosclerosis, 32
  fibrinolytic properties, 73
  formulas containing, 117
  for hypertension, 41
  as mast cell stabilizer, 144t
  PAF inhibiting properties, 77
  for respiratory infections, 130
  specific indications, 157
  sulfur donating qualities, 143
  for tuberculosis, 155
  for varicosities, 81
  for vascular support, 31, 34
Allium sativum (garlic), *34*
  for altered breath sounds, 118
  for arrhythmias, 55, 56
  for atherosclerosis, 32
  for cerebral vascular insufficiency, 62
  for COPD, 148, 149t
  for cor pulmonale, 155
  fibrinolytic properties, 73
  formulas containing
    arrhythmias, 53
    asthma, 147
    bronchitis, 132, 133
    cardiopulmonary disease, 85
    COPD, 150, 151
    cystic fibrosis, 135
    endocarditis, 72
    hyperlipidemia, 35, 36, 37, 47
    hypertension, 42, 44, 45, 46
    peripheral vascular insufficiency, 58
    phlebitis, 81
    pleurisy, 139
    pneumonia, 141, 142
    upper respiratory infections, 119
    venous congestion, 76
  for hyperlipidemia, 34, 35, 36, 37, 47
  for hypertension, 41, 44, 46
  for opportunistic lung infections, 134
  PAF inhibiting properties, 77
  for pleurisy, 138
  for poor circulation, 58
  preparation of cloves, 33
  for respiratory infections, 121, 130
  specific indications, 90, 157
  sulfur donating qualities, 143
  for tuberculosis, 155
  for varicosities, 81
  for vascular infections, 72
  for vascular support, 31, 34

allopathic, definition of, 202
all-purpose vascular support, 27–31
Aloe vera (aloe), 83
Alpinia galanga (Thai galangal), 156
Alstonia scholaris (dita wood), 108t, 157
alteratives
  cooling remedies, 21t
  definition of, 202
  for hypertension and hyperlipidemia, 35
  for poor circulation, 63
  role in treating respiratory conditions, 10
altered breathing, 114–123
Althaea officinalis (marshmallow), 157
amala. See *Phyllanthus emblica*
  (Indian gooseberry)
amla. See *Phyllanthus emblica*
  (Indian gooseberry)
Ammi visnaga (khella), *146*
  for airway inflammation, 146
  for altered breath sounds, 118
  for angina, 38
  formulas containing
    asthma, 145, 146, 147
    cardiopulmonary disease, 85
    coughs, 119
    dyspnea, 125
  for hyperlipidemia, 34
  for hypertension, 46
  as mast cell stabilizer, 144t
  specific indications, 90, 157–58
  for vascular support, 29
amrita. See *Tinospora cordifolia* (guduchi)
analgesic, definition of, 202
Ananas comosus (pineapple)
  fibrinolytic properties, 73
  specific indications, 158
  for varicosities, 81
  for vascular support, 31
anaphrodisiac, definition of, 202
Andrographis paniculata (king of bitters)
  for COPD, 149t
  formulas containing
    COPD, 150, 151
    cor pulmonale, 155
  for hypertension, 42
  immune modulating properties, 23
  specific indications, 158
  for vascular infections, 72
anemia
  Arctium lappa for, 90
  Camellia sinensis for, 91
  dyspnea with, 124
  formulas for, 68–70
  Mentha pulegium for, 99
  overview, 68
  Stellaria media for, 104
  Terminalia arjuna for, 104
Anemopsis californica (yerba mansa), 158
anesthetic, definition of, 202
Anethum graveolens (dill), 29

Angelica spp.
  for atherosclerosis, 32
  for hypertension, 46
  for vascular support, 29
Angelica archangelica (garden angelica), 56, 80
Angelica dahurica (bai zhi), 144t, 158
Angelica sinensis (dong quai), *59*
  for arrhythmias, 55
  for atherosclerosis, 32
  cardiovascular benefits, 59
  for congestive heart failure, 48t
  for cystic fibrosis, 135
  flavonoids in, 33
  formulas containing
    angina, 38, 39
    asthma, 145
    congestive heart failure, 50
    coronary artery disease, 40
    endocarditis, 72
    hyperlipidemia, 37
    hypertension, 44, 45
    peripheral vascular insufficiency, 58, 61
    phlebitis, 81
    Raynaud's syndrome, 67
    varicosities, 75
    vascular support, 27, 29, 30
    vasculitis, 82
    venous congestion, 76
  for hypertension, 42, 44, 45
  PAF inhibiting properties, 77
  for poor circulation, 58
  specific indications, 90, 158
  for stasis ulcers, 60
  for strokes, 65
  for varicosities, 74, 81
  for vascular headaches, 27
  for vascular support, 27, 29, 30
angiitis. See vasculitis
angina
  Ammi visnaga for, 90
  Angelica sinensis for, 90
  Ardisia japonica for, 158
  Arnica montana for, 91
  Asclepius tuberosa for, 159
  Bryonia dioica for, 160
  Cinnamomum camphora for, 161
  Eupatorium perfoliatum for, 165
  formulas for, 38–40
  herbs for, 38
  Iris versicolor for, 168
  Lobelia inflata for, 98
  nitroglycerin patches for, 113
  overview, 38
  Petasites hybridus for, 175
  Pueraria montana var. *lobata* for, 101
  Terminalia arjuna for, 104
  Viscum album for, 106
anhydrotic, definition of, 202
anise. See *Pimpinella anisum* (anise)
an lu (*Artemisia keiskeana*), 144t

anodyne, definition of, 202
antacid, definition of, 202
antagonist, definition of, 202
anthelmintic, definition of, 202
Anthoxanthum odoratum (sweet vernal grass), 79, 90
anticoagulants, herbal, 73, 79, 80
antidote, definition of, 202
antiemetic, definition of, 202
antigalactogogue, definition of, 202
antihemorrhagic, definition of, 202
anti-inflammatory, definition of, 202
   See also inflammation
antilithic, definition of, 202
antimicrobials, 21t
   See also infections
antioxidant, definition of, 202
antiperiodic, definition of, 202
antiphlogistic, definition of, 202
antipyretic, definition of, 202
   See also fever
antiscorbutic, definition of, 202
antiseptic, definition of, 202
antisialagogue, definition of, 202
antispasmodic, definition of, 202
antitussives
   definition of, 202
   herbal, 117
   See also coughs
anxiety
   Actaea racemosa for, 157
   Albizia lebbeck for, 157
   with dyspnea, 125
   Elaeocarpus serratus for, 163
   Eschscholzia californica for, 95
   hyperventilation with, 140
   Lavandula spp. for, 97
   Melissa officinalis for, 99, 172
   Nepeta cataria for, 173
   Passiflora incarnata for, 175
   Tilia x *europaea* for, 105
   Withania somnifera for, 184
aperient, definition of, 202
aphrodisiac, definition of, 202
Apiaceae herbs
   for arrhythmias, 56
   for atherosclerosis, 32
   coumarins in, 78
   for hyperlipidemia, 34
   PAF inhibiting properties, 77
   seeds from, 29
apigenin, 113
Apium graveolens (celery)
   formulas containing
      arrhythmias, 56
      cerebral vascular insufficiency, 65
      hyperlipidemia, 35
      hypertension, 42
   for hyperlipidemia, 34, 35
   for hypertension, 42, 46

   specific indications, 90, 158
   for vascular support, 29, 31
Apocynum cannabinum (dogbane)
   cardiac glycosides from, 49
   for congestive heart failure, 48
   formulas containing
      congestive heart failure, 50
      endocarditis, 72
   specific indications, 90
appetite, observing, 11
apple. *See Malus domesticus* (apple)
apple pectin, 39, 99
Arachis hypogaea (peanut), *31*
arctic rose. *See Rhodiola rosea* (arctic rose)
Arctium lappa (burdock)
   formulas containing, 27
   iron from, 70
   for restless insomnia, 17
   specific indications, 90
   for vascular support, 27, 35
Ardisia japonica (marlberry)
   formulas containing, 153
   specific indications, 158
ARDS. *See* acute respiratory distress syndrome (ARDS)
arjuna. *See Terminalia* spp.
Armillaria fungus (gastrodia mushroom), 95
Armoracia rusticana (horseradish)
   for altered breath sounds, 118
   as counterirritant, 132
   formulas containing
      allergic rhinosinusitis, 128
      coughs, 117
      cystic fibrosis, 135
      emphysema, 137
      stridor, 121
   specific indications, 158–59
Arnica montana (leopard's bane)
   for angina, 38
   for bruising, 66
   formulas containing, 38, 123
   specific indications, 90–91
aromatic, definition of, 202
arrhythmias
   Adonis vernalis for, 89
   Convallaria majalis for, 93
   Crataegus spp. for, 93
   Digitalis purpurea for, 94
   formulas for, 52–57
   herbs for, 55, 56–57
   Leonurus cardiaca for, 98
   Lycopus virginicus (bugleweed), 98
   overview, 52–53
   Viscum album for, 106
Artemisia annua (sweet Annie), 159
Artemisia caerulescens (artemisia), 108t
Artemisia keiskeana (an lu), 144t
Artemisia vulgaris (mugwort), 70
artemisinin
   in *Artemisia annua*, 159

   for tuberculosis, 155, 156, 159
arterial insufficiency. *See* vascular insufficiency
arteritis. *See* vasculitis
artesunate, 155, 159
artichoke. *See Cynara scolymus* (artichoke)
art of herbal formulation, 9–24
   acne sample case, 18, *18*
   asking the right questions, 9, 10–11
   components of a formula, 13
   importance of symptoms, 9–10
   insomnia sample case, 16–17, *16*, *17*
   mastering the actions of herbs, 13–16
   narrowing down herbs to use, 23–24
   pharmacologic vs. physiologic therapy, 15
   rheumatoid arthritis sample case, 18–19, *19*
   supporting vitality vs. opposing disease, 14
   toxic herbs use, 20
   Triangle philosophy, 11–13, *11*, 23–24
   types of herbal preparations, 21–23
   warming and cooling formulas, 20–21, 21t
Asclepius tuberosa (pleurisy root)
   for altered breath sounds, 118
   for bronchitis, 131
   formulas containing
      coughs, 117, 119, 121
      pleurisy, 139
   for pleurisy, 138, 139
   specific indications, 159
ashwagandha. *See Withania somnifera* (ashwagandha)
Asian knotweed. *See Polygonum cuspidatum* (Japanese knotweed)
Asian water plantain. *See Alisma plantago-aquatica* (Asian water plantain)
Aspidosperma quebracho (quebracho), 131, 159
asthma
   Actaea racemosa for, 157
   Albizia lebbeck for, 157
   Ammi visnaga for, 157, 158
   Apium graveolens for, 158
   Aspidosperma quebracho for, 159
   Astragalus membranaceus for, 159
   Bacopa monnieri for, 160
   Cissampelos sympodialis for, 161
   Cnidium monnieri for, 161
   Coleus forskohlii for, 177
   Curcuma longa for, 162
   Datura stramonium for, 163
   Digitalis purpurea for, 163
   dyspnea with, 124
   Eclipta prostrata for, 163
   Elaeocarpus serratus for, 163
   Ephedra sinica for, 163
   Eriodictyon californicum for, 164
   Eucalyptus globulus for, 164
   Foeniculum vulgare for, 165
   formulas for, 142–47
   Grindelia spp. for, 167
   Hemidesmus indicus for, 167
   Inula helenium for, 168

asthma (*continued*)
  Juniperus communis for, 169
  Justicia adhatoda for, 169
  Lobelia inflata for, 170
  Marrubium vulgare for, 171
  Medicago sativa for, 171
  Mikania glomerata for, 172
  Neopicrorhiza scrophulariiflora for, 172
  Oenothera biennis for, 173
  overview, 142–43
  Passiflora incarnata for, 175
  Pelargonium sidoides for, 175
  Petasites hybridus for, 175
  Petasites spp. for, 129
  Phyllanthus emblica for, 176
  Picrorhiza kurroa for, 176
  Pinellia ternata for, 176
  Pinus pinaster for, 176
  Polygala spp. for, 177
  Raphanus sativus var. *niger* for, 178
  Rosmarinus officinalis for, 179
  Saussurea costus for, 179
  Scutellaria baicalensis for, 180
  Stemona spp. for, 181
  Sticta pulmonaria for, 181
  Tephrosia purpurea for, 182
  Thymus vulgaris for, 182
  Tylophora asthmatica for, 183
asthma weed. *See Euphorbia* spp.
Astragalus membranaceus (milk vetch)
  for bronchitis, 132
  for COPD, 148
  for cystic fibrosis, 135
  formulas containing
    COPD, 150, 153
    cystic fibrosis, 135
    heart stress at high altitudes, 52
    hyperlipidemia, 36
    lipid health, 33
    pneumonia, 142
    vascular support, 29
  immune modulating properties, 23
  for insomnia with exhaustion, 16
  for lipid health, 32
  PAF inhibiting properties, 77
  specific indications, 91, 159
  for strokes, 65
astringents
  cooling remedies, 21*t*
  definition of, 202
atherosclerosis
  Cinnamomum spp. for, 92
  Coptis chinensis for, 93
  coronary artery disease progression
    from, 38
  formulas for, 32–37
  Gymnema sylvestre for, 96
  Hibiscus sabdariffa for, 96
  Olea europaea for, 99
  overview, 32, 61

  Punica granatum for, 101
  Tanacetum parthenium for, 104
  Terminalia arjuna for, 104
Atractylodes macrocephala (atractylodes)
  formulas containing, 150
  specific indications, 91
  for strokes, 65
atrial fibrillation, formulas for, 53
Atropa belladonna (belladonna), *152*
  in calming formulas, 20
  formulas containing, 44
  safety concerns, 20, 44
  specific indications, 159
  transdermal delivery of, 109
  tropane alkaloids from, 152
atropine
  for bronchospasms, 159, 167
  for lung conditions, 152
  transdermal delivery of, 109
Avena sativa (oats)
  formulas containing, 33, 43
  for hyperventilation, 140
  for restless insomnia, 16, 17
  specific indications, 91
Azadirachta indica (neem)
  as mast cell stabilizer, 144*t*
  specific indications, 159–160

B

Bacopa monnieri (brahmi)
  for asthma, 144
  formulas containing, 147
  as mast cell stabilizer, 144*t*
  specific indications, 160
bacterial infections
  Armoracia rusticana for, 158
  Pinus pinaster for, 176
  Rheum palmatum for, 178
  secondary, 120, 130, 133, 140
  See also specific types
bagflower (*Clerodendron serratum*), 108*t*, 161
bahupatra (*Phyllanthus amarus*), 100
bai bu (*Stemona* spp.), 180–81
baical. *See Scutellaria baicalensis* (huang qin)
bai zhi. *See Angelica dahurica* (bai zhi)
bai zhu. *See Atractylodes ovata*
  (white atractylodes)
base herbs, 11, *11*, 12, 13
bayberry. *See Myrica cerifera* (wax myrtle)
beeswax
  in plaster base, 110
  topical protocols, 112
belladonna. *See Atropa belladonna* (belladonna)
bentonite clay, for varicosities, 75
Berberis aquifolium (Oregon grape)
  for cystic fibrosis, 135
  formulas containing
    cystic fibrosis, 135
    for hyperlipidemia, 35
    hypertension, 42

  phlebitis, 81
  vascular support, 27
  for heart infections, 72
  for poor circulation, 63
  specific indications, 91, 160
  for stasis ulcers, 60
  for varicosities, 74
berries
  flavonoids in, 33
  formulas containing, 35
beta blockers, 47, 152
Beta vulgaris (beet), 31
bhadrasey. *See Elaeocarpus serratus* (bhadrasey)
bhringraj. *See Eclipta alba* (bhringraj)
Biancaea sappan. See Caesalpinia sappan (su mu)
bibhitaki. *See Terminalia* spp.
bilberry. *See Vaccinium myrtillus* (bilberry)
biofilms, 137
biscuitroot. *See Lomatium dissectum* (biscuitroot)
bishop's hat. *See Epimedium brevicornu*
  (horny goatweed)
bitter almond (*Prunus dulcis*), 154
bitter orange. *See Citrus aurantium* (bitter orange)
bitters
  cooling remedies, 21*t*
  definition of, 202
blackbean tree. *See Castanospermum australe*
  (blackbean tree)
blackberry. *See Rubus* spp.
blackboard tree. *See Alstonia scholaris* (dita wood)
black cohosh. *See Actaea racemosa* (black cohosh)
black currant (*Ribes nigrum*), 43
blackhaw. *See Viburnum prunifolium* (blackhaw)
black mustard. *See Brassica nigra* (black mustard)
black pepper. *See Piper nigrum* (black pepper)
black Spanish radish. *See Raphanus sativus* var.
  niger (black Spanish radish)
black walnut. *See Juglans nigra* (black walnut)
bladderwrack. *See Fucus vesiculosus*
  (bladderwrack)
bleeding. *See* hemorrhage
blessed thistle (*Cnicus benedictus*), 92
blood clots
  nattokinase for, 80
  tea for, 81
  Trifolium pratense for, 105
blood movers
  Angelica sinensis as, 59, 61, 90
  for angina, 39
  cardiovascular benefits, 32, 46, 61
  for cold extremities, 76
  definition of, 202
  for Raynaud's disease, 67
  for stasis ulcers, 60
  for weak circulation, 58
blood pressure, high. *See* hypertension
blood pressure, low. *See* hypotension
bloodroot. *See Sanguinaria canadensis* (bloodroot)
blue flag. *See Iris versicolor* (blue flag)
boneset. *See Eupatorium perfoliatum* (boneset)

Borago officinalis (borage), 43, 91
brahmi. *See Bacopa monnieri* (brahmi)
Brassica nigra (black mustard), 111, 160
breathing, altered, 114–123
brezel wood. *See Caesalpinia sappan* (su mu)
bromelain
    from *Ananas comosus*, 158
    for bruising, 66
    formulas containing
        asthma, 147
        bronchitis, 132–33
        emphysema, 138
    PAF inhibiting properties, 77
    for varicosities, 81
bronchitis
    Achillea millefolium for, 156
    Aconitum napellus for, 156
    Albizia lebbeck for, 157
    Allium sativum for, 157
    Ananas comosus for, 158
    Anemopsis californica for, 158
    Apium graveolens for, 158
    Aspidosperma quebracho for, 159
    Cinnamomum verum for, 161
    Cissampelos sympodialis for, 161
    Commiphora myrrha for, 162
    Curcuma longa for, 162
    Eucalyptus globulus for, 164
    formulas for, 130–34
    Grindelia spp. for, 167
    Hydrastis canadensis for, 167
    Hyoscyamus niger for, 168
    Hyssopus officinalis for, 168
    Inula helenium for, 168
    Justicia adhatoda for, 169
    Ligusticum porteri for, 169
    Lomatium dissectum for, 170
    Lonicera japonica for, 170
    Lophatherum gracile for, 170
    Marrubium vulgare for, 171
    Melilotus suaveolens for, 172
    overview, 130–31
    Panax ginseng for, 174
    Pelargonium sidoides for, 175
    Polygala spp. for, 177
    Prunus serotina for, 177
    Raphanus sativus var. *niger* for, 178
    Rosmarinus officinalis for, 179
    Scutellaria baicalensis for, 180
    Stellaria media for, 104
    Sticta pulmonaria for, 181
    Stillingia sylvatica for, 181
    Thymus vulgaris for, 182
bronchodilators, herbal, 108*t*, 143*t*
bronchodilators, pharmaceutical, 142, 143*t*
bronchospasms
    Ammi visnaga for, 157
    Atropa belladonna for, 159
    Datura stramonium for, 163
    Foeniculum vulgare for, 166

    Hyoscyamus niger for, 167
    Lepidium meyenii, 169
    Mentha piperita for, 172
    Petroselinum crispum for, 175
    Picrorhiza kurroa for, 176
    Pimpinella anisum for, 176
    Prunus serotina for, 177
    Vitex negundo for, 184
bruising
    herbs for, 66*t*
    Hypericum perforatum for, 97, 168
    therapies for, 66
Bryonia alba (white bryony)
    formulas containing, 139, 141
    specific indications, 160
Bryonia dioica (bryony)
    for altered breath sounds, 118
    formulas containing
        costochondritis, 123
        coughs, 121, 123
    for pleurisy, 138
    specific indications, 160
buckwheat (*Fagopyrum esculentum*), 79
Bufei Yishen granules, 153
bugleweed. *See Lycopus virginicus* (bugleweed)
burdock. *See Arctium lappa* (burdock)
butcher's broom (*Ruscus aculeatus*), 102
butterbur. *See Petasites hybridus* (butterbur)
butterfly weed. *See Asclepius tuberosa*
    (pleurisy root)

C

cacao. *See Theobroma cacao* (cacao)
Cactus grandiflorus. See Selenicereus grandiflorus
Caesalpinia sappan (su mu), 131, 160
Calendula officinalis (pot marigold)
    formulas containing
        capillary fragility, 66, 67
        endocarditis, 72
        hemorrhoids, 78
        phlebitis, 81
        stasis ulcers, 60
        varicosities, 75
    for heart infections, 72
    for microvascular fragility, 66*t*
    specific indications, 91
    for stasis ulcers, 60
    for vascular infections, 72
    for vascular support, 65
California poppy. *See Eschscholzia californica*
    (California poppy)
Camellia sinensis (green tea)
    for COPD, 148, 150, 151
    for cor pulmonale, 155
    formulas containing
        COPD, 150, 151
        emphysema, 138
        hypertension, 42
        vascular support, 28, 29

    for hemochromatosis, 71
    for hypertension, 42
    iron chelation properties, 71
    iron from, 70
    as mast cell stabilizer, 144*t*
    for respiratory conditions, 108
    specific indications, 91, 160
camphorated castor oils, 112
camphor crystals, 112
camphor tree (*Cinnamomum camphora*), 161
candleberry (*Myrica cerifera*), 133, 172
capillary fragility, 65–67
capillary wormwood. *See Artemisia capillaris*
    (capillary wormwood)
Capsicum spp.
    formulas containing, 119, 134
    liniments containing, 111
Capsicum annuum (cayenne)
    for cor pulmonale, 155
    as counterirritant, 132
    for emphysema, 136
    fibrinolytic properties, 73
    formulas containing
        emphysema, 137
        hyperlipidemia, 37
        peripheral vascular insufficiency, 61
    plasters containing, 110
    for respiratory conditions, 115
    specific indications, 160
Capsicum frutescens (cayenne), 77
capsules
    dosage strategy, 23
    pros, cons, and indications, 22
caraway (*Carum carvi*), 160
cardamom (*Elettaria cardamomum*), 29
cardiac cough
    Digitalis purpurea for, 163
    formula for, 85
    Lobelia inflata for, 98
    Urginea maritima for, 105
cardiac glycosides
    for cardiomyopathy, 49
    for congestive heart failure, 47–50
    from *Convallaria majalis*, 51
cardiac inflammation
    Adonis vernalis for, 89
    Ammi visnaga for, 90
cardiac muscle spasms, *Lobelia* for, 13
cardiomyopathy
    Apocynum cannabinum for, 90
    Astragalus membranaceus for, 91
    benefits of early intervention, 7
    Borago officinalis for, 91
    cardiac glycosides for, 49
    Castanospermum australe for, 92
    Convallaria majalis for, 93
    Crataegus spp. for, 93, 162
    Digitalis purpurea for, 94
    formulas for, 47–51, 52
    Glycyrrhiza glabra for, 96

cardiomyopathy (*continued*)
   Ligusticum porteri for, 169
   overview, 47
   Paeonia lactiflora for, 100, 174
   Panax ginseng for, 174
   Stevia rebaudiana for, 104
   Terminalia arjuna for, 104
   Viscum album for, 106
   Ziziphus spinosa for, 106
cardiopulmonary disease
   Desmodium styracifolium for, 94
   formulas for, 83–85
   Gastrodia elata for, 95
   herbs for, 89–106
   Ligusticum striatum for, 98
   Withania somnifera for, 106
cardiotonic, definition of, 202
cardiovascular diseases
   Allium sativum for, 90
   Andrographis paniculata for, 158
   Angelica sinensis for, 90
   Berberis aquifolium for, 91
   Coptis chinensis for, 93
   Crataegus spp. for, 93
   Digitalis purpurea for, 94
   Echinacea spp. for, 94
   Equisetum spp. for, 95
   herbs for, 89–106
   Hibiscus sabdariffa for, 96
   Ligusticum striatum for, 98
   Linum usitatissimum for, 98
   Lycopus virginicus (bugleweed), 98
   Olea europaea for, 99
   overview, 25–27
   Panax ginseng for, 100
   Pueraria montana var. *lobata* for, 177
   Punica granatum for, 101
   Rheum palmatum for, 102
   Rhodiola rosea for, 102
   Rosa canina for, 102
   Schisandra chinensis for, 180
   Scutellaria baicalensis for, 103
   Ziziphus spinosa for, 106
   See also specific diseases
carminative, definition of, 202
Carum carvi (caraway), 160
Carum copticum (ajwain), 105
cascara. *See Rhamnus purshiana* (cascara)
Cassia spp. *See Senna* spp.
Castanospermum australe (blackbean tree), 92
castor oil. *See Ricinus communis* (castor oil)
caterpillar fungus. *See Cordyceps sinensis*
   (caterpillar fungus)
cathartic, definition of, 203
catnip. *See Nepeta cataria* (catnip)
cat's claw. *See Uncaria tomentosa* (uña de gato)
Caulophyllum thalictroides (blue cohosh), 75
caustic, definition of, 203
cayenne. *See Capsicum annuum* (cayenne)
Ceanothus americanus (red root)

formulas containing, 58, 81
   for poor circulation, 58
   specific indications, 92
celery. *See Apium graveolens* (celery)
cellulitis
   Echinacea spp. for, 94
   Equisetum spp. for, 95
   Phytolacca decandra for, 100
Centella asiatica (gotu kola), 62
   for altered breath sounds, 118
   for bronchitis, 132
   for cardiopulmonary disease, 83
   for COPD, 148, 151
   for fibrosis, 62
   flavonoids in, 33
   formulas containing
      capillary fragility, 66
      cardiopulmonary disease, 84, 85
      cerebral vascular insufficiency, 62–63, 64
      COPD, 151
      coughs, 120
      emphysema, 138
      endocarditis, 72
      hemoptysis, 126
      lymphedema, 83
      phlebitis, 81
      stasis ulcers, 60
      varicosities, 74, 75
      vasculitis, 82
   for heart infections, 72
   specific indications, 92, 161
   for stasis ulcers, 60
   for vascular support, 65
   for venous insufficiency, 74
cerebral ischemia, 61
cerebral vascular insufficiency
   formulas for, 61–65
   Ginkgo biloba for, 95
Cereus grandiflorus. See Selenicereus grandiflorus
chamomile. *See Matricaria chamomilla*
   (chamomile)
chanca piedra (*Phyllanthus amarus*), 100
chest pain. *See* angina
chickweed. *See Stellaria media* (chickweed)
Chimaphila umbellata (pipsissewa), 123
Chinese chaste tree. *See Vitex negundo*
   (Chinese chaste tree)
Chinese cucumber. *See Trichosanthes kirilowii*
   (Chinese cucumber)
Chinese date (*Ziziphus jujuba*), 106
Chinese indigo (*Persicaria tinctoria*), 131, 175
Chinese knotweed. *See Polygonum multiflorum*
   (fo ti, he shou wu)
Chinese licorice. *See Glycyrrhiza uralensis*
   (Chinese licorice)
Chinese privet. *See Ligustrum lucidum*
   (Chinese privet)
Chinese rhubarb. *See Rheum officinale*
   (Chinese rhubarb)
chi tonics, 21*t*, 203

cholagogue, definition of, 203
choleretic, definition of, 203
cholesterol, high. *See* high cholesterol
choline, for strokes, 63
chopi. *See Zanthoxylum piperitum*
   (Japanese pepper)
chronic bronchitis, 131
   See also bronchitis
chronic conditions
   cooling remedies for, 21
   definition of, 203
   types of formulas to use, 22
   warming stimulants for, 20
   See also specific conditions
chronic obstructive pulmonary disorder (COPD)
   Aspidosperma quebracho for, 159
   from bronchitis, 132
   Equisetum spp. for, 164
   formulas for, 84, 147–153
   Grindelia spp. for, 167
   herbs for, 149*t*
   Magnolia officinalis for, 171
   Melilotus suaveolens for, 172
   Neopicrorhiza scrophulariiflora for, 172
   Origanum vulgare for, 174
   overview, 147–48
   Paeonia lactiflora for, 174
   Petroselinum crispum for, 175
   Scutellaria baicalensis for, 180
   Stemona spp. for, 181
   Syzygium aromaticum for, 181
   vascular protectants for, 83
chuanxiong. *See Ligusticum striatum*
   (ligusticum)
cilantro. *See Coriandrum sativum* (coriander)
Cimicifuga foetida. See Actaea cimicifuga
   (shengma)
Cimicifuga racemosa. See Actaea racemosa
   (black cohosh)
Cinnamomum spp.
   formulas containing
      hemoptysis, 126
      hyperlipidemia, 35
      peripheral vascular insufficiency, 58, 61
   for hypertension, 46
   specific indications, 92
Cinnamomum camphora (camphor tree), 161
Cinnamomum cassia (cassia cinnamon), 38, 130
Cinnamomum verum (cinnamon)
   formulas containing
      hemoptysis, 126
      hyperlipidemia, 36
      lipid health, 33
      phlebitis, 81
      Raynaud's syndrome, 67
      vascular support, 27, 29
      venous congestion, 76
   for hypertension, 46
   for poor circulation, 58
   specific indications, 161

circulation, poor
  Aesculus hippocastanum for, 89
  Alisma plantago-aquatica for, 90
  Arnica montana for, 91
  Caesalpinia sappan for, 160
  Ceanothus americanus for, 92
  Cnicus benedictus for, 92
  Cornus officinalis for, 162
  Crataegus spp. for, 93
  Desmodium styracifolium for, 94
  Digitalis purpurea for, 94
  Echinacea spp. for, 94
  Epimedium brevicornu for, 95
  Gastrodia elata for, 95
  herbs for, 58, 63
  Hydrastis canadensis for, 97
  Juniperus communis for, 97
  Lepidium latifolium for, 98
  Medicago sativa for, 171
  Paeonia lactiflora for, 100
  Polygonum cuspidatum for, 101
  Polygonum multiflorum for, 102
  Sanguinaria canadensis for, 179
  Urtica dioica for, 105
  Vinca minor for, 106
  Zingiber officinale for, 106
Cirsium japonicum (Japanese thistle), 32–33
Cissampelos sympodialis (cissampelos), 161
citicoline, for strokes, 63
Citrus spp.
  flavonoids in, 26
  rutin from, 79
  specific indications, 161
  for vascular support, 29, 30
Citrus aurantium (bitter orange)
  for COPD, 151
  for coughs, 117
Citrus paradisi (grapefruit)
  specific indications, 161
  zest of, 29
Clerodendron serratum (glorybower), 108*t*, 161
cloves. *See Syzygium aromaticum* (cloves)
Cnicus benedictus (blessed thistle), 92
Cnidium monnieri (snow parsley)
  for asthma, 144
  formulas containing, 87
  for hypertension, 42
  as mast cell stabilizer, 144*t*
  specific indications, 92, 161
  for vascular support, 29, 87
Cochlearia armoracia. See Armoracia rusticana (horseradish)
cocklebur. *See Xanthium cavanillesii* (cocklebur)
cocoa powder
  formulas containing, 33
  for vascular support, 30
coconut flakes, for lipid health, 33
coconut oil
  formulas containing, 35
  for vascular support, 31

cod liver oil, 35
coenzyme Q10 (CoQ10), 37, 48*t*
cognitive decline
  Huperzia serrata for, 97
  Ligusticum striatum for, 98
  Panax ginseng for, 100
  Vinca minor for, 106
cold, common. *See* upper respiratory infections (URIs)
Coleus forskohlii (coleus)
  bronchodilating properties, 108*t*
  for congestive heart failure, 47–48, 50
  formulas containing
    arrhythmias, 55
    congestive heart failure, 50
    hyperlipidemia, 37
    hypotension, 86
    impotence, 87
  as mast cell stabilizer, 144*t*
  for poor circulation, 63
  specific indications, 101, 177
Collinsonia canadensis (stoneroot)
  formulas containing, 75, 76
  specific indications, 93, 161–62
coltsfoot. *See Tussilago farfara* (coltsfoot)
comfrey. *See Symphytum officinale* (comfrey)
Commiphora mukul (guggul)
  for altered breath sounds, 118
  for cerebral vascular insufficiency, 62
  formulas containing
    arrhythmias, 55
    coughs, 119
    for hyperlipidemia, 35
    hyperlipidemia, 36–37, 47
    hypertension, 45
    impotence, 87
    stridor, 121
    vascular support, 28
  for hemorrhoids, 74
  for hyperlipidemia, 32, 34
  PAF inhibiting properties, 77
  for poor circulation, 63
  specific indications, 93
Commiphora myrrha (myrrh), 121, 162
common cold. *See* upper respiratory infections (URIs)
compresses
  for hemorrhoids, 78
  for stasis ulcers, 60
coneflower. *See Echinacea* spp.
congestion. *See specific types*
congestive heart failure
  Aspidosperma quebracho for, 159
  Cytisus scoparius for, 94
  folkloric herbs for, 50
  formulas for, 47–51
  herbs for, 48*t*
  Lepidium meyenii for, 98
  overview, 47
  Terminalia arjuna for, 104

  Thuja occidentalis for, 105
  Viscum album for, 106
connective tissues
  Centella asiatica for, 92
  Curcuma longa for, 162
  Fucus vesiculosus for, 95
  Medicago sativa for, 171
  Rosa canina for, 179
constitution considerations, 11
continuous positive airway pressure (CPAP) devices, 88
Convallaria majalis (lily of the valley), *51*
  for altered breath sounds, 118
  for arrhythmias, 56
  cardiac glycosides from, 49
  for congestive heart failure, 48*t*, 50, 51
  for cor pulmonale, 155
  formulas containing
    cardiopulmonary disease, 84, 85
    congestive heart failure, 50
    hypotension, 86
  specific indications, 93
cooling remedies, 20–21, 21*t*
COPD. *See* chronic obstructive pulmonary disorder (COPD)
Coptis chinensis (goldthread)
  formulas containing, 43, 53
  for heart infections, 72
  for lipid health, 32
  specific indications, 93
Cordyceps sinensis (caterpillar fungus), 16, 148
Coriandrum sativum (coriander), 29, 123
Cornus officinalis (Japanese cornel), 153, 162
coronary artery disease
  Aconitum carmichaelii for, 89
  formulas for, 38–40
  Ginkgo biloba for, 95
  overview, 38
  Punica granatum for, 101
  Salvia miltiorrhiza for, 102, 103
cor pulmonale, formulas for, 85, 154–55
corrigent, definition of, 203
Corydalis cava (corydalis)
  formulas containing, 46, 123, 125
  specific indications, 162
cosmopolitan political herbalism, 3–5
costochondritis, formulas for, 121, 123
cough reflex, 115
coughs
  Achillea millefolium for, 156
  Actaea racemosa for, 157
  Albizia lebbeck for, 157
  Allium cepa for, 157
  Althaea officinalis for, 157
  Ammi visnaga for, 157
  Angelica sinensis for, 158
  Ardisia japonica for, 158
  Asclepius tuberosa for, 159
  with bronchitis, 130

coughs (*continued*)
  Capsicum annuum for, 160
  Cinnamomum camphora for, 161
  Cissampelos sympodialis for, 161
  Citrus spp. for, 161
  Collinsonia canadensis for, 161, 162
  Corydalis cava for, 162
  Dracocephalum rupestre for, 163
  Epimedium brevicornu for, 164
  Eriodictyon californicum for, 164
  Eschscholzia californica for, 164
  Foeniculum vulgare for, 165
  formulas for, 114–123
  Grindelia spp. for, 167
  Hemidesmus indicus for, 167
  Hydrangea macrophylla for, 167
  Hydrastis canadensis for, 167
  Hyoscyamus niger for, 167–68
  Hyssopus officinalis for, 168
  Justicia adhatoda for, 169
  Ligusticum striatum for, 169
  Lobelia inflata for, 122, 170
  Lomatium dissectum for, 170
  Lonicera japonica for, 170
  Lophatherum gracile for, 170
  Lycopus virginicus for, 170
  Magnolia officinalis for, 171
  Marrubium vulgare for, 171
  Matricaria chamomilla for, 171
  Mentha piperita for, 172
  Mikania glomerata for, 172
  Neopicrorhiza scrophulariiflora for, 172
  Nepeta cataria for, 173
  nitroglycerin patches for, 113
  Ophiopogon japonicus for, 174
  overview, 114–16
  Passiflora incarnata for, 175
  Pelargonium sidoides for, 175
  Petasites hybridus for, 175
  Phyllanthus emblica for, 176
  Pimpinella anisum for, 176
  Piper methysticum for, 177
  Polygala spp. for, 177
  Prunus serotina for, 177
  Rosmarinus officinalis for, 179
  Salvia officinalis for, 179
  Schisandra chinensis for, 180
  Selaginella uncinata for, 180
  Stemona spp. for, 180
  Sticta pulmonaria for, 181
  Stillingia sylvatica for, 181
  Syzygium aromaticum for, 181
  Tanacetum parthenium for, 182
  Thymus vulgaris for, 182
  Tussilago farfara for, 183
  Ulmus fulva for, 183
  Usnea barbata for, 183
  Verbascum thapsus for, 184
  Vitex negundo for, 184
  Xanthium cavanillesii for, 184

coumarins
  from *Angelica sinensis*, 59
  from *Anthoxanthum odoratum*, 90
  from Apiaceae plants, 78
  in calming formulas, 87
  lack of anticoagulant properties, 79
  from *Melilotus officinalis*, 99
  potential toxicity of, 78
counterirritants
  Armoracia rusticana as, 121
  definition of, 203
  Juniperus communis as, 97
  Lobelia inflata as, 122
  in respiratory formulas, 119, 122, 132, 134
  Thuja occidentalis as, 58
cow-itch (*Mucuna pruriens*), 71
cow plant. *See Gymnema sylvestre* (sugar
  destroyer, cow plant)
CPAP (continuous positive airway pressure)
  devices, 88
crackling breath sounds, herbs for, 118
crampbark (*Viburnum opulus*), 13
Crataegus spp.
  for arrhythmias, 55
  for congestive heart failure, 48*t*
  for cor pulmonale, 155
  for cystic fibrosis, 135
  flavonoids in, 33
  formulas containing
    angina, 38
    arrhythmias, 53, 55
    capillary fragility, 67
    cardiopulmonary disease, 84, 85
    congestive heart failure, 50
    coronary artery disease, 40
    cystic fibrosis, 135
    for hyperlipidemia, 35
    hypertension, 43
    peripheral vascular insufficiency, 58, 61
    vascular support, 29
    vasculitis, 82
  heart health benefits, 40
  for hypertension, 41–42
  for lipid health, 32
  specific indications, 162
  for varicosities, 75
  for vascular infections, 72
  for vascular support, 65
Crataegus monogyna (common hawthorn)
  for arrhythmias, 56
  formulas containing
    hypertension, 44, 46
    telangiectasias, 66
  for microvascular fragility, 66*t*
  specific indications, 93
  for stasis ulcers, 60
Crataegus oxyacantha (hawthorn), 26
  for angina, 38
  for arrhythmias, 56
  formulas containing

    cardiopulmonary disease, 85
    congestive heart failure, 50
    coronary artery disease, 40
    endocarditis, 72
    hemochromatosis, 71
    hypotension, 86
    telangiectasias, 66
    vasculitis, 82
  for hypertension, 44
  for microvascular fragility, 66*t*
  for poor circulation, 58
  specific indications, 93
  for stasis ulcers, 60
  for vasculature protection, 26
Crinum glaucum (swamp lily), 93
cromolyn sodium, 146
Croton lechleri (dragon's blood)
  formulas containing, 135
  specific indications, 162
croup
  Drosera rotundifolia for, 120, 163
  Polygala spp. for, 177
  Sticta pulmonaria for, 181
  Stillingia sylvatica for, 113, 181
crow dipper. *See Pinellia ternata*
  (crow dipper)
cultural appropriation, 2, 3
Curcuma longa (turmeric)
  for altered breath sounds, 118
  antiallergy qualities, 128
  for bronchitis, 132
  for cardiopulmonary disease, 83
  for COPD, 148, 150
  for cor pulmonale, 155
  for cystic fibrosis, 135
  fibrinolytic properties, 73
  formulas containing
    asthma, 147
    bronchitis, 134
    cardiopulmonary disease, 84, 85
    cerebral vascular insufficiency,
      62–63, 64
    COPD, 150
    costochondritis, 123
    coughs, 120
    cystic fibrosis, 135
    emphysema, 138
    endocarditis, 72
    hemochromatosis, 71
    hemoptysis, 126
    hyperlipidemia, 37
    phlebitis, 81
    stridor, 121
    telangiectasias, 66
    tuberculosis, 156
    vasculitis, 82
    venous congestion, 76
  for heart infections, 72
  for hemochromatosis, 71
  for hypertension, 46

as mast cell stabilizer, 144*t*
PAF inhibiting properties, 77
for poor circulation, 58, 63
for respiratory conditions, 108
for restless insomnia, 17
specific indications, 93–94, 162
for varicosities, 74
for vascular infections, 72
for vascular support, 31, 65
curcumin
for COPD, 148
for emphysema, 136, 138
formulas containing, 138
for hemochromatosis, 71
currant. *See Ribes* spp.
Cynara scolymus (artichoke)
formulas containing, 80
for lipid health, 33
specific indications, 94
for varicosities, 74
cystic fibrosis
Croton lechleri for, 162
formulas for, 134–35
Origanum vulgare for, 174
Cytisus scoparius (Scotch broom), 56, 94

D

dacryagogue, definition of, 203
dampness
Clerodendron serratum for, 161
definition of, 203
Desmodium styracifolium for, 94
formula for, 58
Pinellia ternata for, 176
Poria cocos for, 101
Schisandra chinensis for, 103
dandelion. *See Taraxacum officinale*
(dandelion)
dan shen. *See Salvia miltiorrhiza* (dan shen, red sage)
dan zhu ye gen (*Lophatherum gracile*), 170
Datura stramonium (jimson weed)
specific indications, 163
transdermal delivery of, 109
tropane alkaloids from, 152
Daucus carota (wild carrot), 29, 46
decoctions, defined, 22
deep vein thrombosis (DVT), 73, 80
deficient, definition of, 203
demulcents, 21*t*, 203
depurant, definition of, 203
Desmodium styracifolium (desmodium), 94
devil's club. *See Oplopanax horridus*
(devil's club)
diabetes
Andrographis paniculata for, 158
Borago officinalis for, 91
Ceanothus americanus for, 92
Cinnamomum spp. for, 92
Commiphora mukul for, 93

Coptis chinensis for, 93
Curcuma longa for, 94
Gymnema sylvestre for, 96
Hibiscus sabdariffa for, 96
Hypericum perforatum for, 97
impotence with, 87
Melilotus suaveolens for, 172
Oplopanax horridus for, 99
Panax ginseng for, 100
poor circulation with, 58
Rheum palmatum for, 102
stasis ulcers with, 59, 60
Stevia rebaudiana for, 104
Syzygium spp. for, 104
diaphoretics
Asclepius tuberosa as, 159
Collinsonia canadensis as, 162
definition of, 203
Eupatorium perfoliatum as, 165
Inula helenium as, 168
Lycopus virginicus as, 170
Marrubium vulgare as, 171
Polygala spp. as, 177
Sambucus canadensis as, 179
warming stimulants, 21*t*
Dicoumarol, 79
Digitalis purpurea (foxglove), 49
for altered breath sounds, 118
for arrhythmias, 56
cardiac glycosides from, 49
for congestive heart failure, 48*t*, 50
for cor pulmonale, 155
formulas containing, 126
safety concerns, 20, 94
specific indications, 94, 163
digoxin, 49, 94
dill (*Anethum graveolens*), 29
diminished breath sounds, herbs for, 118
Dioscorea villosa (wild yam), 13, 74
dita wood (*Alstonia scholaris*), 108*t*, 157
dittander (*Lepidium latifolium*), 98
diuretics, herbal
for congestive heart failure, 48*t*
cooling remedies, 21*t*
definition of, 203
for hypertension, 44
diuretics, pharmaceutical, 47
dizziness
Apium graveolens for, 90
Polygonum cuspidatum for, 101
Polygonum multiflorum for, 102
dogbane. *See Apocynum cannabinum* (dogbane)
dong quai. *See Angelica sinensis* (dong quai)
dosages
acute vs. chronic formulas, 7
toxic herbs, 7, 20
doshas system, in Ayurveda, 14
Dracocephalum rupestre (dragon's head), 163
dragon's blood. *See Croton lechleri*
(dragon's blood)

dragon's head. *See Dracocephalum rupestre*
(dragon's head)
Drimia maritima (sea squill), 48*t*, 105
Drosera rotundifolia (sundew)
for coughs, 117
formulas containing, 120, 146
specific indications, 163
drying agents, for mucous production, 133
du xing cai (*Lepidium apetalum*), 153
DVT (deep vein thrombosis), 73, 80
dwarf lilyturf. *See Ophiopogon japonicus*
(mai men dong)
dyspnea
Aconitum napellus for, 156
Ammi visnaga for, 157
Angelica sinensis for, 158
Aspidosperma quebracho for, 159
Cinnamomum camphora for, 161
Coleus forskohlii for, 177
Ephedra sinica for, 163
Euphorbia spp. for, 165
formulas for, 124–25
Hemidesmus indicus for, 167
Ligusticum porteri for, 169
Lobelia inflata for, 98, 170
Magnolia officinalis for, 171
Neopicrorhiza scrophulariiflora for, 172
overview, 124
Phytolacca americana for, 176
Selenicereus grandiflorus for, 103
Tanacetum parthenium for, 182
Terminalia chebula for, 182

E

ecbolic, definition of, 203
Echinacea spp.
formulas containing
COPD, 151
coughs, 119, 123
lymphedema, 83
for heart infections, 72
need for cultural context of, 4
Echinacea angustifolia (coneflower)
formulas containing, 59, 72
for respiratory infections, 121
specific indications, 94, 163
for stasis ulcers, 60
for vascular infections, 72
Echinacea purpurea (purple coneflower)
formulas containing, 81
specific indications, 94
for stasis ulcers, 60
Eclipta prostrata (eclipta), 144, 163
edema
Adonis vernalis for, 89
Apocynum cannabinum for, 90
Ginkgo biloba for, 95
Juniperus communis for, 97
Petroselinum crispum for, 100
Phytolacca americana for, 176

edema (*continued*)
  Phytolacca decandra for, 100
  Polygonum cuspidatum for, 101
  Quercus robur for, 101
  Selenicereus grandiflorus for, 103
  Taraxacum officinale for, 104, 182
  Urginea maritima for, 105
  Urtica dioica for, 105
Elaeocarpus serratus (bhadrasey), 108t, 163
elderberry. *See Sambucus* spp.
elecampane. *See Inula helenium* (elecampane)
Elettaria cardamomum (cardamom), 29
Eleutherococcus senticosus (Siberian ginseng)
  formulas containing
    arrhythmias, 55
    hypertension, 44
    hyperventilation, 140
  for insomnia with exhaustion, 16
  specific indications, 94
Emblica officinalis. See Phyllanthus emblica
  (Indian gooseberry)
emesis routine, for emphysema, 136–37
emetic, definition of, 203
emmenagogue, definition of, 203
emodin
  for ARDS, 153
  for lung fibrosis, 110
  in *Rheum palmatum*, 178
  in *Senna* spp., 180
emollient, definition of, 203
emphysema
  Aspidosperma quebracho for, 159
  Centella asiatica for, 161
  Equisetum spp. for, 164
  Euphorbia spp. for, 165
  formulas for, 136–38
  Grindelia spp. for, 167
endocarditis
  Adonis vernalis for, 89
  Calendula officinalis for, 91
  Castanospermum australe for, 92
  Coptis chinensis for, 93
  formulas for, 72
  Glycyrrhiza glabra for, 96
  Salvia miltiorrhiza for, 102
endocrine imbalances
  Eleutherococcus senticosus for, 94
  Leonurus cardiaca for, 98
  Lycopus virginicus (bugleweed), 98
  Melissa officinalis for, 99
endothelial inflammation, 26, 41
enemas, for respiratory distress, 154
energetic considerations
  fine-tuning formulas, 14–15
  toxic herbs use, 20
  in Triangle philosophy, 11
energy dispersants, 21t
Ephedra sinica (ma-huang), 128
  antiallergy qualities, 128
  bronchodilating properties, 108t

formulas containing
    allergic rhinosinusitis, 130
    COPD, 153
    vascular support, 30
  simple tea, from twigs of, 128
  specific indications, 163–64
ephedrine, 163–64
epicatechins, 71, 74
epilepsy
  Albizia lebbeck for, 157
  Viscum album for, 106
Epimedium brevicornu (horny goatweed)
  formulas containing, 153
  for hyperlipidemia, 36
  specific indications, 95, 164
Epimedium grandiflorum (horny goatweed), 36
Equisetum spp.
  for altered breath sounds, 118
  for COPD, 148, 150
  for cor pulmonale, 155
  formulas containing
    capillary fragility, 66
    congestive heart failure, 50
    COPD, 150
    coughs, 120
    emphysema, 138
    hemoptysis, 126
    hemorrhoids, 78
    stasis ulcers, 60
  specific indications, 95, 164
Equisetum arvense (horsetail), 59, 81, 84, 138
Equisetum hyemale (horsetail), 59, 84
erectile dysfunction. *See* impotence
Eriodictyon californicum (yerba santa)
  for bronchitis, 131
  for COPD, 149t
  formulas containing, 135, 148
  specific indications, 164
errhine, definition of, 203
escharotic, definition of, 203
Eschscholzia californica (California poppy)
  formulas containing, 46, 125
  for restless insomnia, 17
  specific indications, 95, 164
essential fatty acids (EFAs), 43
essential hypertension, 41
  See also hypertension
essential oils
  definition of, 204
  nebulized form, 137
  for pneumonia, 141
estrogen, heart disease and, 40
Eucalyptus globulus (eucalyptus), 112
  for altered breath sounds, 118
  for coughs, 117
  formulas containing
    asthma, 146
    bronchitis, 133
    COPD, 151
    costochondritis, 123

dyspnea, 125
    emphysema, 138
    pleurisy, 139
    pneumonia, 141
  liniments containing, 111
  for lung conditions, 112
  nebulized form, 137
  for respiratory infections, 121, 130
  specific indications, 164–65
Eupatorium perfoliatum (boneset)
  formulas containing, 127
  as mast cell stabilizer, 144t
  specific indications, 165
Euphorbia spp.
  for asthma, 144
  as counterirritant, 132
  formulas containing, 147
  specific indications, 165
Euphrasia officinalis (eyebright)
  formulas containing, 119, 128
  specific indications, 165
evidence-based medicine, need for cultural
  competence, 4–5
excess, definition of, 203
excitant, definition of, 203
exhilarant, definition of, 203
expectorant, definition of, 203
  See also coughs; phlegm
eyebright. *See Euphrasia officinalis* (eyebright)

F
Fagopyrum esculentum (buckwheat), 79
Fallopia japonica. See Polygonum cuspidatum
  (Japanese knotweed)
fang feng (*Saposhnikovia divaricata*), 150
febrifuge, definition of, 203
  See also fever
fennel. *See Foeniculum vulgare* (fennel)
fenugreek. *See Trigonella foenum-graecum*
  (fenugreek)
ferulic acid, 32, 59, 67, 71
fever
  Alstonia scholaris for, 157
  Ardisia japonica for, 158
  Artemisia annua for, 159
  Asclepius tuberosa for, 159
  Digitalis purpurea for, 163
  Eupatorium perfoliatum for, 165
  Euphorbia spp. for, 165
  Hemidesmus indicus for, 167
  Hydrangea macrophylla for, 167
  Lonicera japonica for, 170
  Melilotus suaveolens for, 171
  Morinda citrifolia for, 172
  Ocimum sanctum for, 173
  Rheum palmatum for, 178
  Tinospora cordifolia for, 182
  Xanthium cavanillesii for, 184
feverfew. *See Tanacetum parthenium* (feverfew)
fibrin, 73, 80

fibrinolytics
  botanicals, 73
  definition of, 203
  Desmodium styracifolium as, 94
  Ginkgo biloba as, 69
  nattokinase as, 80
fibrosis
  Centella asiatica for, 62
  myocardial, 70
  Senna spp. for, 180
  Vaccinium myrtillus for, 183
five flavor fruit. *See Schisandra chinensis*
  (magnolia vine)
flavonoids
  for COPD, 148
  for hemochromatosis, 71
  iron chelation properties, 71
  for lipid health, 32, 33
  for respiratory conditions, 108
  for vascular support, 65, 67
  for vasculature protection, 26
  for venous insufficiency, 74
flax. *See Linum usitatissimum* (flax)
flor de Jamaica. *See Hibiscus sabdariffa* (hibiscus)
fluid retention. *See* edema
Foeniculum vulgare (fennel)
  formulas containing
    asthma, 145
    bronchitis, 134
    coughs, 117, 119, 120
    dyspnea, 125
    stridor, 121
  for hypertension, 46
  nebulized form, 137
  specific indications, 165–66
  for vascular support, 29
folkloric herbs
  for coughs, 114
  for emphysema, 137
  for heart failure, 50
  for hypertension, 44
  for respiratory disorders, 121, 129, 131
formula types, 21–22
forskolin, 47, 177
fo ti. *See Polygonum multiflorum* (fo ti,
  he shou wu)
foundational herbs, 7
four-elements theory, 14
foxglove. *See Digitalis purpurea* (foxglove)
French Paradox, 31
friction rubs
  formula for, 120–21
  herbs for, 118
  with pleurisy, 138
Fritillaria cirrhosa (chuan bei mu), 153
Fucus vesiculosus (bladderwrack)
  formulas containing, 28, 87
  for poor circulation, 63
  specific indications, 95
fu zhi (*Aconitum carmichaelii*), 89

G

gagging sensations, from *Lobelia*, 122
galactogogue, definition of, 203
Galphimia glauca (calderona amarilla), 77, 108*t*
Ganoderma lucidum (reishi), *28*
  as chi tonic, 16
  immune modulating properties, 23
  for vascular support, 28
Gardenia latifolia (gardenia), 108*t*
garlic. *See Allium sativum* (garlic)
Gastrodia elata (tian ma), 61, 95
gastrodia mushroom (*Armillaria* fungus), 95
gastrodin, 61
Gelsemium sempervirens (yellow jessamine), 56
Gentiana lutea (gentian), 20
Geranium spp., 137
ginger. *See Zingiber officinale* (ginger)
gingerol, 77
Ginkgo biloba (ginkgo), *69*
  for angina, 38
  for arrhythmias, 53, 55, 56
  bronchodilating properties, 108*t*
  for cerebral vascular insufficiency,
    62–63, 64, 65
  for congestive heart failure, 50
  for COPD, 148
  for cor pulmonale, 155
  for cystic fibrosis, 135
  fibrinolytic properties, 73
  flavonoids in, 33
  formulas containing
    angina, 38
    arrhythmias, 53, 55
    asthma, 145, 146, 147
    bronchitis, 134
    cardiopulmonary disease, 84, 85
    cerebral vascular insufficiency,
      62–63, 64, 65
    congestive heart failure, 50
    COPD, 150
    coughs, 120
    cystic fibrosis, 135
    dyspnea, 124
    emphysema, 138
    heart stress at high altitudes, 52
    hyperlipidemia, 36
    hypotension, 86
    impotence, 87
    peripheral vascular insufficiency, 58, 59, 61
    pneumonia, 142
    Raynaud's syndrome, 67
    stridor, 121
    varicosities, 74
    vascular support, 29
    vasculitis, 82
  for hypertension, 44
  as mast cell stabilizer, 144*t*
  PAF inhibiting properties, 77
  for poor circulation, 58, 69
  specific indications, 95

for stasis ulcers, 60
for varicosities, 81
for vascular support, 27, 65
for vasculature protection, 26
ginseng. *See Panax ginseng* (ginseng)
glorybower. *See Clerodendron serratum*
  (glorybower)
glossy privet. *See Ligustrum lucidum*
  (Chinese privet)
Glycine max (soy)
  formulas containing, 35
  for hyperlipidemia, 34, 35
  specific indications, 95
Glycyrrhiza glabra (licorice)
  formulas containing
    asthma, 145, 146
    bronchitis, 133
    COPD, 151
    coughs, 117, 120
    cystic fibrosis, 135
    dyspnea, 125
    emphysema, 138
    endocarditis, 72
    heart stress at high altitudes, 52
    hemoptysis, 126
    for hyperlipidemia, 35
    hyperventilation, 140
    hypotension, 86
    pneumonia, 141
    stridor, 121
    tuberculosis, 156
    vascular support, 27
    vasculitis, 82
    venous congestion, 80
  for hyperventilation, 140
  PAF inhibiting properties, 77
  specific indications, 96
Glycyrrhiza uralensis (Chinese licorice)
  formulas containing, 38, 130
  for viral infections, 131
gobo root. *See Arctium lappa* (burdock)
goji (*Lycium barbarum*), 170
goldenseal. *See Hydrastis canadensis*
  (goldenseal)
goldthread. *See Coptis* spp.
gotu kola. *See Centella asiatica* (gotu kola)
grapefruit (*Citrus paradisi*), 29, 161
green tea. *See Camellia sinensis* (green tea)
Grindelia spp.
  for bronchitis, 131
  as counterirritant, 132
  formulas containing
    asthma, 147
    bronchitis, 133
    coughs, 119
    emphysema, 137
Grindelia hirsutula (gumweed), 167
Grindelia robusta (gumweed)
  formulas containing, 135
  specific indications, 167

Grindelia squarrosa (gumweed)
    formulas containing, 134, 135
    specific indications, 167
guaco (*Mikania glomerata*), 108t, 172
guduchi. *See Tinospora cordifolia* (guduchi)
guggul. *See Commiphora mukul* (guggul)
guggulsterones, 32, 34
gumweed. *See Grindelia* spp.
Gymnema sylvestre (sugar destroyer, cow plant)
    formulas containing, 35, 36
    specific indications, 96

H

hacking coughs, 114
Hamamelis virginiana (witch hazel)
    formulas containing
        capillary fragility, 67
        hemorrhoids, 78
        phlebitis, 81
        varicoceles, 75
        varicosities, 75
        vasculitis, 82
        venous congestion, 76
    for microvascular fragility, 66t
    specific indications, 96
    for vascular support, 65
hawthorn. *See Crataegus* spp.
hay fever
    Achillea millefolium for, 156
    Angelica sinensis for, 158
    Armoracia rusticana for, 159
    Curcuma longa for, 162
    Euphorbia spp. for, 165
    Euphrasia officinalis for, 165
    formula for, 128
    Juniperus communis for, 169
    Petasites hybridus for, 175
    Sticta pulmonaria for, 181
    Tanacetum parthenium for, 182
    Thymus vulgaris for, 182
    Tinospora cordifolia for, 182
    Xanthium cavanillesii for, 184
headaches
    Angelica dahurica for, 158
    Apium graveolens for, 90
    Cnidium monnieri for, 92
    herbs for, 27
    Huperzia serrata for, 97
    with hypertension, formula for, 44
    Ligusticum porteri for, 169
    Ligusticum striatum for, 98, 169
    Magnolia officinalis for, 171
    Melilotus officinalis for, 99
    Melilotus suaveolens for, 171
    Petasites hybridus for, 175
    Tanacetum parthenium for, 182
    Tinospora cordifolia for, 182
    Vinca minor for, 106
healing, complexity of, 9
healing crises, 10

heart, weak. *See* weak heart
heart disease. *See* cardiomyopathy
heart failure
    Aspidosperma quebracho for, 159
    Coleus forskohlii for, 101
    cor pulmonale, 85, 154–55
    Digitalis purpurea for, 94, 163
    dyspnea with, 124
    hypotension with, 86
    lack of research on formulas for, 7
    Panax ginseng for, 100
    Petroselinum crispum for, 100
    Rhodiola rosea for, 102
    Salvia miltiorrhiza for, 102
    Selenicereus grandiflorus for, 103
    Theobroma cacao for, 105
    See also congestive heart failure
heart infections and inflammation,
    herbs for, 72
heat stroke, *Lobelia inflata* for, 98
Helichrysum italicum (immortelle), 81
hematic, definition of, 203
Hemidesmus indicus (hemidesmus)
    for asthma, 145, 147
    formulas containing
        asthma, 147
        hemoptysis, 126
        tuberculosis, 156
    specific indications, 167
hemming symptoms, 114
hemochromatosis, 70–71, 94
hemoptysis
    Achillea millefolium for, 156
    Digitalis purpurea for, 163
    formulas for, 125–27
    Hamamelis virginiana for, 96
    Lycopus virginicus for, 170
    Melilotus suaveolens for, 171
    overview, 125–26
    Sticta pulmonaria for, 181
    Usnea barbata for, 183
hemorrhage
    Achillea millefolium for, 89
    Cinnamomum spp. for, 92
    Hamamelis virginiana for, 96
    nosebleeds, 99
hemorrhagic strokes, nutraceuticals for, 63
hemorrhoids
    Collinsonia canadensis for, 93
    formulas for, 78
    overview, 74
    Quercus robur for, 101
hemostatics
    Achillea millefolium as, 89
    Cinnamomum spp. as, 92
    definition of, 203
    Gardenia latifolia as, 166
henbane (*Hyoscyamus niger*), 152, 167–68
hepatic, definition of, 203
herbal teas, 21–22

Hercules' club/Hercules' herb. *See Zanthoxylum clava-herculis* (southern prickly ash)
hereditary hemorrhagic telangiectasia, topical
    protocol for, 127
he shou wu. *See Polygonum multiflorum* (fo ti,
    he shou wu)
Hibiscus rosa-sinensis (hibiscus)
    flavonoids in, 33
    formulas containing, 35, 43
    for poor circulation, 58
Hibiscus sabdariffa (hibiscus)
    flavonoids in, 33
    formulas containing, 27, 36, 42
    for hypertension, 42
    specific indications, 96
    for vascular infections, 72
    for vascular support, 30
high cholesterol
    Cinnamomum spp. for, 92
    Cynara scolymus for, 94
    Polygonum cuspidatum for, 101
    See also hyperlipidemia
Hippocrates, 9
Hippophae rhamnoides (sea buckthorn)
    formulas containing, 75, 81
    seed oil from, 43
    specific indications, 96
hispidulin, 167, 179
hives
    Cnidium monnieri for, 92
    Tanacetum parthenium for, 104
hoelen (*Poria cocos*), 65, 101
holy basil. *See Ocimum sanctum* (holy basil)
honeysuckle (*Lonicera japonica*), 170
horehound. *See Marrubium vulgare*
    (horehound)
horny goatweed. *See Epimedium pubescens*
    (horny goatweed)
horse chestnut. *See Aesculus hippocastanum*
    (horse chestnut)
horseradish. *See Armoracia rusticana*
    (horseradish)
horsetail. *See Equisetum arvense* (horsetail)
hot packs
    for pleurisy, 139
    for venous congestion, 80–81
huaco (*Mikania glomerata*), 108t, 172
huang qin. *See Scutellaria baicalensis* (huang qin)
humid asthma, 146–47, 164
Huperzia serrata (Chinese club moss), 62, 96–97
huperzines, 97
hurricane weed (*Phyllanthus amarus*), 100
hu zhang. *See Polygonum cuspidatum*
    (Japanese knotweed)
hydragogue, definition of, 203
Hydrangea macrophylla (hydrangea), 144t, 167
Hydrastis canadensis (goldenseal)
    as drying agent, 133
    specific indications, 97, 167
    for stasis ulcers, 60

hydrotherapy
  for emphysema, 136–37
  for pneumonia, 141
Hyoscyamus niger (henbane), 152, 167–68
hyperglycemia
  Avena sativa for, 91
  Castanospermum australe for, 92
  Cinnamomum spp. for, 92
  Crataegus spp. for, 93
  Cynara scolymus for, 94
  formulas for, 35
  Glycine max for, 95
  Nigella sativa for, 99
  Ophiopogon japonicus for, 99
  Opuntia ficus-indica for, 100
  Raphanus sativus var. *niger* for, 101
  role in cardiovascular disease, 26
  Syzygium spp. for, 104
  Urtica dioica for, 105
  Zingiber officinale for, 106
Hypericum perforatum (St. Johnswort)
  for bruising, 66
  flavonoids in, 33
  formulas containing
    arrhythmias, 55
    capillary fragility, 67
    costochondritis, 123
    hemorrhoids, 78
    lymphedema, 83
    pleurisy, 139
    telangiectasias, 66
    varicosities, 74, 75
    venous congestion, 76
  for heart infections, 72
  for pleurisy, 138
  for restless insomnia, 17
  specific indications, 97, 168
hyperlipidemia
  Atractylodes spp. for, 91
  Avena sativa for, 91
  Borago officinalis for, 91
  Commiphora mukul for, 93
  Coptis chinensis for, 93
  Crataegus spp. for, 93
  Curcuma longa for, 94
  Cynara scolymus for, 94
  formulas for, 32–37, 47
  Glycine max for, 95
  Hemidesmus indicus for, 167
  Hippophae rhamnoides for, 96
  hypolipidemic agents for, 34
  Iris versicolor for, 97
  Linum usitatissimum for, 98
  Malus domesticus for, 99
  Ophiopogon japonicus for, 99
  Opuntia ficus-indica for, 100
  overview, 32
  Panax ginseng for, 100
  Raphanus sativus var. *niger* for, 101
  Rheum palmatum for, 102

  role in cardiovascular disease, 26
  Salvia miltiorrhiza for, 103
  Schisandra chinensis for, 103
  Scutellaria baicalensis for, 103
  Silybum marianum for, 103
  Taraxacum officinale for, 104
  Theobroma cacao for, 105
  Trachyspermum ammi for, 105
  Zingiber officinale for, 106
  See also high cholesterol
hypertension
  Achillea millefolium for, 89
  Allium sativum for, 90
  Ammi visnaga for, 90
  Angelica sinensis for, 90
  Apium graveolens for, 90
  Avena sativa for, 91
  Camellia sinensis for, 91
  Castanospermum australe for, 92
  Cinnamomum spp. for, 92
  Cnidium monnieri for, 92
  Coleus forskohlii for, 177
  Coptis chinensis for, 93
  Crataegus spp. for, 93
  Crinum glaucum for, 93
  Desmodium styracifolium for, 94
  Eleutherococcus senticosus for, 94
  energetic considerations, 14–15
  Eschscholzia californica for, 95
  folkloric herbs for, 44
  formulas for, 41–47
  Ginkgo biloba for, 95
  Glycine max for, 95
  Gymnema sylvestre for, 96
  Hibiscus sabdariffa for, 96
  Iris versicolor for, 97
  Lavandula spp. for, 97
  Leonurus cardiaca for, 98
  Lepidium latifolium for, 98
  Matricaria chamomilla for, 99
  Melissa officinalis for, 99
  Nigella sativa for, 99
  Oplopanax horridus for, 99
  overview, 41–42
  Panax ginseng for, 100
  Passiflora incarnata for, 100
  Pinus pinaster for, 100
  Plantago psyllium for, 101
  Pueraria montana var. *lobata* for, 101
  role in cardiovascular disease, 25–26
  Salvia miltiorrhiza for, 102
  Schisandra chinensis for, 103
  Scutellaria baicalensis for, 103
  Selenicereus grandiflorus for, 103
  Senna spp. for, 103
  Silybum marianum for, 103
  Taraxacum officinale for, 104
  Theobroma cacao for, 105
  Tilia x *europaea* for, 105
  Trachyspermum ammi for, 105

  Uncaria tomentosa for, 105
  Urtica dioica for, 105
  Valeriana officinalis for, 106
  Ziziphus spinosa for, 106
hyperthyroidism, 53
hyperventilation, dyspnea with, 124
hyperventilation syndrome, treating, 140
hypolipidemic agents, 34
hypotension
  Aconitum carmichaelii for, 89
  Adonis vernalis for, 89
  Coleus forskohlii for, 101
  Crataegus spp. for, 93
  formulas for, 85–86
  Fucus vesiculosus for, 95
  Gymnema sylvestre for, 96
  Strophanthus hispidus for, 104
hypothyroidism, *Fucus vesiculosus* for, 95
Hyssopus officinalis (hyssop), 133, 168

I

immune-supporting herbs
  Astragalus membranaceus, 159
  Coleus forskohlii, 177
  Eleutherococcus senticosus, 94
  Glycyrrhiza glabra, 96
  Morinda citrifolia, 172
  Neopicrorhiza scrophulariiflora, 172
  Oenothera biennis, 173
  Panax ginseng, 174
  Persicaria tinctoria, 175
  Piper longum, 177
  Piper methysticum, 177
  Saussurea costus, 179
  Schisandra chinensis, 180
  Selaginella uncinata, 180
  Withania somnifera, 184
impotence
  Cnidium monnieri for, 87, 92
  Epimedium spp. for, 95
  formulas for, 87
  Fucus vesiculosus for, 95
Indian gooseberry. *See Phyllanthus emblica*
  (Indian gooseberry)
Indian ipecac. *See Tylophora asthmatica*
  (Indian ipecac)
Indian snakeroot. *See Rauvolfia serpentina*
  (Indian snakeroot)
indigenous knowledge, honoring of, 1–8
indirubin, 131, 175
individualized formulas, importance of, 5, 7
infections
  Albizia lebbeck for, 157
  Alstonia scholaris for, 157
  Andrographis paniculata for, 158
  Astragalus membranaceus for, 159
  Berberis aquifolium for, 160
  Caesalpinia sappan for, 160
  Echinacea spp. for, 94, 163
  Hyssopus officinalis for, 168

infections (*continued*)
  Isatis tinctoria for, 168
  Juniperus communis for, 168–69
  Lycium barbarum for, 170
  Ocimum sanctum for, 173
  Panax ginseng for, 174
  Polygonum multiflorum for, 102
  role of terrain in addressing, 9–10
  Rosmarinus officinalis for, 179
  Uncaria tomentosa for, 105, 183
  See also specific types
inflammation
  Alstonia scholaris for, 157
  Camellia sinensis for, 91
  Centella asiatica for, 92
  Glycyrrhiza glabra for, 96
  Hemidesmus indicus for, 167
  Hypericum perforatum for, 168
  Linum usitatissimum for, 169
  Magnolia officinalis for, 171
  Morinda citrifolia for, 172
  Ocimum sanctum for, 173
  Oenothera biennis for, 173
  Olea europaea for, 173
  Paeonia lactiflora for, 174
  Passiflora incarnata for, 175
  Perilla frutescens for, 175
  Persicaria tinctoria for, 175
  Petroselinum crispum for, 175
  Pinus pinaster for, 176
  Piper longum for, 177
  Piper methysticum for, 177
  Polygonum cuspidatum for, 178
  Polygonum multiflorum for, 102
  Poria cocos for, 101
  Rheum palmatum for, 178
  Rosa canina for, 179
  Rosmarinus officinalis for, 179
  Saussurea costus for, 179
  Scutellaria baicalensis for, 103, 180
  Selaginella uncinata for, 180
  Silybum marianum for, 103, 180
  Stemona spp. for, 181
  Taraxacum officinale for, 104
  Tephrosia purpurea for, 182
  Terminalia chebula for, 182
  Tinospora cordifolia for, 182
  Trichosanthes kirilowii for, 182
  Tylophora asthmatica for, 183
  Uncaria tomentosa for, 183
  Vaccinium myrtillus for, 183
  Vitis vinifera for, 184
  Zingiber officinale for, 106, 184
  See also specific types
influenza
  Achillea millefolium for, 156
  Actaea racemosa for, 157
  Eupatorium perfoliatum for, 165
  formulas for, 123
  hemoptysis with, 127

  herbs for, 131
  Hydrastis canadensis for, 167
  Isatis tinctoria for, 168
  Ligusticum porteri for, 169
  Ligusticum striatum for, 169
  Lomatium dissectum for, 170
  Lonicera japonica for, 170
  Lophatherum gracile for, 170
  Melissa officinalis for, 172
  Nepeta cataria for, 173
  Ocimum sanctum for, 173
  Pelargonium sidoides for, 175
  Persicaria tinctoria for, 175
  Sambucus canadensis for, 179
  Stemona spp. for, 180
  Sticta pulmonaria for, 181
  Taraxacum officinale for, 182
  Terminalia chebula for, 182
infusions, defined, 22
inotropic, definition of, 203
insomnia
  sample cases, 16–17, *16, 17*
  Schisandra chinensis for, 180
  Tilia x *europaea* for, 105
  Ziziphus spinosa for, 106
insulin sensitivity, 99, 104
Inula helenium (elecampane)
  for bronchitis, 131, 133
  formulas containing, 133
  for respiratory infections, 130
  specific indications, 168
Inula racemosa (inula), 144*t*
Iris versicolor (blue flag)
  formulas containing
    emphysema, 137
    hyperlipidemia, 37
    lung conditions, 123
    peripheral vascular insufficiency, 58
    vascular support, 28
  for poor circulation, 63
  specific indications, 97, 168
  for varicosities, 74
iron
  potential toxicity of, 68
  sources of, 68, 70
iron deficiency anemia, formulas for, 68–70
irritant, definition of, 203
Isatis tinctoria (woad), 131, 144*t*, 168

J

jaborandi (*Pilocarpus jaborandi*), 20
jambul. *See Syzygium jambos* (roseapple)
Japanese arrowroot. *See Pueraria montana*
  var. *lobata* (kudzu, gegen)
Japanese bigleaf magnolia (*Magnolia
  obovata*), 144*t*
Japanese cornel. *See Cornus officinalis*
  (Japanese cornel)
Japanese knotweed. *See Polygonum cuspidatum*
  (Japanese knotweed)

Japanese pepper. *See Zanthoxylum piperitum*
  (Japanese pepper)
jimson weed. *See Datura stramonium*
  (jimson weed)
juices
  for capillary fragility, 67
  vegetable, 31
jujube (*Ziziphus jujuba*), 106
Juniperus communis (juniper)
  for congestive heart failure, 50
  formulas containing, 50
  specific indications, 97, 168–69
  for tuberculosis, 156
Justicia adhatoda (malabar nut), 169
Justicia vasica, 108*t*

K

kava kava. *See Piper methysticum* (kava)
khella. *See Ammi visnaga* (khella)
kidney insufficiency. *See* renal insufficiency
kidney stones
  Hydrangea macrophylla for, 167
  Lepidium latifolium for, 98
  Phyllanthus amarus for, 100
king of bitters. *See Andrographis paniculata*
  (king of bitters)
Korean perilla (*Perilla frutescens*), 127, 175
kudzu. *See Pueraria* spp. (kudzu)
kutki. *See Picrorhiza kurroa* (kutki)

L

laryngitis
  Achillea millefolium for, 156
  Aconitum napellus for, 156
  Cinnamomum camphora for, 161
  Collinsonia canadensis for, 161
  Dracocephalum rupestre for, 163
  Eupatorium perfoliatum for, 165
  Rumex crispus for, 179
  Sticta pulmonaria for, 181
  Stillingia sylvatica for, 181
Lavandula spp. (lavender)
  formulas containing
    phlebitis, 81
    pneumonia, 141
    stasis ulcers, 60
  for hypertension, 41
  specific indications, 97
laxative, definition of, 203
leafless mistletoe (*Viscum articulatum*), 48
lectins, 54, 80, 132, 178
lemon balm. *See Melissa officinalis* (lemon balm)
Leonurus cardiaca (motherwort)
  for angina, 38
  for arrhythmias, 53, 55, 56–57
  fibrinolytic properties, 73
  formulas containing
    arrhythmias, 53, 55
    hypertension, 42, 44, 45–46, 47
  specific indications, 98

leopard's bane. *See Arnica montana* (leopard's bane)
Lepidium apetalum (du xing cai), 153
Lepidium latifolium (rompe piedras), 98
Lepidium meyenii (maca)
 for arrhythmias, 55
 for asthma, 145
 bronchodilating properties, 108*t*
 for congestive heart failure, 50
 formulas containing
  arrhythmias, 55
  asthma, 145
  dyspnea, 124
  heart stress at high altitudes, 52
  hypotension, 86
  impotence, 87
  peripheral vascular insufficiency, 58, 61
 for poor circulation, 58, 63
 specific indications, 98, 169
lesser periwinkle. *See Vinca minor* (lesser periwinkle)
licorice. *See Glycyrrhiza glabra* (licorice)
lidocaine, 114, 115
Ligusticum spp.
 for COPD, 148
 for opportunistic lung infections, 134
 for pleurisy, 138
Ligusticum chuanxiong. See Ligusticum striatum (ligusticum)
Ligusticum officinale. See Cnidium officinale (snow parsley)
Ligusticum porteri (osha)
 for bronchitis, 131
 formulas containing
  arrhythmias, 53
  bronchitis, 132
  COPD, 151
  coughs, 117
  pneumonia, 141
 for respiratory infections, 121
 specific indications, 169
Ligusticum striatum (ligusticum)
 for cerebral vascular insufficiency, 61, 62
 formulas containing
  coronary artery disease, 40
  hemochromatosis, 71
 for hemochromatosis, 71
 for respiratory infections, 130
 specific indications, 98, 169
 for strokes, 65
 for viral infections, 131
ligustrazine, 61
lily of the valley. *See Convallaria majalis* (lily of the valley)
linden. *See Tilia x europaea* (linden)
liniments, 111, 113
Linum usitatissimum (flax)
 in oatmeal, 33
 PAF inhibiting properties, 77
 for respiratory conditions, 108

seed oil from, 43
 specific indications, 98, 169
 for vascular support, 31
lipids, elevated. *See* hyperlipidemia
lipotropic, definition of, 204
lithotriptic, definition of, 203
liver congestion
 Raphanus sativus var. *niger* for, 101
 Silybum marianum for, 103
 Taraxacum officinale for, 104
liver function
 Apium graveolens for, 90
 Cnicus benedictus for, 92
 Coptis chinensis for, 93
 Cynara scolymus for, 94
 Nigella sativa for, 99
 Schisandra chinensis for, 103
 in TCM, 17
liver herbs
 for cystic fibrosis, 135
 for hypertension and hyperlipidemia, 35
 for poor circulation, 63
 for varicosities, 74
Lobelia inflata (pukeweed), *122*
 for altered breath sounds, 118
 for angina, 38
 antispasmodic qualities, 13
 bronchodilating properties, 108*t*
 in calming formulas, 20
 for COPD, 149*t*
 for coughs, 117, 130
 as counterirritant, 132
 for emphysema, 136, 137
 formulas containing
  angina, 39
  asthma, 145, 146
  bronchitis, 133
  coughs, 117, 119, 120, 123
  cystic fibrosis, 135
  dyspnea, 125
  emphysema, 137
  pleurisy, 139
 for hemoptysis, 127
 liniments containing, 111, 113
 for lung conditions, 109, 122
 plasters containing, 110
 for pleurisy, 139
 for respiratory conditions, 115
 safety concerns, 20
 specific indications, 98, 170
lobeline, 98, 109, 111, 122
Lomatium dissectum (biscuitroot)
 for altered breath sounds, 118
 for COPD, 148
 formulas containing
  COPD, 148
  coughs, 117, 119, 121
  pleurisy, 139
  pneumonia, 141
 for opportunistic lung infections, 134

 for pleurisy, 138
 for respiratory infections, 121, 130
 specific indications, 170
 for viral infections, 131
long pepper. *See Piper longum* (long pepper)
Lonicera japonica (honeysuckle), 170
Lophatherum gracile (dan zhu ye gen), 131, 170
lung cancer
 apigenin for, 113
 Camellia sinensis for, 160
 Hemidesmus indicus for, 167
 lack of formulas for, 7
 Origanum vulgare for, 174
 Saussurea costus for, 179
 Scutellaria baicalensis for, 180
 Silybum marianum for, 180
lung conditions
 Andrographis paniculata for, 158
 Angelica dahurica for, 158
 Brassica nigra for, 160
 Camellia sinensis for, 160
 Centella asiatica for, 161
 Cnidium monnieri for, 161
 Corydalis cava for, 162
 Curcuma longa for, 162
 Digitalis purpurea for, 163
 Dracocephalum rupestre for, 163
 Echinacea angustifolia for, 163
 Epimedium brevicornu for, 164
 Equisetum spp. for, 164
 Eriodictyon californicum for, 164
 Euphorbia spp. for, 165
 Foeniculum vulgare for, 166
 herbs for, 89–106
 Inula helenium for, 168
 Lepidium meyenii, 169
 Ligusticum porteri for, 169
 Ligusticum striatum for, 169
 Magnolia officinalis for, 171
 Marrubium vulgare for, 171
 Matricaria chamomilla for, 171
 Melaleuca alternifolia for, 171
 Melilotus suaveolens for, 171–72
 Melissa officinalis for, 172
 Morella cerifera for, 172
 Morinda citrifolia for, 172
 Neopicrorhiza scrophulariiflora for, 172
 Ophiopogon japonicus for, 174
 Origanum vulgare for, 174
 Paeonia lactiflora for, 174
 Panax ginseng for, 174
 Passiflora incarnata for, 175
 Phyllanthus emblica for, 176
 Piper nigrum for, 101
 Polygonum cuspidatum for, 178
 Pueraria montana var. *lobata* for, 177
 Rhamnus purshiana for, 178
 Selaginella uncinata for, 180
 Silybum marianum for, 180
 Stellaria media for, 104

lung conditions (*continued*)
 Stillingia sylvatica for, 181
 Syzygium aromaticum for, 181
 Taraxacum officinale for, 182
 topical protocols for, 109–14
 Trichosanthes kirilowii for, 182
 Verbascum thapsus for, 184
 Zingiber officinale for, 184
 See also specific disorders
lung pain, topical protocols for, 114
luteolin, 108, 144, 148
Lycium barbarum (goji), 153, 170
Lycium chinense. See Lycium barbarum (goji)
Lycopus virginicus (bugleweed)
 formulas containing
 arrhythmias, 53
 hemoptysis, 127
 hypertension, 47
 varicoceles, 75
 venous congestion, 76
 specific indications, 98, 170
lymphatic system
 Iris versicolor for, 97
 Phytolacca decandra for, 100
 Quercus robur for, 101
lymphedema
 Echinacea spp. for, 94
 formula for, 82–83
 Phytolacca decandra for, 100

M

maca. *See Lepidium meyenii* (maca)
Magnolia obovata (Japanese bigleaf
 magnolia), 144*t*
Magnolia officinalis (magnolia), 144*t*, 171
Mahonia aquifolium. See Berberis aquifolium
 (Oregon grape)
ma-huang. *See Ephedra sinica* (ma-huang)
maidenhair tree. *See Ginkgo biloba* (ginkgo)
mai men dong. *See Ophiopogon japonicus*
 (mai men dong)
malabar nut (*Justicia adhatoda*), 169
malaria
 Alstonia scholaris for, 157
 Hydrangea macrophylla for, 167
Malus domesticus (apple), 39, 99
maritime pine. *See Pinus pinaster* (maritime pine)
marlberry (*Ardisia japonica*)
 formulas containing, 153
 specific indications, 158
Marrubium vulgare (horehound)
 for bronchitis, 131
 formulas containing, 133
 specific indications, 171
marshmallow (*Althaea officinalis*), 157
mast cell stabilizers, herbal, 144*t*
material dose, definition of, 204
materia medica
 specific herb choice and, 12
 as term, 2

Matricaria chamomilla (chamomile)
 formulas containing, 53
 for restless insomnia, 17
 specific indications, 99, 171
 topical protocols, 114
 for vascular headaches, 27
Medicago sativa (alfalfa)
 formulas containing
 cardiopulmonary disease, 84
 COPD, 150
 dyspnea, 124
 specific indications, 171
Melaleuca alternifolia (tea tree)
 formulas containing, 60
 liniments containing, 113
 specific indications, 171
Melilotus officinalis (sweet clover), 79, 99
Melilotus suaveolens (sweet clover)
 for ARDS, 154
 formulas containing, 85, 155
 specific indications, 171
Melissa officinalis (lemon balm)
 formulas containing
 arrhythmias, 53
 dyspnea, 125
 hypertension, 46, 47
 hyperventilation, 140
 for hypertension, 41, 46, 47
 for hyperventilation, 140
 for restless insomnia, 17
 specific indications, 99, 172
Meniere's disease, 101
menopausal symptoms
 Actaea racemosa for, 157
 Glycine max for, 95
 heart palpitations, 53
 Leonurus cardiaca for, 98
 Pueraria montana var. *lobata* for, 101
menses, heavy, 96
Mentha spp.
 formulas containing
 COPD, 151
 pneumonia, 141
 varicosities, 75
 topical protocols using, 112–13
Mentha piperita (peppermint)
 formulas containing
 anemia, 69
 asthma, 146
 bronchitis, 133
 capillary fragility, 66
 cardiopulmonary disease, 84
 congestive heart failure, 50
 hemorrhoids, 78
 pleurisy, 139
 liniments containing, 111
 as mast cell stabilizer, 144*t*
 nebulized form, 137
 specific indications, 172
Mentha pulegium (pennyroyal), 70, 99

Mentha spicata (spearmint), 137, 145
menthol, 112–13
metabolic acidosis, dyspnea with, 124
metabolic syndrome
 Hibiscus sabdariffa for, 96
 Opuntia ficus-indica for, 100
 Stevia rebaudiana for, 104
migraines, *Tanacetum parthenium* for, 104
Mikania glomerata (huaco), 108*t*, 172
milk thistle. *See Silybum marianum* (milk thistle)
milk vetch. *See Astragalus membranaceus*
 (milk vetch)
milkwort (*Polygala* spp.), 177
mimosa (*Albizia julibrissin*), 157
miotic, definition of, 204
mistletoe. *See Viscum album* (mistletoe)
mocktails, 30
molasses, formulas containing, 68, 69, 70
Monascus purpureus (red yeast), 37
mondograss. *See Ophiopogon japonicus*
 (mai men dong)
Morella cerifera (bayberry), 172
Morinda citrifolia (noni), 172
motherwort. *See Leonurus cardiaca* (motherwort)
mucilage, extraction of, 23
mucolytic herbs, 88, 114, 120, 136
mucous membranes
 Anemopsis californica for, 158
 Berberis aquifolium for, 160
 Collinsonia canadensis for, 93
 Glycyrrhiza glabra for, 166
 Hamamelis virginiana for, 96
 Hydrastis canadensis for, 167
 Hyssopus officinalis for, 168
 Marrubium vulgare for, 171
 Morella cerifera for, 172
 overview, 120
 Prunus serotina for, 177
 Quercus robur for, 101
 Salvia officinalis for, 179
 Sanguinaria canadensis for, 179
 sulfur gasotransmitters and, 143
 Symphytum officinale for, 181
Mucuna pruriens (velvet bean), 71
mugwort (*Artemisia vulgaris*), 70
mullein. *See Verbascum thapsus* (mullein)
mustard poultices, 109, 111, 160
mydriatic, definition of, 204
myocardial infarction
 Aconitum carmichaelii for, 89
 Calendula officinalis for, 91
 Centella asiatica for, 92
 Olea europaea for, 99
 Schisandra chinensis for, 103
 Scutellaria baicalensis for, 103
myocardial ischemia
 Punica granatum for, 101
 Schisandra chinensis for, 103
myotic, definition of, 204
Myrica cerifera (wax myrtle), 133, 172

myrobalan. *See Terminalia chebula* (myrobalan)

myrrh (*Commiphora myrrha*), 121, 162

N

N-acetylcysteine (NAC)
   expectorant properties, 143
   formulas containing, 133, 138, 150

narcotic, definition of, 204

naringenin, 29

nattokinase, 46, 80

nebulized medications, 137

neem. *See Azadirachta indica* (neem)

Neopicrorhiza scrophulariiflora (picrorhiza)
   for COPD, 149*t*
   picrosides in, 149*t*, 154
   specific indications, 172–73
   as substitute for *Picrorhiza kurroa*, 176

Nepeta cataria (catnip), 70, 127, 173

nervine herbs
   definition of, 204
   for hypertension, 41, 42, 43, 46
   for hyperventilation, 125, 140

nervous disorders
   Leonurus cardiaca for, 98
   Matricaria chamomilla for, 99
   Viscum album for, 106
   Withania somnifera for, 106

nettle. *See Urtica* spp.

neuropathy
   Borago officinalis for, 91
   flavonoids for, 26
   Hypericum perforatum for, 97

New Jersey tea. *See Ceanothus americanus* (red root)

Nigella sativa (black cumin), 43, 99

night-blooming cactus. See Selenicereus grandiflorus

nitric oxide, 41, 42

nitroglycerin patches, 113

noni (*Morinda citrifolia*), 172

nosebleeds, 99

nourishing herbs, 11, 12

number of herbs in a formulary, 7, 12

nutraceuticals, for ischemic and hemorrhagic stroke, 63

nutrient, definition of, 204

O

oats. *See Avena sativa* (oats)

obesity
   Commiphora mukul for, 93
   Hemidesmus indicus for, 167
   Rheum palmatum for, 102
   Stellaria media for, 104

obstructive sleep apnea, treating, 88–89

Ocimum sanctum (holy basil)
   bronchodilating properties, 108*t*
   for hypertension, 41
   as mast cell stabilizer, 144*t*
   specific indications, 173

Oenothera biennis (evening primrose)
   for respiratory conditions, 108
   seed oil from, 43
   specific indications, 173

old man's beard. *See Usnea barbata* (old man's beard)

Olea europaea (olive)
   for respiratory infections, 121
   seed oil from, 43
   specific indications, 99, 173–74

omega-3 fatty acids
   cardiovascular benefits, 32, 43, 63
   in *Linum usitatissimum*, 169
   lipid metabolism benefits, 34
   in *Oenothera biennis*, 173
   in *Perilla frutescens*, 175

onion. *See Allium cepa* (onion)

Ophiopogon japonicus (mai men dong)
   for congestive heart failure, 48
   formulas containing, 53, 153
   specific indications, 99, 174

Oplopanax horridus (devil's club), 16, 63, 99–100

opportunistic infections
   Andrographis paniculata for, 155, 158
   herbs for, 134, 138
   Origanum vulgare for, 174
   pulmonary disease and, 151

Opuntia ficus-indica (prickly pear)
   flavonoids in, 33
   formulas containing, 35, 36
   for poor circulation, 58
   specific indications, 100
   for vascular support, 30, 31

oregano. *See Origanum vulgare* (oregano)

Oregon grape. *See Berberis aquifolium* (Oregon grape)

Origanum vulgare (oregano)
   for COPD, 148
   formulas containing, 119, 148
   for opportunistic lung infections, 134
   specific indications, 174

osha. *See Ligusticum porteri* (osha)

osthol
   in *Angelica archangelica*, 80
   for arrhythmias, 56
   as calcium channel blocker, 46
   in *Cnidium monnieri*, 87, 144, 161
   hypolipidemic properties, 34

osthole. *See Cnidium monnieri* (snow parsley)

oxytocic, definition of, 204

P

Paeonia anomala, 153

Paeonia emodi (Himalayan peony), 149*t*

Paeonia lactiflora (white/red peony)
   for COPD, 149*t*
   formulas containing
      allergic rhinosinusitis, 130
      arrhythmias, 53
      dyspnea, 124

   specific indications, 100, 174
   for strokes, 65

paeoniflorin, 108

palpitations
   Eleutherococcus senticosus for, 94
   Leonurus cardiaca for, 98
   Melilotus officinalis for, 99
   Tilia x *europaea* for, 105
   Ziziphus spinosa for, 106

Panax ginseng (ginseng)
   for altered breath sounds, 118
   for arrhythmias, 55
   for cerebral vascular insufficiency, 62
   for congestive heart failure, 48
   for COPD, 148
   formulas containing
      arrhythmias, 53, 55
      asthma, 145
      bronchitis, 133
      cardiopulmonary disease, 84
      COPD, 148, 153
      coughs, 120
      dyspnea, 125
      emphysema, 138
      hypertension, 44
      impotence, 87
      peripheral vascular insufficiency, 58, 61
      pneumonia, 142
   for hyperlipidemia, 36
   for hyperventilation, 140
   for insomnia with exhaustion, 16
   PAF inhibiting properties, 77
   for poor circulation, 63
   specific indications, 100, 174

panic attacks
   dyspnea with, 124
   formulas for, 125
   hyperventilation with, 140

paraffin, in plaster base, 110

parsley. *See Petroselinum crispum* (parsley)

parturifacient, definition of, 204

partus preparator, definition of, 204

Passiflora incarnata (passionflower)
   bronchodilating properties, 108*t*
   formulas containing, 42, 43
   for hypertension, 41
   for restless insomnia, 16, 17
   specific indications, 100, 174–75

peanut (*Arachis hypogaea*), 31

Pelargonium sidoides (African geranium), 131, 175

pennyroyal (*Mentha pulegium*), 70, 99

pennywort. *See Centella asiatica* (gotu kola)

peony, tree. *See Paeonia suffruticosa* (tree peony)

peony, white. *See Paeonia lactiflora* (peony)

pepper, black. *See Piper nigrum* (black pepper)

pepper, long. *See Piper longum* (long pepper)

peppermint. *See Mentha piperita* (peppermint)

pepperweed. *See Lepidium latifolium* (rompe piedras)

pericarditis
   Calendula officinalis for, 91
   Centella asiatica for, 92
   Glycyrrhiza glabra for, 96
Perilla frutescens (Korean perilla), 127, 175
peripheral vascular insufficiency
   Angelica sinensis for, 90
   Collinsonia canadensis for, 93
   Crataegus spp. for, 93
   Equisetum spp. for, 95
   formulas for, 57–61
   Ginkgo biloba for, 69
   Hamamelis virginiana for, 96
   herbs for, 89–106
   overview, 57–58
peripheral vasodilators
   Achillea millefolium as, 35, 61, 89
   formulas containing, 61, 76
   Ginkgo biloba as, 61, 69
   warming stimulants, 21*t*
Persicaria tinctoria (Chinese indigo), 131, 175
Petasites hybridus (butterbur), *129*
   antiallergy qualities, 127
   for coughs, 130
   for cystic fibrosis, 135
   formulas containing
      allergic rhinosinusitis, 128
      asthma, 147
      bronchitis, 134
      cystic fibrosis, 135
      emphysema, 138
      upper respiratory infections, 119
      vascular support, 29, 30
   for respiratory allergies, 129
   specific indications, 175
   for vascular headaches, 27
Petasites japonicus (butterbur), 70, 129
Petroselinum crispum (parsley)
   formulas containing, 50, 59
   specific indications, 100, 175
   for vascular support, 29
Petroselinum sativum. See Petroselinum crispum (parsley)
pharmaceuticals
   for asthma, herbal alternatives for, 143*t*
   diuretics, 47
   physiologic therapy vs., 15
   transdermal drug delivery, 109
pharyngitis
   Hydrastis canadensis for, 167
   Pelargonium sidoides for, 175
   Salvia officinalis for, 179
   Tinospora cordifolia for, 182
phlebitis
   Centella asiatica for, 92
   Coptis chinensis for, 93
   formulas for, 81
   nattokinase for, 80
   overview, 74
   Phytolacca decandra for, 100

phlegm
   Magnolia officinalis for, 171
   Ophiopogon japonicus for, 174
   Perilla frutescens for, 175
   Pinellia ternata for, 176
   Poria cocos for, 101
   Trichosanthes kirilowii for, 183
   warming stimulants for, 20
Phoradendron chrysocladon (mistletoe)
   for angina, 38
   for arrhythmias, 57
   for hypertension, 44
Phyllanthus amarus (bahupatra, hurricane weed), 100
Phyllanthus emblica (Indian gooseberry)
   for COPD, 149*t*
   formulas containing, 126, 156
   specific indications, 176
phytic acid, 68
Phytolacca decandra (pokeroot)
   for altered breath sounds, 118
   formulas containing
      lymphedema, 83
      peripheral vascular insufficiency, 59
      pneumonia, 141
      stridor, 121
   as lymph mover, 20
   specific indications, 100, 176
   for stasis ulcers, 60
picrorhiza. *See Neopicrorhiza scrophulariiflora* (picrorhiza)
Picrorhiza kurroa (kutki)
   bronchodilating properties, 108*t*
   formulas containing, 148
   as mast cell stabilizer, 144*t*
   PAF inhibiting properties, 77
   specific indications, 176
Picrorhiza scrophulariiflora. See Neopicrorhiza scrophulariiflora (picrorhiza)
picrosides, 149*t*, 154, 176
pills, use of, 22
Pilocarpus jaborandi (jaborandi), 20
Pimpinella anisum (anise)
   formulas containing, 146
   liniments containing, 111
   nebulized form, 137
   specific indications, 176
   for vascular support, 29
pineapple. *See Ananas comosus* (pineapple)
Pinellia ternata (crow dipper), 130, 176
Pinus pinaster (maritime pine)
   Pycnogenol from, 127
   specific indications, 100, 176
Piper longum (long pepper)
   formulas containing, 126, 156
   specific indications, 177
Piper methysticum (kava)
   antispasmodic qualities, 13
   formulas containing, 46, 123
   for hypertension, 46

   for restless insomnia, 16, 17
   specific indications, 177
Piper nigrum (black pepper)
   formulas containing, 29, 40
   specific indications, 101
pippali. *See Piper longum* (long pepper)
Plantago psyllium (plantain), 31, 101
plasters, 110–11
platelet antiaggregators, 73, 77, 92
Plectranthus forskohlii. See Coleus forskohlii (coleus)
pleurisy
   Asclepius tuberosa for, 159
   formulas for, 138–39
   herbs for, 118
   overview, 138
   Prunus serotina for, 177
   Stemona spp. for, 180
pleurisy root. *See Asclepius tuberosa* (pleurisy root)
pleuritis
   herbs for, 118
   overview, 138
pneumonia
   Allium sativum for, 157
   Ananas comosus for, 158
   Asclepius tuberosa for, 159
   from biofilms, 137
   from bronchitis, 130
   Curcuma longa for, 162
   dyspnea with, 124
   formulas for, 140–42
   Hyoscyamus niger for, 168
   Juniperus communis for, 169
   Ligusticum porteri for, 169
   Lomatium dissectum for, 170
   Lycopus virginicus for, 170
   Melilotus suaveolens for, 172
   Origanum vulgare for, 174
   overview, 140–41
   Panax ginseng for, 174
   Pelargonium sidoides for, 175
   Polygala spp. for, 177
   Prunus serotina for, 177
   Stemona spp. for, 181
   Thymus vulgaris for, 182
pokeroot. *See Phytolacca decandra* (pokeroot)
polydatin, 108
Polygala spp. (milkwort), 177
Polygala karensium (snakeroot), 131
Polygonum cuspidatum (Japanese knotweed)
   emodin in, 110
   formulas containing, 151
   for hyperlipidemia, 36
   specific indications, 101, 178
Polygonum multiflorum (fo ti, he shou wu)
   for hyperlipidemia, 36
   specific indications, 101–2
Polygonum tinctorium. See Persicaria tinctoria (Chinese indigo)
pomegranate. *See Punica granatum* (pomegranate)
poor circulation. *See* circulation, poor

Poria cocos (hoelen), 65, 101
portal congestion
  Angelica sinensis for, 90
  Berberis aquifolium for, 91
  formulas for, 79–80
  Iris versicolor for, 97
pot marigold. *See Calendula officinalis*
  (pot marigold)
PQRST (provocation, quality, radiation,
  severity, time) questions, 10–11
prickly ash. *See Zanthoxylum clava-herculis*
  (prickly ash)
prince's pine. *See Chimaphila umbellata*
  (pipsissewa)
proanthocyanidins
  in *Crataegus* spp., 26
  natural iron chelation and, 71
  in *Punica granatum*, 30
  for venous insufficiency, 74
propranolol, for hereditary hemorrhagic
  telangiectasia, 127
Prunus spp.
  for altered breath sounds, 118
  formulas containing, 117, 119
Prunus dulcis (bitter almond), 154
Prunus serotina (cherry)
  for bronchitis, 131
  for capillary fragility, 67
  for coughs, 117, 130
  formulas containing, 146
  specific indications, 177
Pueraria montana var. *lobata* (kudzu, gegen)
  for cystic fibrosis, 135
  formulas containing, 27, 80
  for hyperlipidemia, 36
  specific indications, 101, 177
  for vascular support, 27
pukeweed. *See Lobelia inflata* (pukeweed)
pulmonary congestion
  Ammi visnaga for, 90
  Aspidosperma quebracho for, 159
  Eriodictyon californicum for, 164
  formula for, 126
  Lepidium meyenii, 169
  Lomatium dissectum for, 170
pulmonary hypertension and fibrosis
  cor pulmonale from, 154–55
  formulas for, 84–85
  Rhamnus purshiana for, 178
  Salvia miltiorrhiza for, 102
  vascular protectants for, 83
pulmonary inflammation
  Ammi visnaga for, 90
  Hypericum perforatum for, 168
pulse, observing, 11
Punica granatum (pomegranate)
  for capillary fragility, 67
  flavonoids in, 33
  as mast cell stabilizer, 144*t*
  for microvascular fragility, 66*t*

specific indications, 101, 178
  for vascular support, 30, 65
purgative, definition of, 204
purple coneflower. *See Echinacea purpurea*
  (purple coneflower)
Pycnogenol, 100, 127, 176

Q

qian ceng ta. *See Huperzia serrata*
  (Chinese club moss)
quebracho (*Aspidosperma quebracho*), 131, 159
queen's root. *See Stillingia sylvatica* (queen's root)
quercetin
  for COPD, 148
  for emphysema, 136
  in herbal bronchodilators, 108*t*
  as mast cell stabilizer, 144
  for natural iron chelation, 71
Quercus spp. (oak)
  formulas containing
    hemorrhoids, 78
    venous congestion, 76, 80
  for varicosities, 74
  for venous insufficiency, 74
Quercus robur (common oak), 81, 101

R

radish, black Spanish. *See Raphanus sativus* var.
  niger (black Spanish radish)
Raphanus sativus var. *niger* (black Spanish
  radish), 101, 178
Rauvolfia serpentina (Indian snakeroot), 45
  for arrhythmias, 57
  formulas containing, 44
  for hypertension, 42, 44, 45
  for vascular headaches, 27
Ravensara aromatica (clove nutmeg), 137
rayasanas, 156
Raynaud's syndrome, 67, 95
red clover. *See Trifolium pratense* (red clover)
red peony. *See Paeonia lactiflora*
  (white/red peony)
red root. *See Ceanothus americanus* (red root)
red sage. *See Salvia miltiorrhiza* (dan shen,
  red sage)
red yeast rice, for hyperlipidemia, 37
refrigerant, definition of, 204
reishi (*Ganoderma lucidum*), 28
  as chi tonic, 16
  immune modulating properties, 23
  for vascular support, 28
renal insufficiency
  formula for, 58
  Juniperus communis for, 97
  Phyllanthus amarus for, 100
  Thuja occidentalis for, 105
ren dong teng (*Lonicera japonica*), 170
ren shen. *See Panax ginseng* (ginseng)
reserpine
  in *Rauvolfia serpentina*, 45, 57

in *Vinca minor*, 106
respiratory allergies
  Astragalus membranaceus for, 159
  formulas for, 127–130
  See also specific types
respiratory disorders
  Aconitum napellus for, 156
  Actaea racemosa for, 157
  Andrographis paniculata for, 158
  Angelica sinensis for, 158
  Apocynum cannabinum for, 90
  Aspidosperma quebracho for, 159
  Brassica nigra for, 160
  Centella asiatica for, 161
  Citrus spp. for, 161
  Clerodendron serratum for, 161
  Collinsonia canadensis for, 93, 161–62
  Corydalis cava for, 162
  Crataegus spp. for, 162
  Eriodictyon californicum for, 164
  Eucalyptus globulus for, 164
  Eupatorium perfoliatum for, 165
  Euphorbia spp. for, 165
  Grindelia spp. for, 167
  Hemidesmus indicus for, 167
  Iris versicolor for, 168
  Ligusticum striatum for, 98
  Lobelia inflata for, 13, 98, 170
  Lycium barbarum for, 170
  Lycopus virginicus for, 170
  Mikania glomerata for, 172
  Morella cerifera for, 172
  Morinda citrifolia for, 172
  Ocimum sanctum for, 173
  overview, 107–8
  Petasites hybridus for, 175
  Phyllanthus emblica for, 176
  Pinellia ternata for, 176
  Piper methysticum for, 177
  Polygala spp. for, 177
  Raphanus sativus var. *niger* for, 178
  Schisandra chinensis for, 180
  Silybum marianum for, 180
  Symplocarpus foetidus for, 181
  Syzygium aromaticum for, 181
  Thymus vulgaris for, 182
  Tinospora cordifolia for, 182
  Vaccinium myrtillus for, 183–84
  Withania somnifera for, 184
  Zanthoxylum clava-herculis for, 184
  Zingiber officinale for, 184
  See also specific disorders
respiratory distress syndrome. *See* acute
  respiratory distress syndrome (ARDS)
respiratory infections
  Achillea millefolium for, 156
  Actaea cimicifuga for, 157
  Allium cepa for, 157
  Alstonia scholaris for, 157
  Althaea officinalis for, 157

respiratory infections (*continued*)
  Armoracia rusticana for, 158
  Asclepius tuberosa for, 159
  Berberis aquifolium for, 160
  Bryonia dioica for, 160
  Citrus spp. for, 161
  Commiphora myrrha for, 162
  Cornus officinalis for, 162
  Curcuma longa for, 162
  Dracocephalum rupestre for, 163
  Echinacea angustifolia for, 163
  Eucalyptus globulus for, 164
  herbs for, 121
  Inula helenium for, 168
  Ligusticum porteri for, 169
  Ligusticum striatum for, 169
  Lomatium dissectum for, 170
  Lonicera japonica for, 170
  Lycopus virginicus for, 170
  Magnolia officinalis for, 171
  Matricaria chamomilla for, 171
  Melaleuca alternifolia for, 171
  Melissa officinalis for, 172
  Morinda citrifolia for, 172
  Nepeta cataria for, 173
  Olea europaea for, 173
  Ophiopogon japonicus for, 174
  Panax ginseng for, 174
  Pelargonium sidoides for, 175
  Picrorhiza kurroa for, 176
  Pinus pinaster for, 176
  Polygala spp. for, 177
  Punica granatum for, 178
  Rosmarinus officinalis for, 179
  Sambucus canadensis for, 179
  Sanguinaria canadensis for, 179
  Schisandra chinensis for, 180
  Scutellaria baicalensis for, 180
  Stemona spp. for, 180
  Sticta pulmonaria for, 181
  Terminalia chebula for, 182
  Thymus vulgaris for, 121, 130, 134
  Trichosanthes kirilowii for, 182
  Ulmus fulva for, 183
  Verbascum thapsus for, 184
  See also upper respiratory infections (URIs)
respiratory syncytial virus (RSV)
  Actaea spp. for, 123, 131
  Lophatherum gracile for, 131, 170
  overview, 130
  Selaginella uncinata for, 131, 180
  Terminalia chebula for, 131, 182
resveratrol
  antioxidant and anti-inflammatory
    properties, 184
  for COPD, 148, 150
  for emphysema, 136, 138
  for excessive CFTR activity, 134
  flavonoids in, 33
  formulas containing

    asthma, 147
    COPD, 150
    emphysema, 138
  for hemochromatosis, 71
  for vascular support, 31
revulsive, definition of, 204
Reynoutria japonica. See Polygonum cuspidatum
  (Japanese knotweed)
Reynoutria multiflora. See Polygonum
  multiflorum (fo ti, he shou wu)
Rhamnus purshiana (cascara)
  emodin in, 110
  formulas containing, 151
  as intestinal smooth-muscle mover, 20
  specific indications, 178
rheumatoid arthritis sample case, 18–19, *19*
Rheum emodi, 110
Rheum officinale (Chinese rhubarb)
  for ARDS, 153, 154
  emodin in, 110
  formulas containing, 85, 153
Rheum palmatum (turkey rhubarb)
  emodin in, 110
  for hyperlipidemia, 36
  specific indications, 102, 178
rhinosinusitis, allergic. *See* allergic rhinosinusitis
Rhodiola spp.
  as chi tonic, 16
  for cystic fibrosis, 134
Rhodiola kirilowii (rhodiola), 135
Rhodiola rosea (arctic rose)
  for congestive heart failure, 50
  formulas containing
    cerebral vascular insufficiency, 64
    coronary artery disease, 40
    heart stress at high altitudes, 52
    hypotension, 86
    impotence, 87
  for poor circulation, 58, 63
  specific indications, 102
rhonchi, 118, 120
rhubarb. *See Rheum* spp.
Ribes spp., 102
Ribes nigrum (black currant), 43
Ricinus communis (castor oil)
  formulas containing
    hemorrhoids, 78
    pleurisy, 139
    venous congestion, 80
  in plaster base, 110
  plasters containing, 111
  specific indications, 178
  in topical protocols, 112, 113
rompe piedras (*Lepidium latifolium*), 98
Rosa canina (dog rose)
  flavonoids in, 33
  formulas containing
    arrhythmias, 56
    capillary fragility, 66
    emphysema, 138

    endocarditis, 72
    hypertension, 43
    vascular support, 27
    vasculitis, 82
  hips of, 26
  for lipid health, 32
  specific indications, 102, 178–79
  for varicosities, 75
  for vascular infections, 72
roselle. *See Hibiscus sabdariffa* (hibiscus)
Rosmarinus officinalis (rosemary)
  antiallergy qualities, 127
  bronchodilating properties, 108*t*
  as drying agent, 133
  formulas containing, 62–63, 64
  specific indications, 179
RSV. *See* respiratory syncytial virus (RSV)
rubefacient, definition of, 204
Rubus spp., *31*, 69, 82
rue (*Ruta graveolens*), 79
Rumex spp.
  emodin in, 110
  for iron deficiency anemia, 68
Rumex crispus (yellow dock)
  for poor circulation, 63
  specific indications, 179
ruscogenins, 102, 174
Ruscus aculeatus (butcher's broom), 102
Ruta graveolens (rue), 79
rutin, 79

S

safety concerns
  Aconitum napellus, 89, 156
  Adonis vernalis, 89
  Apocynum cannabinum, 90
  Atropa belladonna, 159
  Digitalis purpurea, 20, 94
  Hyoscyamus niger, 168
  Lobelia inflata, 170
  Strophanthus hispidus, 104
  toxic herb dosages, 7, 20
  Urginea maritima, 105
sage. *See Salvia* spp.
St. Johnswort. *See Hypericum perforatum*
  (St. Johnswort)
salivant, definition of, 204
Salix spp., 65
Salvia miltiorrhiza (dan shen, red sage)
  for arrhythmias, 57
  for atherosclerosis, 32
  for cardiopulmonary disease, 83
  for cerebral vascular insufficiency, 62
  formulas containing
    arrhythmias, 53, 56
    cerebral vascular insufficiency, 64, 65
    congestive heart failure, 50
    endocarditis, 72
    heart stress at high altitudes, 52
    hemochromatosis, 71

hyperlipidemia, 37
hypertension, 45
peripheral vascular insufficiency, 58
phlebitis, 81
vascular support, 27, 30
venous congestion, 76, 80
for heart infections, 72
for hemochromatosis, 70–71
for hyperlipidemia, 36
specific indications, 102–3
for varicosities, 81
for vascular infections, 72
for vascular support, 27
Salvia officinalis (sage)
for altered breath sounds, 118
as drying agent, 133
formulas containing, 62–63, 64
specific indications, 179
Sambucus spp.
formulas containing, 117, 119
for respiratory infections, 130
Sambucus canadensis (elderberry)
formulas containing, 82
specific indications, 179
Sambucus nigra (elderberry)
formulas containing, 123, 132
for respiratory infections, 121
for viral infections, 131
sample cases
acne variations, 18, *18*
insomnia variations, 16–17, *16*, *17*
rheumatoid arthritis variations, 18–19, *19*
sandalwood, nebulized form, 137
sangre de drago. *See Croton lechleri*
(dragon's blood)
Sanguinaria canadensis (bloodroot)
for bronchitis, 131
formulas containing, 123, 133
specific indications, 179
as tissue mover, 20
Saposhnikovia divaricata (fang feng), 150
Saussurea costus (saw wort)
formulas containing, 126, 156
specific indications, 179
Schisandra chinensis (magnolia vine)
for congestive heart failure, 48
formulas containing
allergic rhinosinusitis, 130
arrhythmias, 53
COPD, 153
venous congestion, 80
specific indications, 103, 180
Scotch broom. *See Cytisus scoparius*
(Scotch broom)
Scutellaria baicalensis (huang qin), *150*
for COPD, 148
for cystic fibrosis, 135
formulas containing
asthma, 145
bronchitis, 134

COPD, 148
cystic fibrosis, 135
hemochromatosis, 71
hypertension, 43
stridor, 121
for hemochromatosis, 71
for hyperlipidemia, 36
for respiratory conditions, 150
for sleep difficulties, 16, 17
specific indications, 103, 180
for viral infections, 131
Scutellaria lateriflora (skullcap)
formulas containing, 140
for hyperventilation, 140
for sleep difficulties, 16, 17
sea almond. *See Terminalia* spp. (sea almond)
sea buckthorn. *See Hippophae rhamnoides*
(sea buckthorn)
sea squill. *See Drimia maritima* (sea squill)
secretory stimulants, 21*t*
sedative, definition of, 204
Seedy Sea Salt recipe, 33
Selaginella uncinata (spikemoss), 131, 180
Selenicereus grandiflorus (night-blooming
cactus), *51*
for arrhythmias, 57
for congestive heart failure, 51
for cor pulmonale, 155
formulas containing
arrhythmias, 55, 83
cardiopulmonary disease, 83
congestive heart failure, 50
coronary artery disease, 40
cor pulmonale, 155
specific indications, 103
Senna spp.
emodin in, 110
specific indications, 103, 180
Senna alata (candle bush), 144*t*
shengma. *See Actaea cimicifuga* (shengma)
Shin'iseihaito (xin yi qing fei tang), 174
shiso (*Perilla frutescens*), 127, 175
sialagogue, definition of, 204
Siberian ginseng. *See Eleutherococcus senticosus*
(Siberian ginseng)
silk tree (*Albizia julibrissin*), 157
Silybum marianum (milk thistle)
for cystic fibrosis, 135
formulas containing
cardiopulmonary disease, 85
cystic fibrosis, 135
heart stress at high altitudes, 52
hyperlipidemia, 37, 47
for hemochromatosis, 71
for lipid health, 32–33
as mast cell stabilizer, 144*t*
for restless insomnia, 17
specific indications, 103, 180
for varicosities, 74
for vascular support, 35

simple, definition of, 204
sinusitis
Albizia lebbeck for, 157
Allium sativum for, 157
Angelica dahurica for, 158
Angelica sinensis for, 158
Armoracia rusticana for, 158, 159
Berberis aquifolium for, 160
Curcuma longa for, 162
Ligusticum porteri for, 169
Linum usitatissimum for, 169
Lomatium dissectum for, 170
Magnolia officinalis for, 171
Panax ginseng for, 174
Petasites hybridus for, 175
Xanthium cavanillesii for, 184
siris. *See Albizia lebbeck* (siris)
sitz baths, 78
skunk cabbage (*Symplocarpus foetidus*), 117, 181
sleep apnea, 88–89
sleep difficulties
Corydalis cava for, 162
Eschscholzia californica for, 95, 164
Nepeta cataria for, 173
observing, 11
Withania somnifera for, 184
See also insomnia
slippery elm. *See Ulmus* spp.
Smilax ornata (sarsaparilla), 63
snakeroot (*Polygala* spp.), 177
snowball bush (*Viburnum opulus*), 13
snow parsley. *See Cnidium officinale* (snow parsley)
sore throat
Althaea officinalis for, 157
Anemopsis californica for, 158
Berberis aquifolium for, 160
Capsicum annuum for, 160
Cinnamomum camphora for, 161
Ligusticum porteri for, 169
Ligusticum striatum for, 169
Lonicera japonica for, 170
Marrubium vulgare for, 171
Mentha piperita for, 172
Morella cerifera for, 172
Phytolacca americana for, 176
Polygala spp. for, 177
Rumex crispus for, 179
Salvia officinalis for, 179
Syzygium aromaticum for, 181
soybean. *See Glycine max* (soy)
Spanish black radish. *See Raphanus sativus*
var. *niger* (black Spanish radish)
spearmint. *See Mentha spicata* (spearmint)
specific herbs
definition of, 204
role in formulas, *11*, 12, 13
in Triangle exercise, *11*, 12
specific indications
choosing herbs based on, 5, 12
as term, 5

spider veins. *See* telangiectasias (spider veins)

spikemoss. *See Selaginella uncinata* (spikemoss)

spurge. *See Euphorbia* spp.

squill. *See Drimia maritima* (sea squill)

standardized medicines, indications for, 22

staphylokinase, 64

stasis ulcers
  Berberis aquifolium for, 91
  Centella asiatica for, 92
  Echinacea spp. for, 94
  formulas for, 59, 60
  herbs for, 60
  Hydrastis canadensis for, 97
  Hypericum perforatum for, 97

steam inhalation
  for emphysema, 136–37
  for pneumonia, 141

steeping herbs, 22–23

Stellaria media (chickweed)
  formulas containing, 70
  iron from, 70
  specific indications, 104

Stemona spp. (bai bu), 180–81

sternutatory, definition of, 204

Stevia rebaudiana (stevia, sweet leaf)
  formulas containing, 36, 43
  specific indications, 104
  for vascular support, 30

Sticta pulmonaria (lungwort)
  for bronchitis, 131
  for coughs, 117
  formulas containing, 141
  specific indications, 181

Stillingia sylvatica (queen's root)
  for coughs, 117
  as counterirritant, 132
  formulas containing
    hemoptysis, 126
    lung conditions, 123
    stridor, 121
  liniments containing, 113
  specific indications, 181

stinging nettle. *See Urtica dioica* (stinging nettle)

stoneroot. *See Collinsonia canadensis* (stoneroot)

streptokinase, 64

stress
  hyperventilation with, 140
  Lavandula spp. for, 97
  Passiflora incarnata for, 175
  Tilia x *europaea* for, 105
  Withania somnifera for, 184

stridor
  formulas for, 121
  herbs for, 118
  overview, 115–16

strokes
  Angelica sinensis for, 90
  Astragalus membranaceus for, 91
  Atractylodes spp. for, 91
  Epimedium brevicornu for, 95

Ligusticum striatum for, 98

nutraceuticals for, 63

Ophiopogon japonicus for, 99

overview, 61

Paeonia lactiflora for, 100

Poria cocos for, 101

Scutellaria baicalensis for, 103

thrombolytic agents for, 64

traditional Chinese herbs for, 65

Ziziphus spinosa for, 106

Strophanthus hispidus (strophanthus)
  for angina, 38
  for arrhythmias, 57
  for congestive heart failure, 48t
  specific indications, 104

styptics
  for bleeding from the lungs, 126
  definition of, 204

suan zao ren. *See Ziziphus spinosa* (suan zao ren)

sudorific, definition of, 204

sugar destroyer. *See Gymnema sylvestre* (sugar destroyer, cow plant)

sulfur gasotransmitters, 143

su mu. *See Caesalpinia sappan* (su mu)

sundew. *See Drosera rotundifolia* (sundew)

swamp lily (*Crinum glaucum*), 93

sweet Annie. *See Artemisia annua* (sweet Annie)

sweet clover (*Melilotus officinalis*), 79

sweet leaf. *See Stevia rebaudiana* (stevia, sweet leaf)

sweet vernal grass (*Anthoxanthum odoratum*), 79, 90

Symphytum officinale (comfrey)
  for bruising, 66
  for coughs, 117
  formulas containing, 60
  for heart infections, 72
  specific indications, 181
  for stasis ulcers, 60

Symplocarpus foetidus (skunk cabbage), 117, 181

symptoms
  choosing herbs based on, 5, 9–10, 114
  concomitant, 11
  of healing crises, 10
  importance of, 9–10
  totality of, 7

synergist herbs
  definition of, 204
  role in formulas, *11*, *12*, *13*
  in Triangle exercise, *11*, *12*

Syzygium spp., specific indications, 104

Syzygium aromaticum (cloves)
  for COPD, 149t
  liniments containing, 111
  for poor circulation, 58
  specific indications, 181–82

Syzygium jambos (roseapple)
  formulas containing, 35
  PAF inhibiting properties, 77
  for poor circulation, 63

Szechuan lovage. *See Ligusticum striatum* (ligusticum)

T

tablets, use of, 22

tachycardia
  formula for, 53
  herbs for, 55
  Senna spp. for, 103
  Strophanthus hispidus for, 104
  Viscum album for, 106

taenicide, definition of, 204

Tanacetum parthenium (feverfew)
  antiallergy qualities, 127
  formulas containing
    asthma, 145
    pneumonia, 142
    Raynaud's syndrome, 67
    vascular support, 29, 30
  as mast cell stabilizer, 144t
  specific indications, 104, 182
  for vascular headaches, 27

Taraxacum officinale (dandelion)
  in broth for electrolyte imbalance, 86
  for cor pulmonale, 155
  formulas containing
    anemia, 68, 69, 70
    cardiopulmonary disease, 84
    congestive heart failure, 50
    cor pulmonale, 155
    venous congestion, 76
  iron from, 70
  for poor circulation, 63
  for restless insomnia, 17
  specific indications, 104, 182
  for varicosities, 74
  for viral infections, 131

taurine, for congestive heart failure, 48t

TCM. *See* Traditional Chinese Medicine (TCM)

tea tree. *See Melaleuca alternifolia* (tea tree)

telangiectasias (spider veins)
  formulas for, 65–67
  herbs for, 66t
  overview, 65

Tephrosia purpurea (wild indigo)
  bronchodilating properties, 108t
  as mast cell stabilizer, 144t
  specific indications, 182

Terminalia spp., 134

Terminalia arjuna (arjuna)
  formulas containing, 40
  specific indications, 104

Terminalia chebula (myrobalan)
  as mast cell stabilizer, 144t
  specific indications, 182
  for viral infections, 131

terrain, 9–10

testicular varicocele, tincture for, 75

Thai galangal (*Alpinia galanga*), 156

Thea sinensis. See Camellia sinensis (green tea)

Theobroma cacao (cacao), 104–5
thrombolytic agents
  nattokinase as, 80
  for strokes, 64
Thuja occidentalis (northern white cedar)
  formulas containing, 58
  specific indications, 105
Thuja plicata (western red cedar), 50
Thymus vulgaris (thyme)
  for altered breath sounds, 118
  for COPD, 148
  for coughs, 117
  as drying agent, 133
  formulas containing
    allergic rhinosinusitis, 128
    asthma, 145, 146, 147
    bronchitis, 132, 133
    COPD, 151
    coughs, 117, 119, 120, 121
    dyspnea, 125
    emphysema, 138
    hemoptysis, 126, 127
    pneumonia, 141
    upper respiratory infections, 119
  nebulized form, 137
  for opportunistic infections, 134, 151
  for pleurisy, 138
  for respiratory infections, 121, 130, 134
  specific indications, 182
tian ma. *See Gastrodia elata* (tian ma)
TIAs (transient ischemic attacks), 61
Tilia x *europaea* (linden)
  formulas containing, 42, 46
  for hypertension, 44
  specific indications, 105
tinctures
  dosage strategy, 23
  pros, cons, and indications, 22
Tinospora cordifolia (guduchi)
  formulas containing, 126, 156
  as mast cell stabilizer, 144*t*
  specific indications, 182
Tongsai granules, 153
tongue, observing, 11
tonic, definition of, 204
tonsillitis
  Cinnamomum camphora for, 161
  Commiphora myrrha for, 162
  Hydrastis canadensis for, 167
  Pelargonium sidoides for, 175
  Phytolacca americana for, 176
  Salvia officinalis for, 179
topical protocols
  for COPD, 151–52
  for hemorrhoids, 78
  for hereditary hemorrhagic telangiectasia, 127
  liniments, 111, 113
  for lung conditions, 109–14
  for phlebitis, 81
  plasters, 110–11

  for pleurisy, 138, 139
  for pneumonia, 141
  for stasis ulcers, 60
  for varicosities, 75
toxic herbs. *See* safety concerns
toxicity, definition of, 204
Trachyspermum ammi (ajwain), 105
Traditional Chinese Medicine (TCM)
  energetic state of patient, 14
  evidence-based formulas from, 6–7
  for hyperlipidemia, 36
  liver symptoms, 17
  royal terms in herbal formula triangle, 12
  stroke recovery herbs, 65
transdermal drug delivery, 109, 115
transient ischemic attacks (TIAs), 61
Triangle exercise, 11–13, *11*
Triangle philosophy, 23–24
Trichosanthes kirilowii (Chinese cucumber)
  for ARDS, 154
  specific indications, 182–83
Trifolium pratense (red clover)
  formulas containing, 81
  for hypertension, 42
  iron from, 70
  specific indications, 105
Trigonella foenum-graecum (fenugreek)
  formulas containing, 35
  for poor circulation, 58
triterpene glycosides, 157
tropane alkaloids
  for asthma, 143
  for coughs, 167
  for dyspnea, 159
  for lung conditions, 152
  transdermal delivery of, 109
tuberculosis
  Albizia lebbeck for, 157
  Artemisia annua for, 159
  Azadirachta indica for, 160
  formulas for, 155–56
  hemoptysis from, 126
  Inula helenium for, 168
  Juniperus communis for, 169
  Lycopus virginicus for, 170
  Morinda citrifolia for, 172
  Stemona spp. for, 180
  Sticta pulmonaria for, 181
  Xanthium cavanillesii for, 184
tulsi. *See Ocimum sanctum* (holy basil)
turkey corn. *See Corydalis* spp.
turkey rhubarb. *See Rheum palmatum* (turkey rhubarb)
turmeric. *See Curcuma longa* (turmeric)
Tussilago farfara (coltsfoot)
  for altered breath sounds, 118
  for coughs, 117, 130
  formulas containing
    asthma, 146
    bronchitis, 133

    coughs, 117, 120
    hemoptysis, 126, 127
    pneumonia, 141
    stridor, 121
  specific indications, 183
Tylophora asthmatica (Indian ipecac)
  formulas containing, 147
  as mast cell stabilizer, 144*t*
  specific indications, 183
Tylophora indica, bronchodilating properties, 108*t*, 183

U

ulcers
  Hemidesmus indicus for, 167
  Hydrastis canadensis for, 97
  See also stasis ulcers
Ulmus fulva (slippery elm)
  formulas containing, 121
  specific indications, 183
umbel family. *See Apiaceae* herbs
Uncaria tomentosa (uña de gato)
  for COPD, 149*t*
  specific indications, 105, 183
upper respiratory infections (URIs)
  Achillea millefolium for, 156
  Allium cepa for, 157
  Allium sativum for, 157
  Andrographis paniculata for, 158
  Anemopsis californica for, 158
  Angelica sinensis for, 158
  Armoracia rusticana for, 158, 159
  Berberis aquifolium for, 160
  Dracocephalum rupestre for, 163
  Echinacea angustifolia for, 163
  Eucalyptus globulus for, 164
  Foeniculum vulgare for, 165, 166
  formulas for, 119
  herbs for, 121
  Hydrastis canadensis for, 167
  Hyssopus officinalis for, 168
  Juniperus communis for, 169
  Lomatium dissectum for, 170
  Lophatherum gracile for, 170
  Lycopus virginicus for, 170
  Matricaria chamomilla for, 171
  Mentha piperita for, 172
  Morella cerifera for, 172
  Nepeta cataria for, 173
  Ocimum sanctum for, 173
  Ophiopogon japonicus for, 174
  Panax ginseng for, 174
  Pelargonium sidoides for, 175
  Perilla frutescens for, 175
  Persicaria tinctoria for, 175
  Prunus serotina for, 177
  Punica granatum for, 178
  Rosmarinus officinalis for, 179
  Sambucus canadensis for, 179
  Schisandra chinensis for, 180

Tussilago farfara (coltsfoot) (*continued*)
  Sticta pulmonaria for, 181
  Thymus vulgaris for, 182
  Tinospora cordifolia for, 182
  Usnea barbata for, 183
  Verbascum thapsus for, 184
Urginea maritima (sea squill), 48*t*, 105
urinary conditions
  Juniperus communis for, 97
  Piper methysticum for, 13
  Thuja occidentalis for, 105
URIs. *See* upper respiratory infections (URIs)
Urtica spp.
  for congestive heart failure, 50
  for COPD, 148
  for cor pulmonale, 155
  formulas containing
    cardiopulmonary disease, 84
    congestive heart failure, 50
    COPD, 151
    peripheral vascular insufficiency, 59
Urtica dioica (stinging nettle), 84, 105
Urtica urens (nettle), 76, 84
Usnea spp., 141
Usnea barbata (old man's beard)
  formulas containing, 126
  for respiratory infections, 121
  specific indications, 183
Usnea hirta (old man's beard), 121

V

Vaccinium myrtillus (bilberry)
  for capillary fragility, 67
  flavonoids in, 33
  formulas containing
    cardiopulmonary disease, 84, 85
    hyperlipidemia, 36
  for microvascular fragility, 66*t*
  PAF inhibiting properties, 77
  specific indications, 105, 183–84
  for varicosities, 75
  for vascular support, 65
Valeriana officinalis (valerian)
  formulas containing
    angina, 39
    arrhythmias, 55
    hypertension, 44, 46
  for hypertension, 41, 44, 46
  for restless insomnia, 16, 17
  specific indications, 106
Valeriana sitchensis (Sitka valerian), 44
valve disease
  Adonis vernalis for, 89
  Convallaria majalis for, 93
  Selenicereus grandiflorus for, 103
  Viscum album for, 106
varicoceles, 73, 96
varicosities
  Aesculus hippocastanum for, 89
  Collinsonia canadensis for, 93

formulas for, 73–81
  Hamamelis virginiana for, 96
  with phlebitis, supportive measures for, 81
  Ruscus aculeatus for, 102
  Silybum marianum for, 103
vascular congestion
  Angelica sinensis for, 90
  Anthoxanthum odoratum for, 90
  Cnidium monnieri for, 92
  Collinsonia canadensis for, 93, 161
  Hamamelis virginiana for, 96
vascular dementia
  Huperzia serrata for, 97
  Ligusticum striatum for, 98
  Panax ginseng for, 100
  Salvia miltiorrhiza for, 102
vascular infections
  Echinacea spp. for, 94
  formulas for, 72
vascular inflammation
  Achillea millefolium for, 89
  Angelica sinensis for, 90
  Astragalus membranaceus for, 91
  Atractylodes spp. for, 91
  Cornus officinalis for, 162
  Curcuma longa for, 93
  Equisetum spp. for, 95
  Glycyrrhiza glabra for, 96
  Hippophae rhamnoides for, 96
  Hypericum perforatum for, 97
  Punica granatum for, 178
  Ribes spp. for, 102
  Rosa canina for, 102
  Salvia miltiorrhiza for, 102–3
vascular insufficiency
  Ginkgo biloba for, 95
  Hypericum perforatum for, 97
vascular smooth muscle proliferation, 41
vascular support
  Arctium lappa for, 90
  Astragalus membranaceus for, 91
  Calendula officinalis for, 91
  Castanospermum australe for, 92
  Coleus forskohlii for, 101
  Cornus officinalis for, 162
  Crataegus spp. for, 93
  Cynara scolymus for, 94
  flavonoids for, 26
  formulas for, 27–31
  Ginkgo biloba for, 95
  Gymnema sylvestre for, 96
  Huperzia serrata for, 97
  Hydrastis canadensis for, 97
  Hypericum perforatum for, 97
  Leonurus cardiaca for, 98
  Melilotus suaveolens for, 171
  Nigella sativa for, 99
  Olea europaea for, 99
  Ophiopogon japonicus for, 174
  Ruscus aculeatus for, 102

Salvia miltiorrhiza for, 102
Stevia rebaudiana for, 104
Tanacetum parthenium for, 104
Terminalia arjuna for, 104
Trachyspermum ammi for, 105
Trifolium pratense for, 105
Urtica dioica for, 105
Vinca minor for, 106
vasculitis
  Ammi visnaga for, 90
  formulas for, 82
  Ruscus aculeatus for, 102
vasoconstrictor, definition of, 204
vasodepressant, definition of, 204
vasodilators
  Achillea millefolium as, 89
  Ammi visnaga as, 90
  for angina, 38
  for cold extremities, 76
  definition of, 204
  Valeriana officinalis as, 106
  warming stimulants, 21*t*
  for weak circulation, 61
vegetable juices, 31
vegetable mercury. *See Iris versicolor* (blue flag)
venous congestion, formulas for, 73–81
venous insufficiency
  Ginkgo biloba for, 95
  overview, 57–58
  Pinus pinaster for, 100
  Ruscus aculeatus for, 102
Veratrum spp. (hellebore), 20
Veratrum viride (false hellebore), 44
Verbascum thapsus (mullein)
  for coughs, 117
  formulas containing
    anemia, 69, 70
    asthma, 146
    hemoptysis, 126
    pneumonia, 141
  iron from, 70
  specific indications, 184
vermifuge, definition of, 204
vertigo
  Ligusticum striatum for, 98
  Vinca minor for, 106
vesicant, definition of, 204
Viburnum opulus (crampbark), 13
Viburnum prunifolium (blackhaw)
  as antispasmodic, 13
  formulas containing, 46
  for hypertension, 44
Vinca minor (lesser periwinkle)
  formulas containing, 64
  specific indications, 106
viral infections
  Azadirachta indica for, 160
  Glycyrrhiza glabra for, 96
  Hypericum perforatum for, 168
  Isatis tinctoria for, 168

Lomatium dissectum for, 170
Lonicera japonica for, 170
Lophatherum gracile for, 170
Melissa officinalis for, 172
Nepeta cataria for, 173
Pelargonium sidoides for, 175
Pinus pinaster for, 176
Scutellaria baicalensis for, 180
Terminalia chebula for, 182
Trichosanthes kirilowii for, 182
Viscum album (mistletoe), 54
  for arrhythmias, 55
  for atherosclerosis, 32
  for congestive heart failure, 48
  formulas containing
    angina, 38
    arrhythmias, 55
    congestive heart failure, 50
  for heart disease, 54
  specific indications, 106
Viscum articulatum (leafless mistletoe), 48
Viscum flavens. See Phoradendron chrysocladon
vitality, supporting, 14
vitamin B_{12} deficiency anemia, formulas for,
  68–70
Vitex negundo (Chinese chaste tree), 108t, 184
Vitis vinifera (grape), 31
  for capillary fragility, 67
  iron chelation properties, 71
  as mast cell stabilizer, 144t
  PAF inhibiting properties, 77
  specific indications, 184
volatile oils. *See* essential oils

W

warming stimulants, 20, 21t
wax myrtle. *See Myrica cerifera* (wax myrtle)
weak circulation. *See* circulation, poor
weak heart
  Adonis vernalis for, 89
  Convallaria majalis for, 93
  Crinum glaucum for, 93
  Glycyrrhiza glabra for, 96
  Panax ginseng for, 100
  Selenicereus grandiflorus for, 103
  Strophanthus hispidus for, 104
Weiss, Rudolf F., 44
Western herbalism
  four-elements theory, 14

as term, 1–2
western red cedar (*Thuja plicata*), 50
wheezing
  Atropa belladonna for, 159
  Euphorbia spp. for, 165
  Foeniculum vulgare for, 165
  herbs for, 118
  Justicia adhatoda for, 169
  Magnolia officinalis for, 171
  overview, 115–16
  Passiflora incarnata for, 175
  Stemona spp. for, 180
  Tanacetum parthenium for, 182
white peony. *See Paeonia lactiflora*
  (white/red peony)
whooping cough
  Digitalis purpurea for, 163
  Drosera rotundifolia for, 163
  Inula helenium for, 168
  Stemona spp. for, 181
  Sticta pulmonaria for, 181
wild carrot. *See Daucus carota*
  (wild carrot)
wild indigo. *See Tephrosia purpurea*
  (wild indigo)
wild iris. *See Iris versicolor* (blue flag)
witch hazel. *See Hamamelis virginiana*
  (witch hazel)
Withania somnifera (ashwagandha)
  formulas containing
    hemoptysis, 126
    hypertension, 42, 46
    tuberculosis, 156
  specific indications, 106, 184
  for tuberculosis, 156
woad. *See Isatis tinctoria* (woad)
wolfberry. *See Lycium barbarum* (goji)
wolfsbane. *See Aconitum napellus* (aconite)
wu wei zi. *See Schisandra chinensis*
  (magnolia vine)

X

Xanthium cavanillesii (cocklebur), 144t, 184
Xanthoxylum americanum, 123
xin yi qing fei tang (Shin'iseihaito), 174
Xuanbai Chengqi decoction, 153, 154

Y

yarrow. *See Achillea millefolium* (yarrow)

yellow clover. *See Melilotus officinalis*
  (sweet clover)
yellow dock. *See Rumex crispus* (yellow dock)
yellow jessamine (*Gelsemium sempervirens*), 56
yerba mansa (*Anemopsis californica*), 158
yerba santa. *See Eriodictyon californicum*
  (yerba santa)
yin yang huo. *See Epimedium brevicornu*
  (horny goatweed)
Yu Ping Feng San, 150

Z

Zanthoxylum clava-herculis (southern
  prickly ash), 132, 184
Zingiber officinale (ginger)
  for altered breath sounds, 118
  for atherosclerosis, 32
  for COPD, 148
  for cor pulmonale, 155
  fibrinolytic properties, 73
  formulas containing
    allergic rhinosinusitis, 130
    angina, 38
    asthma, 147
    bronchitis, 134
    COPD, 150
    coughs, 117
    cystic fibrosis, 135
    emphysema, 138
    for hyperlipidemia, 35
    hyperlipidemia, 47
    peripheral vascular insufficiency, 58, 61
    phlebitis, 81
    Raynaud's syndrome, 67
    vascular support, 27, 28, 29, 30
    venous congestion, 76
  for hypertension, 46
  for insomnia with exhaustion, 16
  PAF inhibiting properties, 77
  for poor circulation, 58, 63
  for respiratory conditions, 108
  specific indications, 106, 184
  strong energy of, 20
  for vascular support, 31
Ziziphus jujuba (Chinese date), 106
Ziziphus spinosa (suan zao ren)
  formulas containing, 53
  specific indications, 106
  for strokes, 65

— ABOUT THE AUTHOR —

Shelly Fry of Battle Ground

Dr. Jill Stansbury is a naturopathic physician with 30 years of clinical experience. She served as the chair of the Botanical Medicine Department of the National University of Natural Medicine in Portland, Oregon, for more than 20 years. She remains on the faculty, teaching herbal medicine and medicinal plant chemistry and leading ethnobotany field courses in the Amazon. Dr. Stansbury presents numerous original research papers each year and writes for health magazines and professional journals. She serves on scientific advisory boards for several medical organizations. She is the author of *Herbal Formularies for Health Professionals*, Volume I and Volume II, and is the coauthor of *The PCOS Health and Nutrition Guide and Herbs for Health and Healing*. Dr. Stansbury lives in Battle Ground, Washington, and is the medical director of Battle Ground Healing Arts. She also runs an herbal apothecary offering the best quality medicines from around the world, featuring many of her own custom tea formulas, blends, powders, and medicinal foods.

Visit www.naturopathicce.com/instructor/jillian-stansbury for continuing education coursework taught by Dr. Stansbury.

www.healingartsapothecary.org

HERBAL FORMULARIES FOR HEALTH PROFESSIONALS

This comprehensive five-volume set by Dr. Jill Stansbury serves as a practical and necessary reference manual for herbalists, physicians, nurses, and allied health professionals everywhere. This set is organized by body system, and each volume includes hundreds of formulas to treat common health conditions, as well as formulas that address specific energetic or symptomatic presentations.

— VOLUME 1 —

DIGESTION AND ELIMINATION

INCLUDING THE GASTROINTESTINAL SYSTEM, LIVER AND GALLBLADDER, URINARY SYSTEM, AND THE SKIN

9781603587075
Hardcover • $59.95

— VOLUME 3 —

ENDOCRINOLOGY

INCLUDING THE ADRENAL AND THYROID SYSTEMS, METABOLIC ENDOCRINOLOGY, AND THE REPRODUCTIVE SYSTEMS

9781603588553
Hardcover • $44.95
Available Spring 2019

— VOLUME 4 —

NEUROLOGY, PSYCHIATRY, AND PAIN MANAGEMENT

INCLUDING COGNITIVE AND NEUROLOGIC CONDITIONS AND EMOTIONAL CONDITIONS

9781603588560
Hardcover
Available Spring 2020

— VOLUME 5 —

IMMUNOLOGY, ORTHOPEDICS, AND OTOLARYNGOLOGY

INCLUDING ALLERGIES, THE IMMUNE SYSTEM, THE MUSCULOSKELETAL SYSTEM, AND THE EYES, EARS, NOSE, AND THROAT

9781603588577
Hardcover
Available Spring 2021

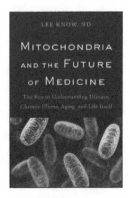
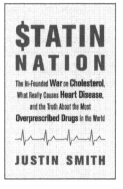
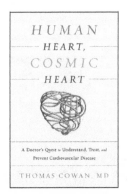